The Large-Scale Structure of Inductive Inference

BSPS Open SERIES

ISSN 2564-3169 (Print) ISSN 2564-3177 (Online)

BSPS Open publishes landmark, cutting edge works that represent the full breadth and diversity of the philosophy of science. Diamond Open Access, all books in this series are available freely to readers everywhere.

Published in collaboration with the British Society for the Philosophy of Science.

No. 1 *The Material Theory of Induction*
John D. Norton

No. 2 *The Large-Scale Structure of Inductive Inference*
John D. Norton

UNIVERSITY OF CALGARY
Press

The Large-Scale Structure of Inductive Inference

JOHN D. NORTON

BSPS | OPEN

BSPS Open SERIES
ISSN 2564-3169 (Print) ISSN 2564-3177 (Online)

University of Calgary Press
2500 University Drive NW
Calgary, Alberta
Canada T2N 1N4
press.ucalgary.ca

Library and Archives Canada Cataloguing in Publication

Title: The large-scale structure of inductive inference / John D. Norton.
Names: Norton, John D., 1960- author.
Description: Series statement: BSPS open series, ISSN 2564-3169 (print), ISSN (ebook) 2564-3177 ; no. 2 | Includes bibliographical references and index.
Identifiers: Canadiana (print) 20240393821 | Canadiana (ebook) 20240393899 | ISBN 9781773855394 (hardcover) | ISBN 9781773855400 (softcover) | ISBN 9781773855424 (PDF) | ISBN 9781773855431 (EPUB) | ISBN 9781773855417 (Open Access PDF)
Subjects: LCSH: Induction (Logic)
Classification: LCC BC91 .N67 2024 | DDC 161—dc23

The University of Calgary Press acknowledges the support of the Government of Alberta through the Alberta Media Fund for our publications. We acknowledge the financial support of the Government of Canada. We acknowledge the financial support of the Canada Council for the Arts for our publishing program.

Alberta Government   Canada   Canada Council for the Arts   Conseil des Arts du Canada

Copyediting by Dallas Harrison
Cover image: Colourbox 7726547
Cover design, page design, and typesetting by Melina Cusano

Contents

List of Figures

List of Tables

Preface

The analyses in this book began as an afterthought in the development of the material theory of induction. My goal with the theory was to resolve once and for all how inductive inference works. Its chief concern was the failure of the many competing accounts of inductive inference already in the literature to do justice to how evidence is actually used in science. The mature development of that project is provided in my earlier work, *The Material Theory of Induction* (Norton 2021b).

When the first sketch of the material theory of induction (Norton 2003) was in a complete first draft, Jim Bogen pointed out to me that the material theory provides an escape from the problem of induction. I added the point to the final version in Section 6 of the paper. I repeated it more briefly in the conclusion to "A Little Survey of Inductive Inference" (Norton 2005). In retrospect, my analysis was too hasty. The basic idea of the escape was sound, but the details were not well developed.

That this was so was brought home at a Philosophy of Science Association symposium in 2008 organized by Peter Achinstein. Papers by John Worrall and Tom Kelly suggested that a version of the problem of induction still troubled the material theory. They were right. The escape as described was not adequately elaborated. I am grateful to them for pressing me. When I worked to clarify the escape from the problem of induction, I saw that it required a clarification of the large-scale structure of relations of inductive support. My first effort to provide a better account was in "A Material Dissolution of the Problem of Induction" (Norton 2014). That paper contains many of the ideas developed in this volume, especially including the nonhierarchical nature of relations of inductive support and the special role of hypotheses.

Although that account was a great improvement on the earlier versions, it contained a weakness. It did not adequately separate the idea of a logic of induction from an epistemology of belief. The problem of induction resides within the first, the logic of induction. It has a presence in the epistemology of belief only indirectly, when we use a logic of induction to reason from belief to belief. In failing to separate them clearly, I conformed to the corresponding failure in much of the present literature on epistemology.

Critiques of the material escape from the problem of induction continued. Nothing is as assured to attract critical responses as the claim of a solution to the problem of induction. Does not everyone know that it cannot be done? Some of them appeared in a volume of *Studies in History and Philosophy of Science* dedicated to *The Material Theory of Induction*. (My replies are in Norton 2021a, in which the original papers can be identified.) Many of these critiques mislocated the material theory of induction within the epistemology of belief and, as a result, conflated issues that should have been kept separate. This alerted me to the need to distinguish the two contexts more clearly.

In the present volume, I have done my best to distinguish the two. The easy way to distinguish them is to note that the two contexts use different relata. The relata of the logic of induction are propositions. Their content and relations are independent of human thoughts and beliefs. The relata of the epistemology of beliefs are beliefs. They are related by psychological processes that might or might not respect a logic. These issues are laid out as clearly as I can in Chapter 6, "The Problem of Induction," and in Chapter 5, "Coherentism and the Material Theory of Induction."

Addressing the problem of induction has been a major stimulus to the ideas developed in this volume. However, tracing a pathway from this origin to these ideas is a poor way of presenting them. The ideas about the large-scale structure of relations of inductive support are more important in their own right. They tell us how all the relations of inductive support fit together when we look at the entirety of science. They would retain this importance even if they had nothing to say about the problem of induction. Once that problem is mentioned, however, it seems to mesmerize many philosophers so that they are unable to see anything else. For this reason, I avoided all mention of the problem of induction in *The Material Theory of Induction* until the epilogue lest it distract readers from the substance laid out in its sixteen chapters. Again for this reason I have delayed discussion of the problem of induction until well into the present work. My hope is that this tactic will induce readers to consider the account developed here of the large-scale structure of inductive inference in its own right and not conceive it as yet another tiresome attempt to solve the problem of induction. No doubt I will fail in these hopes with some readers, but will you, dear reader, not be one of them?

During the writing of this text, I have been helped by colleagues, and I have acknowledged their support in the context of the individual chapters. That identifies their assistance more clearly than would a long, generic list here. However, I do want to thank here the participants in my graduate seminar, HPS 2682 Theories of Confirmation, for their reading of and critical reflections on Chapters 1, 2, 3, 4, and 6 in meetings of March 31 and April 7, 2021, and Youness Ayaita for his reading of Chapters 1 to 6. My thanks also to Marc Lange for his careful reading in 2021 of many chapters and for his copious and helpful comments.

The material in this volume was collected over several years. Some of the chapters were written in their earliest forms when I thought it was possible to include this discussion in the earlier volume, *The Material Theory of Induction*. Other chapters were written subsequently. Many of them were written in the COVID-19 pandemic years of 2020–21. Immersion in writing them provided a little of the comfort and support needed during this awful time of bad news and isolation.

The greater support was provided by my wife, Eve, whose love and companionship brightened each day and to whom this volume is dedicated.

REFERENCES

Norton, John D. 2003. "A Material Theory of Induction." *Philosophy of Science* 70: 647–70.

———. 2005. "A Little Survey of Inductive Inference." In *Scientific Evidence: Philosophical Theories and Applications*, edited by Peter Achinstein, 9–34. Baltimore: Johns Hopkins University Press.

———. 2014. "A Material Dissolution of the Problem of Induction." *Synthese* 191: 671–90.

———. 2021a. "Author's Responses." *Studies in History and Philosophy of Science* 85: 114–26.

———. 2021b. *The Material Theory of Induction*. BSPS Open Series. Calgary: University of Calgary Press.

Introduction

1. The Project of This Volume

According to the material theory of induction, inductive inferences or relations of inductive support are not warranted in a way familiar from accounts of deductive logic. They are not warranted by conformity with some universally applicable schema or template. Rather, each is warranted by background facts peculiar to the domain in which the inference arises. This idea was developed in my earlier monograph, *The Material Theory of Induction* (Norton 2021). A key provision of the theory is that the warranting facts must be facts: that is, truths of the domain. If we seek to sustain an inductive inference by appealing to some warranting proposition in the domain that is false, then we commit the inductive analog of a fallacy. The error is comparable to the deductive fallacy of affirming the consequent as if it were a valid deductive schema.

That warrants must be factual truths places a special burden on us when we assess the inductive inferences or relations of inductive support among the propositions of some science. To establish support fully, we must also establish the truth of the warranting propositions used. Since these warranting propositions are also contingencies of the domain, establishing their truth requires further inductive inferences. Thus, any claim that some particular item of evidence inductively supports some other proposition in a theory is not self-contained. To be sustained to the fullest extent, the truth of these further warranting facts must also be established. Since those facts themselves are contingent propositions, we must establish their truth with still further inductive inferences or relations of inductive support, and we must show that those inductive inferences in turn are warranted by further facts.

And so on. All claims of inductive support, in effect, are claims that concern a large network of contingent propositions within the science of interest and, commonly, extending beyond it.

These considerations define the project of this work. Individual claims of inductive support must be made within a larger ecology of relations of inductive support. How is this larger ecology configured? What is the large-scale structure of inductive inference? What are its problems? Can a cogent account be supplied for it? My goal in this work is to answer these questions.

Some might find this entanglement of inductive support with a larger inductive ecology disquieting and might want to retreat to formal approaches to escape it. Formal approaches that use universal schemas might appear to have an advantage. They can assess the cogency of an inductive inference without engaging a larger ecology. An inductive argument from analogy just has to show that it conforms to the relevant schema. A claim of probabilistic support might just have to show that the associated probabilities relate by Bayes' theorem.

This advantage is illusory. According to the material theory of induction, it is dangerous to assume that each formal schema can be applied unconditionally everywhere. It exposes users to a significant risk of inductive fallacies if the schemas are applied in domains that lack a material warrant. The common remedy by formalists is tacitly to limit the application of the schemas to where they are thought somehow to be appropriate. The remedy is poor since decisions on applicability depend on hunches and intuitions. Here material theorists have the advantage. The question of which inference forms are applicable where is decided by an explicit analysis of the prevailing facts.

Again one might think that a better way to treat the large-scale structure of inductive support is mathematical. We merely need to identify the calculus that applies at this large scale. Questions about the large-scale structure would be answered mathematically by theorems in the calculus. Bayesians in the philosophy of science might believe that the probability calculus already does just this.

Hopes for some universal calculus of inductive inference fail and provably so. In recent work (Norton 2019, 2021, Chapter 12), I have shown the incompleteness of all calculi of inductive inference that meet some minimal conditions. Any such calculus will fail to discern nontrivially the inductive import of any body of evidence unless the computation is supplemented by inductive content supplied externally. The familiar example is that Bayesian analysis always requires prior probabilities. Their stipulation is antecedent to the application of Bayes' theorem, yet their content exerts a strong influence on the outcome of the computations. Efforts have failed to supply Bayes' theorem with vacuous priors that exert no such influence. This incompleteness is not limited to the probability calculus. A form of it will arise in any calculus meeting minimal conditions.

In its briefest form, the answer supplied by the material theory of induction to the question of the large-scale structure of inductive support is this: relations of inductive support within a mature science form a massively entangled network without any clear hierarchical structure. Quine (1951, 39–40), in his celebrated "Two Dogmas of Empiricism," presented a similar structure for beliefs. However, his structure was variously a "fabric" and a "field of force" and later a "web of belief." Its key attribute is its elasticity. A conflict with experience, according to this picture, can always be accommodated. The internal connections, he supposed, are so elastic that there are many ways to do this. This supposition has been responsible for much philosophical mischief. It has encouraged the idea that evidence, even in great measure, is unable to determine the propositions of a science. This indeterminacy is incompatible with our routine experiences of mature science and is not established by Quine's analysis. The elasticity results from reliance on a naive and inadequately weak hypothetico-deductive approach to inductive inference.[1]

The account developed in this volume differs sharply from Quine's (1951) supposition of elasticity. The relations of inductive support in a mature science are better imagined as strong steel cables, not elastic threads. They are connected and interconnected in such a variety of ways that the integrity of the entire structure is threatened if an anomalous experience arises. The affirmation that some ordinary machines can be combined to produce a perpetual motion machine would overturn mechanics. Or consider the discovery of a new mineral not constituted by atoms or not compounded of elements found in the periodic table. It would destabilize chemistry and, after that, the quantum theory that underpins the atomic character of matter and the uniqueness of the elements in the periodic table. Evolutionary theory would fail to accommodate a new species of living beings that spontaneously appears fully formed without any past history of development. The structure

1 Or so I argue (Norton 2008).

of inductive support of mature sciences is not elastic but rigid. A break in one place propagates with revolutionary import far into the structure.

In this volume, I explore and examine this structure. The first chapter is a brief development of the material theory of induction. It does not replace the lengthier elaboration of the theory in *The Material Theory of Induction* (Norton 2021). However, for readers interested in the issues raised in the present work, it will serve well enough as a point of first contact.

Subsequent chapters are divided into two parts. The first part presents general propositions in the philosophy of science concerning the large-scale structure of inductive inference or inductive support. The second part presents historical case studies that provide detailed illustrations of the main claims of the first part and are the source of many of its claims.

2. Part I: General Claims and Arguments

In Chapter 2, I advance four claims, whose support and elaboration occupy the remainder of the text.

1. Relations of inductive support have a nonhierarchical structure.
2. Hypotheses, initially without known support, are used to erect nonhierarchical structures.
3. Locally deductive relations of support can be combined to produce an inductive totality.
4. There are self-supporting inductive structures.

The first claim renounces the idea that inductive support is hierarchical, structured by generality. In this renounced picture, propositions in a science are supported inductively just by propositions of lesser generality. We then would be able to trace a pathway of inductive support from the lowest levels closer to experience, gradually ascending the hierarchy of generality to the most general propositions of the science. The actuality is that relations of inductive support in real science fail to respect any such hierarchy. They cross over in many complicated ways. The very idea of a hierarchy of generality is sustainable only in a crude way, if at all.

The second claim pertains to the practices needed to identify these tangled inductive structures. In the early stages of the development of a new science, inductive inferences commonly can proceed only if we make use of warranting assumptions for which we do not yet have inductive support. They are introduced as hypotheses, and their use is provisional. Their use comes with an obligation to secure their proper inductive support in subsequent investigations. Should that obligation not be met, the original claims of inductive support fail. This role attributed to hypotheses is *not* the traditional role given to them in accounts of hypothetico-deductive confirmation. In this latter case, the hypotheses themselves are confirmed by their success at entailing evidence. Here the hypotheses mediate in establishing inductive support for other propositions. The hypotheses themselves must accrue support by other means in another stage of investigation.

The third claim asserts that it is possible to combine deductive relations of support to produce an overall relation of support that is inductive in character. This is a possibility that, in the abstract, seems to be impossible. Yet, as the examples show, it arises commonly in the actual practice of science. If it can be achieved, then it is a construction to be prized for its reduction in inductive risk. The more familiar construction involves intersecting relations of inductive support that are combined to produce an overall inductive import. An inductive risk is taken, first, in accepting each component relation or inductive inference and, second, in accepting their combined import. When the component inductive relations of support are replaced by deductive relation, that first inductive risk is eliminated.

Finally, the fourth claim is a thesis of completeness. That many inductive inferences are warranted materially is undeniable, or at least so I think after working through the many examples in *The Material Theory of Induction* (Norton 2021). If one is eager to retain general schemas, then it is tempting to suppose that these examples display only a part of the full inductive story. Materially warranted inductive inferences or relations of support alone, one might want to assert, are not enough to sustain all of a science inductively. A full accounting must include general schemas or general rules in some form. This fourth claim asserts otherwise. It is possible for materially warranted propositions to form structures such that every proposition in the structure is inductively supported, without the need for general schemas or other devices outside the material theory of induction.

This completeness is already a corollary of the arguments given for the material theory in Chapter 2 of *The Material Theory of Induction* (Norton 2021) and repeated more briefly in Chapter 1 below. Any general schema must factually expand in some way on the premises supplied to it. This expansion can only be sustained in domains hospitable to the means of the expansion. Any such expansion can fail if the facts of the domain are such as to oppose it. The fact of that hospitality, in the most general terms, is the warranting fact of the inductive inference or relation of inductive support. This argument defeats every attempt to assert the existence of some universal inductive rule. There can be none that escapes it.

A simpler picture does not use universal inductive rules. Each individual proposition of a mature science is inductively well supported. If we are willing to undertake the task of tracing it, we can display the form that support takes and its material character. This is true of each of the propositions of a mature science, taken individually. Their totality is the full material account of the inductive support of the mature science. Nothing further is needed, for no proposition has been left without an account of its inductive support.

These four claims in turn raise further issues that need to be addressed. Relations of inductive support cross over one another in a myriad of ways. Tracing along the pathways of support, we routinely find circles that bring us back to our starting point. Philosophers, brought up in fear of vicious circularity, mistakenly find the mere existence of such circles automatically disqualifying for any system. In Chapter 3, I argue that this disqualification is hasty and mistaken. There are circles throughout our sciences. We routinely consider populations in which the rate of growth of the population is proportional to the size of the population. This is a benign circle of self-reference. It is merely the most convenient definition of exponential growth. When a circularity is uncovered, there can be no default supposition of a systemic failure. Instead, we have a positive obligation to demonstrate that a circularity is harmful, if we seek to represent it as such. Is the circularity vicious and thus leads to a contradiction? Or does it lead to an underdetermination of theories? I argue that the circularities in inductive relations of support within mature scientific theories do neither. They are benign.

In Chapter 4, I address a related issue. Mature sciences, it has been asserted, are inductively self-supporting. The evidence for them is sufficient to sustain relations of inductive support such that every proposition in the science is supported. That leaves open a troubling possibility. Might it be that there are multiple such sciences for a given body of evidence? Then the bearing of evidence would not be univocal, no matter how rich and varied the evidence. Might this be the harm that circularities bring? I argue otherwise in the chapter. Mature sciences are uniquely supported by their evidence. There is only one periodic table of elements supported by the evidence in chemistry and so on for the central claims of mature sciences.

This uniqueness arises from the empirical character of science. Any alternative is only a real alternative if it differs in some factual assertion. Since all such assertions are open to empirical testing, competition among alternatives is transient, if only the evidence that can decide among them is pursued. The material character of inductive inference adds a mechanism that destabilizes any competition. If one theory in the competition gains a small advantage, then the facts thereby secured can serve as warrants for further inductive inferences supporting the theory. The effect is that the advantage of the ascending theory is amplified. When the investigations continue, this amplification is repeated, at the repeated costs of its competitors. If the process continues long enough, then it ends with one theory prevailing over all of its competitors. It is this instability that promotes the uniqueness of mature sciences.

Circularities are a distinctive feature of coherentist accounts of justification. We might hope, as I did originally, that there would be results already developed there of use to the material theory. The differences between the two systems are so great that, it turns out, these expectations are not met. In Chapter 5, I explore these differences. The coherentist account is offered as an alternative to fundamentalist accounts of justification. Coherentists must eschew the fundamentalist supposition that some beliefs are justified primitively by the world. The material theory has no such obligation. It takes observations and experiences of the world to be the foundations upon which inductive structures are built. For coherentists, beliefs are justified by their inclusion in a coherent system. The judgment is essentially global. There is something similar in the material theory. Strong inductive support for a proposition ultimately does depend on the larger-scale integrity of the relations of inductive support. However, that integrity arises from the composition of many individual relations of support. Each of the propositions in the structure must be well supported inductively, and considerable effort is expended in establishing each such individual relation of support. Finally, coherentist justifications concern relations among beliefs: that is, within cognitive states. The material theory is concerned with mind- and belief-independent relations of inductive support among propositions that assert some factual condition in the world.

In Chapter 6, I describe how the material theory of induction dissolves the classic problem of induction. I provide a short history of the problem and show that the problem of induction is specifically a problem for accounts of induction based on universal schemas. Its dissolution by the material theory involves no exotic legerdemain. The material theory of induction does not posit universal schemas. It follows that the problem of induction cannot be set up in it. It is dissolved. Although this claim of dissolution has already attracted considerable attention, it has come with the mistaken claim that the material was devised specifically to solve the problem of induction. As I have related on several occasions, that is not the history of it.[2] My concern is that the claims of the material theory — at both the local scale and the large scale — should be evaluated as an attempt to understand inductive inference better. That can be done independently of whether the theory dissolves the problem of induction. If it does not dissolve that problem, then the failure merely puts it in good company with all of the other failed attempts. The material theory's other results still stand.

Although the material theory's dissolution of the problem of induction is straightforward, a common reaction is to treat it like other claimed solutions to the problem. Under scrutiny, these other solutions prove to depend on unfounded, hidden assumptions, comparable in import to those that produced the problem originally. This reflexive reaction leads to the supposition that the problem must reappear in the material theory in some way in the mutual dependencies of inductive support. The unmet challenge of this reflexive reaction has been to find a way that the problem of induction reappears. Perhaps circularities in the structure are harmful; perhaps there is a fatal regress to warranting propositions of ever greater generality; perhaps, if our starting point is meager, then we have no warranting hypotheses that would allow inductive inferences to be initiated. All of these suppositions fail to identify a problem for the material theory. There is little need to argue the point in great detail since securing the theory against such objections is undertaken in earlier chapters. The theory's circularities are benign, as I argue in Chapter 3. A fatal regress to warranting propositions of ever greater generality requires the presumption of a hierarchical structure that, as I argue in Chapter 2, is

2 My first paper on the material theory (Norton 2003) was already in a complete first draft when Jim Bogen pointed out the possibility of a dissolution of the problem of induction. I added an imperfect sketch of that dissolution as a later section of the paper.

not present in the material theory. Finally, there is no difficulty starting the inductive project. When warranting premises are missing, they are introduced provisionally as hypotheses.

3. Part II: Historical Case Studies

The second part of this volume presents a set of case studies within the history of science. They are detailed and reflect my commitment that an analysis of inductive inferences should be responsible to what actually happens in science. Here the analysis differs from much of what is found in the philosophical literature on inductive inference. There the analysis suffers from adaptation to a few oversimplified examples. We might infer from the observation that some crows are black the conclusion that all crows are black. But such inferences, analyzed in isolation, are oversimplified caricatures of the much more sophisticated inductive inferences of real science. An account designed just to accommodate such oversimplified examples is destined to be woefully oversimplified itself.

Formal accounts of inductive inference in the philosophical literature face the same problem. An erudite formal analysis, no matter how technically clever, is only as good as the assumptions on which it is based. The inductive practice of real science is complicated and messy. Formal systems, if they are to be amenable to mathematical analysis, must be based on a few simple axioms. When they are naive or oversimplified, so, inevitably, is the analysis. These failures are easily overlooked since, commonly, formal accounts are developed without close attention to the actual inductive practices in science. When a formally pretty system is proposed, it is easy to be distracted by the ingenuity of the technical details and beguiled by the lure of the abstract formal puzzles that they pose.

This work takes seriously the obligation to connect its general claims with the actualities of the sciences. It does this by melding general claims in the philosophy of science with detailed historical studies of science. This practice embodies a conception of what it is to do the history and philosophy of science. Theses in the philosophy of science must withstand scrutiny in the history of science. That much is widely accepted as an abstract principle. It is much less widely practiced. The reverse relation is more interesting. I have found repeatedly that investigations in the history of science are a fertile means of identifying powerful and interesting theses in the philosophy of science. The scientists often face daunting inductive challenges. Their ingenuity in meeting the challenges far outstrips the imaginings of the philosophers of science, concerned only with ruminations on abstract principles and ideas. Careful attention to the history can yield ideas that otherwise would not emerge from mere armchair reflection.

Chapters 7 to 14 present case studies selected, I must admit, simply because they are episodes that interest me and, I suspect, will prove to be fertile in supplying general theses for the first part. In almost all of the cases, we find relations of inductive support crossing over one another in a way that violates a hierarchy of generality. That is one of the most important facts provided by the studies. The individual studies typically add extra points of special interest.

In Chapter 7, I recount Hubble's arguments in 1929 for his celebrated "Hubble's law." That law asserts that galaxies recede with a speed proportional to their distance from us. If one does not look at the details of his reporting, then it is all too easy to represent his analysis as a simple act of generalization. Hubble checked that the linear relation held for a sample of galaxies and then just generalized. A little attention to his paper of 1929 shows that his analysis was neither so simple nor that easy. Hubble had distance measurements only for roughly half of the galaxies in his data set. He needed maneuvers of great ingenuity to extend his law to all of the galaxies in his data set. They involved reasoning that inverted the order of inference that one would expect. In one part, they even employed the Hubble law itself as a premise.

In Chapter 8, I recount some of Newton's arguments for his inverse square law of gravity. Newton, we find, was adept at recovering inductive support for his claims by combining deductive relations. Such combinations figure in central portions of the evidential case that he made for his theory of universal gravitation. They arise in his Moon test, which argues for the identification of terrestrial gravity and the force that binds the Moon to the Earth; they arise again in the details of his analysis of the inverse square law of gravity and its relation to the elliptical orbits of the planets.

In Chapter 9, on atomic spectra, I show how the numerical rules governing the series of lines in the hydrogen emission spectrum are supported by multiple relations of inductive support that cross over one another in many ways. Under the warranting authority of Ritz's combination principle, the presence of some lines provided support for the presence of other lines, and entire infinite series of lines provided support for other entire infinite series of lines. A second crossing over of support occurred at a higher level. Ritz's combination principle provided general support for the newly emerging quantum theory. It was the observable manifestation of the fundamental electronic process of Bohr's quantum theory of the atom: the stepwise descent of an excited electron through the allowed orbits of the theory. Soon this relation of support was inverted. The more fully developed quantum theory both entailed the Ritz combination principle and could specify the empirically found circumstances in which it failed.

In Chapter 10, I provide another illustration of the crossing over of relations of support. It arises among two sets of propositions that date historical artifacts. In one set, datings are provided by traditional historical and archaeological methods. In the other, datings are provided by radiocarbon methods. There are uncertainties in both. Historical methods can err when they rely on meager or equivocal clues. Carbon dating can err if the historically varying levels of atmospheric carbon 14 are not accurately known. Then the baseline from which the carbon 14 decay started is uncertain. Each set can be used to correct and calibrate the other set. The calibration curve for historical levels of atmospheric carbon 14 was derived from historical dating methods, including, famously, the counting of tree rings in ancient bristle cone pines. Once well calibrated, carbon 14 dating can correct historical and archaeological datings of artifacts. When the two sets of propositions are in agreement, each mutually supports the other set.

In Chapter 11, I look at the history of the determination of the relative atomic weights of the elements. The task proved to be recalcitrant and strained the resources of chemists for roughly the first half of the nineteenth century. The difficulty was that, after Dalton's introduction of chemical atomism in 1808, chemists were trapped by an incompleteness in his atomic theory. The evidence that 8g of oxygen react with 1g of hydrogen to produce water does not tell us how many atoms of hydrogen combine with how many atoms of oxygen to form water. Was the ratio 1-1, 2-1, 1-2, and so on? We are left uncertain about whether the molecular formula for water is HO, H_2O, HO_2, or something else again. To eliminate the uncertainty, we also need to know the relative weight of each atom of hydrogen and oxygen. But we cannot know those relative weights until we know the correct molecular formulae for water and other related substances.

Chemists struggled for roughly half a century to overcome this incompleteness. Matters were settled only with Cannizzaro's results in 1858 and brought to the notice of chemists through a conference in 1860. Cannizzaro's results depended on a careful selection of fertile hypotheses to break the evidential circle in which Dalton was trapped. The best known is Avogadro's hypothesis on the numbers of molecules in equal volumes of gases. Applying this and other hypotheses to a wide array of elements and compounds, a unique set of atomic weights could be recovered. They emerged from a huge tangle of intersecting relations of support. There were so many that I can sample only a few in the chapter. They extend from intersecting relations of support among the molecular formulae of individual substances to mutual relations of support at the highest levels of abstract theory. The chemists found support for Avogadro's hypothesis in the new physics of the kinetic theory of gases. Conversely, the physicists found support for their new physics in the chemists' adherence to Avogadro's hypothesis.

In Chapter 12, I provide another illustration of the importance of hypotheses in enabling inductive investigations to proceed. Since antiquity, astronomers have sought to determine the distances to the Sun, Moon, and planets. Simple methods of geometric triangulation — called "parallax" when used astronomically — provided only meager results. The angles to be measured were too small for naked eye astronomy to resolve reliably. That changed when telescopic observations became possible in the seventeenth century. The task remained formidable. Attempts to use parallax for this purpose still called for major scientific expeditions as late as the eighteenth and nineteenth centuries.

These observations and simple geometry alone were not enough. Hypotheses were required to warrant inferences from the observations to the distances sought. Distances thus inferred remained provisional until independent support was provided for the hypotheses. Early hypotheses used in these investigations failed to meet the requirement. Ptolemy derived his estimates of the distances to the Sun, Moon, and planets using the hypothesis that space is filled with the spheres of his geocentric cosmology, packed together as closely as possible. His distance estimates collapsed when his geocentric cosmology failed to find the independent support needed. Reliable distance measurements were subsequently recovered only with the mediation of the Copernican hypothesis, in turn further supported by Newton's theory of universal gravitation. These hypotheses did accrue the requisite independent evidence.

The last two chapters provide examples of theories in competition. They are intended to illustrate the claims of the instability of inductive competitions described in Chapter 4. In Chapter 13, I examine the practice of dowsing. Miners in the Harz mountains of Germany in the sixteenth century believed that minerals underground can be detected by the deflections of a hazel twig. Over the centuries, dowsing migrated to the detection of underground water.

The competition recounted was between dowsers and their skeptical critics and how it turned to favor the skeptics. The case for dowsing was mostly secured anecdotally. It lay in repeated accounts of dowsing successes and even the mere existence of a profitable profession of dowsers. The critics were able eventually to challenge successfully the reliability of these accounts. The nineteenth-century identification of ideo-motor effects explained how dowsers erroneously might have come to believe that the effect was real. On the theoretical side, by the rudimentary standards set by the early theories of electric and magnetic attraction, it was plausible that underground minerals might exert an influence above ground. Over the centuries, the growth of theories of electricity and magnetism left no theoretical space for the mechanism of dowsing. The critics' successes in these two strands of phenomena and theory were mutually supporting and came at the cost of proponents of dowsing. By the early twentieth century, dowsing had been reduced to the status of a pseudoscience.

In Chapter 14, I recount a present-day case of systems of prediction in ongoing competition. I recount four systems, all of which are currently applied to predict the future movement of prices on the stock market. They are fundamental analysis, technical analysis ("chartists"), random walk/efficient market analysis, and fractal/scale-free analysis. The competition among the systems is lively. Proponents of each are aware of the competing systems and try to impugn them. I provide a sample of their disagreements. The guiding principles of each system are hypotheses in the sense of Chapter 2. They are proposed provisionally to enable prediction to proceed. However, none has been secured evidentially such that it has found universal acceptance. That follows from the persistence of the disagreements among the proponents of the individual systems. However, these hypotheses are mutually exclusive: at most, one can be true. The evidence that would single it out is available in abundance in the past history of trading on the stock market. Were this evidence to be pursued and evaluated without prejudice, the disputes would be resolved, and at most one system would prevail. Instead, however, we have the curious spectacle of proponents who refuse this task. The disagreement continues in full display, and we can continue to watch how each approach seeks to gain an evidential advantage over the others.

REFERENCES

Norton, John D. 2003. "A Material Theory of Induction." *Philosophy of Science* 70: 647–70.

———. 2008. "Must Evidence Underdetermine Theory?" In *The Challenge of the Social and the Pressure of Practice: Science and Values Revisited*, edited by Martin Carrier, Don Howard, and Janet Kourany, 17–44. Pittsburgh: University of Pittsburgh Press.

———. 2019. "A Demonstration of the Incompleteness of Calculi of Inductive Inference." *British Journal for the Philosophy of Science* 70: 1119–44.

———. 2021. *The Material Theory of Induction*. BSPS Open Series. Calgary: University of Calgary Press.

Quine, Willard Van Orman. 1951. "Two Dogmas of Empiricism." *Philosophical Review* 60: 20–43.

1

The Material Theory of Induction, Briefly

1. Introduction

In this volume, I describe how relations of inductive support are structured on the large scale. I do so in the context of a particular view of inductive inference, the material theory of induction. I elaborated this account of inductive inference extensively earlier in *The Material Theory of Induction* (2021), to which the reader is referred. In this chapter, there is only a brief introduction to the material theory. It is preliminary. I present the main claims of this volume in the next chapter.

Section 2 gives a motivation for, a summary of, and an argument for the material theory of induction. The standard approach to inductive inference characterizes inductive inferences or relations of inductive support formally, by means of schemas or calculi purported to hold universally. They all fail to apply universally, or so I argue. Facts peculiar to each domain determine which are the good inductive inferences or proper relations of inductive support. There is no way to combine these disparate warranting facts into a single, universally applicable system. This is the central claim of the material theory of induction.

The remainder of the chapter illustrates how standard, formal approaches to inductive inference fail and that a material approach can capture what made the formal approach seem to be viable without succumbing to the difficulties of the formal approaches. Since there are so many approaches to inductive inference, I can discuss only a few of them. They are sampled from a survey of accounts of inductive inference in Norton (2005).

This survey divides accounts into three families. The first, "inductive generalization," is based on the principle that we may infer from an instance to a generalization. It includes enumerative induction, discussed in Section 3, and analogical reasoning, discussed in Section 4. The second family, introduced in Section 5, is "hypothetical induction." It is based on the principle that the capacity of a hypothesis to entail the evidence is a mark of its truth. Section 6 reviews one example in which we are to accept the hypotheses that most simply entail the evidence. The third family has accounts in which a calculus governs strengths of inductive support. The probability calculus is overwhelmingly the most popular candidate. Section 7 uses the example of Laplace's rule of succession to sketch some limits of the account and shows how the material approach can escape them.

2. The Material Theory of Induction

2.1. Inductive Inference

Induction and inductive inference are understood here in their broadest senses. They apply to any inference that leads to a conclusion deductively stronger than the premises from which it proceeds. This conception automatically includes traditional forms of ampliative inference, such as enumerative induction. ("This *A* is *B*. Therefore, all *A*s are *B*.") Ampliation is understood in its broadest sense as referring to any expansion of the conclusion beyond the deductive consequences of the premises. The terms "induction" and "inductive inference" will also be taken to encompass what is often called confirmation theory. It applies to accounts in which one does not proceed in the traditional manner of an inference to infer the truth of some conclusion, detached from the premises from which it was derived. Rather, one merely reports a relation of inductive support of such and such a strength between two propositions. The most familiar application is probabilistic analysis. The measure $P(A|B)$ is the strength of support that proposition *A* accrues from proposition *B*.

The account here is restricted to the logical notion of inference. According to it, the relation of inductive support obtains between *A* and *B*, independently of human desires, beliefs, and thoughts. It is not the "psychologized" notion of inference. There, to report an inference from *A* to *B* is merely to report a fact of our psychology. If we hold *A* true, then we will assert *B* as well. Discussions of people inferring from *A* to *B* will appear in the text that follows, especially in the historical narratives. However, I treat them throughout as attempts by the figures in question to conform their thinking to the appropriate logic of inductive inference.

2.2. An Unmet Challenge

Any account of inductive inference must do two things. First, it must provide a means of distinguishing good inductive inferences from bad ones. Second, it must demonstrate that the inferences that it designates as good really are so.

My contention is that all principal accounts of inductive inference so far have failed to meet these challenges. Their failure derives from a pervasive presupposition: they assume that an account of inductive inference must be based on formal rules that can be applied everywhere. In this, they copy a standard approach in deductive inference. Here is a deductive argument schema:

All *A*s are *B*.

Therefore, some *A*s are *B*.

The schema is universally applicable since we can substitute any noun for *A* and any adjective for *B* and end up with a valid inference. The simplest account of inductive inference mimics this approach. Enumerative induction just inverts the order of the sentences in the schema:

Some *A*s are *B*.

Therefore, all *A*s are *B*.

The account is universal in the sense that this schema can be applied everywhere. It is formal in the sense that the schema specifies only the form of valid inferences. It does not constrain the matter in the sense that any nouns and adjectives can be substituted for *A* and *B*. Probabilistic treatments of inductive support are similarly formal and universal. Sentences derived within the probability calculus play the role of universal schema. Consider, for example, the sentence

$$P(\text{not-}A|E) = 1 - P(A|E)$$

in which *P* is the conditional probability of the propositions indicated. It will remain a theorem in the calculus no matter which propositions are substituted for A and E. These two examples reflect the standard practice in the literature. It is to seek schemas that are universal and formal.

The difficulty is that all of these schemas eventually fail somewhere, and, as I shall argue below, the failure is inevitable. The failure of enumerative induction is widely known. Indeed, the schema almost *never* works. One has to choose substitutions for A and B very carefully if one is to recover any acceptable inductive inference. There are similar problems with the sentence in probability theory, although more analysis is required to show them. The sentence is unproblematic if the "P" represents a physical chance. If the chance of outcome A happening given background E is small, say $P(A|E) = 0.01$, then the chance of outcome A not happening is large:

$$P(\text{not-}A|E) = 1 - P(A|E) = 1 - 0.01 = 0.99$$

But now let "P" measure the inductive strength of support for the proposition A from the evidence E, where E is the totality of all evidence available. This last relation precludes the total evidence E from being neutral in its inductive support of A. That would mean that it supplies no support for either A or its negation not-A. We would want that lack of support to be represented by a small or even zero magnitude for both A and not-A. However, if we set $P(A|E)$ to some number close to zero or to zero itself, then the statement in the probability calculus forces us to set $P(\text{not-}A|E)$ close to one or to one itself.[1]

2.3. The Material Solution in Three Slogans

The material theory of induction addresses these problems at their roots: they derive from the presumption that good inductive inferences or relations of support can be identified by a single set of rules or formal schemas applicable universally. That presumption is denied:

There are no universal rules of inductive inference.

Instead, the core claim is this:

All inductive inferences are warranted by facts.

That is, what distinguishes a good inductive inference is its conformity not with some general schema but with background facts of the pertinent domain.

1 Experts will recognize that this consideration is the starting point of a decades-long debate over the representation of the neutrality of support. My view is that it cannot be done satisfactorily using probabilities (see Norton 2008, 2010).

The idea that an inference can be warranted by a fact is familiar from deductive inference. The factual proposition "If *A*, then *B*" is both a mundane fact and a warrant for a deductive inference from *A* to *B*. The warrant derives fully from the meaning of the hypothetical "if . . . then. . . ." To assert "If *A*, then *B*" is also to assert that we can infer from the truth of the antecedent *A* to that of the consequent *B*. In the case of the material theory of induction, a corresponding background fact might be "generally, *A*." Such a proposition authorizes us to conclude *A*. The import of the "generally" is that the inference is inductive. It conveys that there is a small possibility that the conclusion *A* might fail to be true.

Finally, there are no background warranting facts with universal scope. The warranting facts of each domain, in general, will warrant inductive inferences peculiar to that domain. This is expressed in the third slogan:

All inductive inference is local.

There might be similarities in the inductive inferences from different domains. However, these similarities will prove to be superficial and support no general rule. We must always seek the warrant for an inductive inference within the background facts of its domain.

To continue with the oversimplified example of "generally, *A*," it might seem that this fact is somehow to be applied across all domains. However, the meaning of "generally" will vary from domain to domain, so any similarity is superficial. In a probabilistic domain, we would assert "generally, ten successive coin tosses will not all be heads." The "generally" encodes an objective probability of the possibility of failure such that we expect failure on average at a rate of $1/2^{10} = 1/1{,}024$ in many cases of ten successive coin tosses. In particle physics, we might assert "generally, the laws of particle interactions are time reversible." In chemistry, we might assert "generally, metallic elements are solids at room temperature." In these last two cases, we have no possibility of repetition. The laws of particle interactions of the standard model of particle physics are fixed, as is the set of metallic elements. Setting aside dubious contrivances, the "generally" does not lead to a meaningful notion of an expected rate of failure. Once we have scoured the periodic table for metallic elements, there is no other periodic table with different elements in which we can repeat the search anew.

Left open is the extent of the domains in which each specific sort of inductive inference is warranted. A narrowly specific warranting fact might only warrant a few inductive inferences in some narrow domain. A broader warranting fact might warrant a mathematical calculus, which would be applicable across a large range of cases but still be in some limited domain.

In sum, the two challenges for inductive inferences are met as follows. In any domain, the licit inductive inferences are those warranted by the facts of the domain. That they are properly warranted follows from the truth of those facts and is recovered from the meaning of the terms expressing the warranting facts.

2.4. The Background Facts Decide, Not Our Beliefs about Them

Inductive warrants work in the same way as the formal schema of deductive inference. They pick out which are the licit inductive inferences or relations of inductive support, independently of our beliefs. If we reason deductively in accord with the schema *modus ponens*, then we reason validly, even if we know nothing of deductive logic and its schemas. If we reason in accord with the fallacy of affirming the consequent, then we commit a deductive fallacy, even if we mistakenly believe that affirming the consequent is a licit deductive schema.

Correspondingly, we infer well inductively if our inference is warranted by a fact of the domain, independently of whether we know it. We infer poorly inductively if there is no fact of the domain that warrants the inference, even if we believe erroneously that there is such a fact.

In practice, conceived materially, our inductive inferences are guided by our best judgments of which are the prevailing facts in any domain. They are defeasible. Those judgments might prove to be incorrect, and we might be inferring poorly. If we differ in our judgments and arrive at incompatible inductive inferences, then at most one of us is correct. Which of us inferred well is decided by which truly are the facts of the domain.

2.5. The Case for the Material Theory

There are two components of the material theory to be established: first, that facts provide the warrant for inductive inferences; second, that each domain has its own set of warranting facts ("locality").

First, that facts warrant inductive inferences follows from the inevitable failure of accounts of inductive inference that aspire to apply universally. They must fail because of the defining feature of inductive inferences: they are ampliative. That is, they authorize us to infer more than can be deduced from the premises. Thus, there will always be domains, inhospitable to each schema, in which the schema will fail systematically. Characterized most generally, the factual warrant for each inductive inference amounts to the factual contingency that the inference is conducted within a domain hospitable to it.

Standard connective-based deductive inferences are not prone to this mode of failure. Their warrant lies fully within the premises in the meaning of the connectives and is present whatever the domain of the inference.

Domains inhospitable to each formal account can arise in many ways. Philosophy's fabled deceiving demon is a simple if contrived way to see that inhospitable domains are unavoidable in principle. The demon secretly intervenes to frustrate our inferences. The applicability of each account depends on a factual matter: that we are not in the grip of such a demon. Although deceiving demons are fantasies, something close to them is not. Experimentalists must assume that their lab assistants are not disgruntled employees maliciously selecting and suppressing data such as to deceive them into false conclusions. Or they must assume that they are not in the grip of a mechanical equivalent: a loose connection in their cables that introduces enough noise in the results to obscure a regularity or create a spurious one.[2]

These are contrived examples but with the mitigating virtue that they can be expressed tersely. They display the key point. Any account of inductive inference can succeed only if the conditions in the domain are hospitable. That they are so is a factual matter.

Second, the locality of inductive inference follows from there being no universally applicable warranting fact. An old hope, now long abandoned, was that the regularities of the world might be simple enough that they could be expressed in some sort of universal fact that would then underwrite all inductive inference. This was Mill's principle of the uniformity of nature (Mill 1882, Book III, Chapter III, 223): "The universe, so far as known to us, is so constituted that whatever is true in any one case is true in all cases of a certain description; the only difficulty is . . . to find what description." In the abstract, this principle has momentary appeal. Mill himself had already identified the fatal difficulty. For the principle to be something more than idle posturing, there must be a description of it that picks out when we can advance from one case to all cases. Finding it is an intractable problem. Any description

2 In September 2011, the OPERA collaborative reported faster-than-light neutrinos. As Reich (2012) reported, they were misled in part by a loose cable connection.

precise enough to be applied is rife with counterexamples. A description immune to counterexamples can only be so by adopting vagueness to the point of vacuity.[3]

2.6. An Illustration

An example illustrates this general argument. Consider this deductive inference:

> Winters past have been snowy, AND winters future will be snowy.
>
> Therefore, winters past have been snowy.

The warrant for this deductive inference resides entirely within the premises. It comes from the meaning of the connective "and." It can be used only when the truth of the conjunction derives from the truth of each of the conjuncts individually. Hence, we are warranted to infer to each of them individually. Since the entire burden is carried by the connective "and," we can write a general schema for deductive inference, applicable in any domain:

> *A* and *B*.
>
> Therefore, *A*.

Now consider a related inductive inference:

> Winters past have been snowy.
>
> Therefore, winters past have been snowy, AND winters future will be snowy.

The conclusion amplifies the premise. Thus, there will be domains hospitable to the inference, and there will be inhospitable domains in which it fails. An inhospitable domain is one in which there is considerable climate change, including significant warming. A hospitable domain is one in which climate is unchanging. If ours is one of these hospitable domains, then that fact would warrant the inference.

More generally, this fact licenses a schema for inductive inference restricted to a specific domain:

> In domains with unchanging climates,
>
> If climatic fact *A* has always held in the past,

3 For more of this critique, see Salmon (1953).

Then climatic fact *A* will continue to hold.

We can substitute *A* with facts applicable to domains with unchanging climates to recover a licit inductive inference:

In domains with unchanging climates,

If summers past have always been hot and dry,

Then summers past and future will be hot and dry.

This example also illustrates the inherently inductive character of the inference. We can make the warranting fact explicit and even add it to the premises displayed. However, we have not converted the argument into a deductive argument. Climatic conditions concern long-term regularities. An unchanging climate does not preclude a rare anomaly, such as an unusually warm winter among winters that most commonly are snowy. We risk such an anomaly when we employ an inductive inference warranted by the fact of an unchanging climate.

The following sections illustrate at greater length the failure of the universal applicability of some formal accounts of inductive inference. We shall also see how identifying the warranting material facts in some domain helps us to delimit the domains of applicability of each inductive inference.

3. Enumerative Induction

Enumerative inductions — the familiar inferences from "some . . . to all . . . " — are pervasive in science. Just as pervasive in the philosophical literature is a denunciation of the argument form. Francis Bacon's ([1620] 1952, First Book, Section 105) riposte is just the best known of many from antiquity to later times:

> The induction which proceeds by simple enumeration is puerile, leads to uncertain conclusions, and is exposed to danger from one contradictory instance, deciding generally from too small a number of facts, and those only the most obvious.

This poses a puzzle. How is it that these "some-all" inferences are used pervasively in science yet denounced pervasively by philosophers?

The puzzle is readily solved if the some-all inferences are approached materially. The whole problem derives from the mistaken assumption that all of these some-all inferences are warranted by a single formal schema. There is no formal schema that can serve to warrant them all. Efforts to formulate one that works universally collapse. It is that difficulty to which the philosophical literature responds. Rather, insofar as the some-all inference is warranted at all, that warrant derives from facts peculiar to the domain in which each some-all inference is executed. The unity of form of the many some-all inferences in science is superficial. It is not reflected in a unity of the warrants for the inferences.

3.1. Curie's Enumerative Induction

This material solution to the puzzle is illustrated in an enumerative induction of striking scope in Marie Curie's doctoral dissertation, presented to the Faculté des sciences de Paris in June 1903.[4] There Curie reported on years of work with her husband, Pierre Curie. It included the laborious separation of tiny quantities of radium chloride from several tons of uranium ore residue. Mentioned only briefly were the crystalline properties of radium chloride: "The crystals, which form in very acid solution, are elongated needles, those of barium chloride having exactly the same appearance as those of radium chloride" (Curie, 1904, 26). This remark on the crystallographic properties of radium chloride became standard in the new literature that quickly sprang up around the excitement generated by Curie's discovery of radium.

Since the remark is unlimited in scope, it results from an enumerative induction. Indeed, it is one of rather extraordinary scope. Curie initially had prepared just a few tenths of a gram of radium chloride. Subsequent preparations would not have produced large quantities. Yet a general statement on the crystallographic properties of radium chloride was widely accepted without hesitation. Rutherford surveyed what was known of radioactive substances in 1913 and noted without qualification that "radium salts crystallise in exactly the same form as the corresponding salts of barium" (470).

3.2. Failure of Formal Analysis

What can support an induction of such strength from these very few samples of radium chloride? We can see quickly that the universal schema proposed for enumerative induction above falls far short of what is needed:

4 For further details on this example, see Norton (2021, Chapter 1).

Some *A*s are *B*.

Therefore, all *A*s are *B*.

There are simply too many substitutions possible for *A* and *B* that lead to failed inductions:

Some samples of radium chloride were prepared by Marie Curie.

Some samples of radium chloride are in Paris.

Some samples of radium chloride are at 25°C.

Some samples of radium chloride are less than 0.5g.

Some radioactive substances crystallize like barium chloride.

Some substances in Curie's laboratory crystallize like barium chloride.

None of these leads to a credible inference. One might be tempted to propose restrictions on what can be substituted for *A* and *B*. Might we insist that no nouns or adjectives with essentially spatiotemporal character be substituted? That would block the substitution "substances in Curie's lab" for *A* and "in Paris" for *B*. However, it would also block what otherwise would be credible enumerative inductions.

All known kangaroos are indigenous to Australia.

Therefore, all kangaroos are indigenous to Australia.

And

All known moons and planets in our solar system orbit in the same direction as Earth.

Therefore, all moons and planets in our solar system orbit in the same direction as Earth.

The pattern here is evident. For each restriction that we might contemplate on substitutions for *A* and *B*, it takes only a little imagination to find otherwise credible inferences arbitrarily blocked. We must abandon hope for an embellished version of the schema that can serve universally.

3.3. Material Analysis

This failure should not make us pessimistic about the prospects of inductive inferences such as Curie's. It is a vanity of inductive logicians to imagine that

Curie and Rutherford relied on the pronouncements of logicians in forming their inferences. Rather, Curie and Rutherford knew precisely which crystallographic properties of radium chloride could enter into some-all inferences through a century of research in mineralogy on crystals.

Crystals grow in such a bewildering array of shapes that initially it was hard to see that any regularities could be found. If some crystalline sample of a mineral adopted a particular shape, then it would be extraordinary to find another sample with exactly that shape. The problem is reminiscent of the old saw that no two snowflakes are alike. The problems are similar. Which regularities can be found among snowflakes when they all differ?

Figure 1.1. Snowflakes; from Wilson A. Bentley, "Studies among the Snow Crystals during the Winter of 1901–2, with Additional Data Collected during Previous Winters," *Monthly Weather Review* 30, no. 13 (1902): 607–16, Plate XIII

The answer is widely known and easily seen in Figure 1.1. Snowflakes all reflect the same regular hexagonal shape. More abstractly, they exhibit a discrete rotational symmetry. The shapes map back into themselves if we rotate them by 60°.

Essentially, this is the regularity discovered during the nineteenth-century investigation of crystalline forms but promoted from the two-dimensional forms of snowflakes to the three-dimensional forms of most other crystals. Snowflakes are built around one shape, the regular hexagon. The more general three-dimensional theory, however, calls for six[5] crystallographic systems, each with its own fundamental form and symmetries. The most familiar system is the "cubic" system to which sodium chloride, common table salt, belongs. This membership does not mean that all of the crystals of common table salt are just little cubes. Rather, it means that all of them are derived by geometric operations from the basic cubical form, just as all snowflakes derive from the regular hexagon.

By Curie's time, it was a standard result that each crystalline substance generally belongs to a unique crystallographic system. The complication that underwrites the "generally" is that some crystalline substances manifest dimorphism or polymorphism. They might crystallize under different conditions into two ("di-") or more ("poly-") systems. This generally regular association of crystalline substances with one of the six systems is the material fact that warranted Curie's inference. If Curie could identify the crystallographic system to which one sample of radium chloride belongs, then she could infer the crystallographic system of all samples of radium chloride. I distinguish this warranting fact (Norton 2021, 43) as a principle named after René Juste Haüy, an early-nineteenth-century founder of crystallography: "(Weakened Haüy's Principle) *Generally*, each crystalline substance has a single characteristic crystallographic form." The "generally" that weakens the principle ensures that Curie's inference is inductive. Curie took the inductive risk of assuming no polymorphism for radium chloride.

She did not mention by name the monoclinic system to which radium chloride belongs. Rather, she used an indirect locution: radium chloride crystallizes as does barium chloride. That is, the system to which radium chloride belongs is the same as that to which barium chloride belongs. That they should belong to the same system is plausible since the two salts are very similar in their chemical properties, and such similarities often manifest in crystallographic similarities.

What initially appeared as a simple enumerative induction by Curie can now be seen as something richer. The specific generalization that she made on the crystalline form of radium chloride was informed by and warranted by

5 So was the count in Curie's time as provided by Miers (1902, 38).

facts uncovered in a century of research in mineralogy. That research solved the difficult and delicate problem of just which properties of crystals can be generalized in a some-all inference. The warranting fact of the Weakened Haüy's Principle rested in turn on a considerable amount of science. It exploited the atomic theory of matter in picturing crystals as atoms arranged in regular lattices and the mathematics of group theory in discerning how spatial symmetries led specifically to the different crystallographic families. Curie's inference was grounded not in any abstract logical schema but in a considerable range of scientific facts.

4. Analogy

Reasoning by analogy, like enumerative induction, is a long-recognized form of inductive generalization. It, too, was recounted by Aristotle. It asserts in its simplest form that, when some system with property *P* also has property *Q*, this particular fact can be taken as an instance of the generalization that other systems with a similar property *P* will also have a similar property *Q*. The difficulties that analogical reasoning faces are similar to those faced by enumerative induction. Simple schemas for analogical reasoning are not serviceable. A bare schema is too permissive in part through its simplicity and in part through the vagueness of essential terms such as "similar." The obvious repair is to strengthen the schema by careful elaborations, tuned to canonical examples of analogical inference. The results, however, are schemas of increasing complexity that still turn out to be prone to the same troubles. That this should happen is predicted by the material approach. According to it, the best that we can have are different schemas that succeed only in different, factually delimited domains. There is no way to synthesize them into a single coherent schema that applies universally.[6]

There is a curious difference in how philosophers approach analogy and how scientists do. Philosophers treat analogy as a form of inductive inference, and they seek the general rules governing it. Scientists treat analogies as facts that lead to useful results. For them, an analogy itself is an empirical matter subject to normal scientific investigation. If one thinks formally about inductive inference, then this difference is puzzling. It makes perfect sense,

6 Here I discount the trivializing device of simply taking a huge, and likely infinite, disjunction of all the distinct locally applicable schemas and offering it as a single schema.

however, if one approaches inductive inference materially, for the scientists' factual analogy is the material fact that warrants the analogical inference.

4.1. The Bare Formal Schema

In his logical treatise, Joyce (1936, 260) gives a standard schema for analogical inference in its bare form:

> S_1 is *P*.
>
> S_2 resembles S_1 in being *M*.
>
> [Therefore,] S_2 is *P*.

This schema fits many inferences in science. In the eighteenth century, it was noted that electricity resembled gravity in manifesting as a force between bodies that diminishes with distance. The analogy supported the conclusion that electrical forces like gravitational forces diminish with the inverse square of distance. This conclusion was experimentally affirmed by Coulomb.

As with the simple schema for enumerative induction, this bare analogical schema returns good results only when one makes careful substitutions. With little effort, one finds many examples of failed analogical inference. Heat flows like a conserved fluid from hot to cold; however, contrary to the eighteenth-century supposition of the caloric fluid, it is not conserved and not a fluid substance. Perhaps the most famous analogical failure concerns whales. They resemble fish in swimming in the oceans. However, since they are mammals, they neither breathe with gills nor lay eggs, as do fish.

As with enumerative induction, there is a long-standing tradition of deprecation of analogical inference, complete with sage warnings of the dangers of false analogies. Here is one example:

> Even the most successful analogies in the history of science break down at some point. Analogies are a valuable guide as to what facts we may expect, but are never final evidence as to what we shall discover. A guide whose reliability is certain to give out at some point must obviously be accepted with caution. We can never feel certain of a conclusion which rests only on analogy, and we must always look for more direct proof. Also we must examine all our methods of thought carefully, because thinking by analogy is much more extensive than many of us are inclined to suppose. (Thouless 1953, Chapter 12, 91)

4.2. The Two-Dimensional Model

If one thinks formally about analogical inference, then the remedy is to embellish the bare schema in a way that will exclude the plethora of troublesome counterexamples. The dominant approach in the literature develops a two-dimensional account, so named by me because it lends itself to display in a two-dimensional array. It draws from Keynes' (1921, Chapter 19) notion of "positive analogy" and "negative analogy" and was developed by Hesse (1966). The account uses these notions to support inferences about a target system through its analogical relations with a suitable source system. It can be represented in a general tabular schema, provided by Bartha (2010, 15):

Source	Target	
P	P^*	(positive analogy)
A $\sim B$	$\sim A^*$ B^*	(negative analogy)
Q		
	Q^* (plausibly)	

The goal is to infer some as yet unaffirmed property Q^* of the target that corresponds with some property Q of the source. Whether we can do this is decided by the relative strengths of the positive and negative analogies. The positive analogy lies in properties P and P^* of source and target agreeing. The negative analogy lies in the source exhibiting property A but the target lacking the analogous property A^* and conversely with properties B and B^*.

Properties P and Q of the source stand in some relation, which might be causal, explanatory, or something else. If the strength of the positive analogy outweighs the strength of the negative analogy, then that relation can be carried over to the analogous properties P^* and Q^* of the target. We can then affirm that the target system does indeed carry the property Q^*.

Although the bare schema has been enriched considerably, this tabular schema still falls well short of what is needed in a formal account that mechanically can separate the good from the bad analogical reasoning, the true from the false analogy. Rather, it still relies throughout on users of the schema just knowing intuitively when certain relations obtain. They are not given formal specifications that can be applied unambiguously. In the case of the relations laid out vertically in the table, just what is it to be a causal or explanatory relation between *P* and *Q*? And which other relations are admissible? The horizontal relations between *P* and *P**, between *A* and *A**, and so on, are relations of similarity. In formal terms, when are two properties similar? Finally, and most troublesome, how are we to assess the relative strengths of the positive and negative analogies? That balance decides whether we have a true or false analogy overall. These incompletenesses leave sufficient room for us to continue concocting dubious inferences that nonetheless conform to the explicit conditions of the schema.

Joyce's (1936) bare schema for analogical reasoning contained just one term — "resembles" — in need of external, formal specification. In an effort to resolve the bare schema's problems, the two-dimensional account has introduced many more terms and notions. Each in turn is in need of further formal specification. One might, as did Bartha, take this as a challenge to be resolved by still further elaboration. In this vein, Bartha's (2010, Chapter 4) "articulation model" adds considerably more structure to the two-dimensional model. The pattern already established continues. Each elaboration brings new conceptions with it, and each such conception in turn requires a further formal specification.

There is considerably more detail in both Hesse's (1966) two-dimensional model and Bartha's (2010) articulation model than can be presented here. My best effort to provide a richer account of both is in Norton (2021, Chapter 4). However, the overall trend is evident. Each effort to conform the schema better to good and bad examples requires an elaboration that employs new conceptions and artifices that in turn are in need of formal specification. Each effort to repair an inadequate schema does not solve the problems but multiplies it.

4.3. The Material Approach to Analogy

According to the material approach, this mode of failure is inevitable. Each analogical inference is warranted by particular facts peculiar to the domain of the inference. Resemblances among analogical inferences from different domains will be superficial. Efforts to modify a schema to cover more examples of analogical inferences will degenerate into a growing multiplicity of clauses, each responding to particulars of the new examples added. There will be no end to this growth since there will always be new examples.

A material approach accommodates examples of analogical inference on a case-by-case basis. Whether the inference is good is determined by whether there is a background fact in the domain to warrant it. Such facts have an analogical character since they express similarities among systems. They are aptly named the "facts of analogy." There is no requirement for an essential resemblance among different examples of analogical inferences that could be captured in a generally applicable formal schema. The different examples bear superficial resemblances only.

Failed analogical inferences arise when there is no suitable fact of analogy. Relativity theory showed us that we should abandon the absoluteness of motion in favor of the relativity of inertial motion. By analogy, should we abandon the absoluteness of truth and of moral rectitude in favor of their relativity? The analogical argument fails since there is no fact of analogy connecting motion with truth and moral rectitude. The analogical inference attempted depends on a verbal coincidence in the repeated presence of the word *absolute*.

4.4. The Mountains on the Moon

Galileo's ([1610] 1957) *Siderius nuncius* — the *Starry Messenger* — reports an extraordinary finding among his telescopic investigations of the heavens: there are mountains and seas on the Moon. The mountains become manifest when one tracks how the division between light and dark on the Moon grows in a waxing Moon. As the bright edge advances, bright points of light appear ahead of it, grow, and merge with the advancing edge. This is just how mountains on the Earth are illuminated by a rising Sun. Similar observations and analogies support the presence of depressions or "seas" on the Moon.

Galileo's analysis draws from an analogy between the Moon and the Earth. His inference fits the bare schema of analogical inference:

> The Earth (S_1) has mountains and seas (P).
>
> The Moon (S_1) resembles the Earth (S_2) in that both show the same patterns of surface illumination (M).
>
> Therefore, the Moon (S_2) has mountains and seas (P).

The inadequacy of the schema as a warrant is easy to see. Nothing in the schema prevents us from replacing

$$P = \text{“has mountains and seas”}$$

with

P = "has mountains with alpine ski resorts and water-filled seas with submarines."

It is hard to imagine anyone endorsing the resulting inference to ski resorts and submarines on the Moon. The obvious objection is that the presence of ski resorts on earthly mountains plays no role in the formation of patterns of light and dark on the Earth. The analogical inference succeeds only insofar as it uses the right sort of connection between the "M" and the "P" of the schema. With that remark, we have introduced the fact of analogy that warrants the inference:

The process that produces the patterns of light and dark on the Moon is the same as the process that produces them on the Earth.

The similarity to the process on the Earth is inessential to the power of the fact to warrant the inference. What matters is that

The patterns of light and dark on the Moon are produced as shadows in rectilinearly propagating light by opaque bodies.

That is how the patterns on the Earth are produced. In principle, Galileo could proceed entirely by using this reduced form of the fact of analogy. He could demonstrate by some simple geometric constructions that lunar mountains would illuminate in just the patterns that he observed. The Earth need never be mentioned. However, there was a shortcut. Galileo did not need to develop these constructions afresh for his readers. They were already familiar to earthbound observers who experienced a sunrise. It was a rapid expository convenience to recall that experience.

This development oversimplifies Galileo's analysis in that this last warranting fact in conjunction with his observations enables a deductive inference to the presence of mountains on the Moon. The inductive character of his investigation resides in an uncertainty over whether this warranting fact is true. We restore the inductive character of the analysis by inserting the word *likely* into the fact so that it merely asserts "are likely produced." This reflects Galileo's efforts to show that other possible accounts of the origin of the patterns of light and dark are unlikely.[7]

7 For further discussion, see Norton (2021, Chapter 4, Section 8).

5. Hypothetical Induction

5.1. Saving the Appearances

Enumerative induction and analogical reasoning are forms of inductive generalization: we infer from an instance to the generalization. The weakness of this form of inductive inference is that the generalizations are most naturally expressed in the same vocabulary as the instances. That makes it difficult to infer from evidence to hypotheses formulated with a different vocabulary.[8]

Another form of inductive inference that I have called "hypothetical induction" is free from this limitation. According to it, the fact that some hypothesis with suitable adjuncts entails true evidence is a mark of the truth of the hypothesis itself. This form of inductive inference has long been used in science. In ancient Greek astronomy, "saving of the appearances" meant having hypotheses about the motions of celestial bodies whose observable consequences match and correctly predict what is seen in celestial motions. The Copernican planetary system used the astonishing hypothesis of the motion of the Earth to save the appearances of the motions of the planets. This, according to the Copernicans, indicated its truth. Critics of this conclusion, such as Osiander, writing in a preface to Copernicus' work, urged that it merely shows the pragmatic utility of the hypothesis, not its truth.

As scientific theories became more remote from the evidence that supports them, the need for something stronger than mere inductive generalization grew. It was inescapable by the time of Einstein's general theory of relativity. The planetary motions that provide evidence for the theory are expressed in the vocabulary of observational astronomers. It is remote from the vocabulary used to express the core statements of Einstein's theory: metrical and stress-energy tensors, Christoffel symbols, and Riemann's four index symbols (now the curvature tensor). In November 1915, a jubilant Einstein reported the success of his theory with the long-standing astronomical anomaly in the perihelion motion of Mercury. That anomalous motion could be deduced within his theory. It was, to use his term in 1915, "explained."[9] There was no generalization from an instance. Einstein's new

8 It is difficult, but not impossible, as a survey shows (Norton 2005).

9 The title of Einstein's (1915) paper translates to "Explanation of the Perihelion Motion of Mercury by the General Theory of Relativity."

theory saved the appearances, and that was enough to make it one of the revered evidential coups of the twentieth century.

5.2. Limitations of Hypothetical Induction

The strength of hypothetical induction is that it can lead to the confirmation of hypotheses remote from the evidence. That is also its weakness. It can lead to the confirmation of too much. We can keep adding as many epicycles and other devices as we wish to Ptolemy's geocentric system. Do it cleverly enough and we create a suitably adjusted version that can also save the appearances of planetary motion just as well as Copernicus' heliocentric system. Indeed, so too can a Ptolemaic geocentric cosmology, larded with fanciful crystalline spheres, each powered in its rotation by angels. Does that fanciful hypothesis also earn a mark of truth? If saving the appearances is all that matters, then we must answer yes.

The near-universal response is that merely saving the appearances is too permissive. They must be saved in the right way. Selecting this "right way" becomes almost the full substance of the rescued account. Otherwise, the appearances *A* are saved by *every* proposition of the form *A*&*X*, in which *X* can be anything at all. The "right way" is what selects, among this overwhelming infinity of possibilities, just which is best favored by the evidence of the appearance.

A leading candidate is the requirement that the hypothesis must not merely entail the appearances but must also explain them. This notion is the basis of abduction or "inference to the best explanation."[10] It was, according to this account, what distinguished Einstein's treatment of the anomalous motion of Mercury from mere saving the appearances. His theory explained them. As my survey (Norton 2005) recounts, there are other candidates for this "right way" promoted in different sectors of the literature. I will pursue just one here. It is that the favored hypothesis is the one that saves the appearances in a simple and harmonious way.

---

10 Providing a material explication of inference to best explanation is difficult. There are many notions of explanation, so the approach is not univocal. My best efforts are given in Norton (2021, Chapters 8–9). Successful inferences to the best explanations do not draw from any special inductive power of explanation. Rather, their success comes from deprecating alternatives to the favored hypothesis, either as inconsistent with the evidence or as taking on undischarged evidential debts.

One of Copernicus' arguments for his system was based on considerations of simplicity mixed with esthetics. In his preface to *On the Revolutions of the Heavenly Spheres*, Copernicus censured the Ptolemaic geocentric cosmology as monstrous:

> [The geocentric astronomers'] experience was just like some one taking from various places hands, feet, a head, and other pieces, very well depicted, it may be, but not for the representation of a single person; since these fragments would not belong to one another at all, a monster rather than a man would be put together from them. ([1543] 1992, 4)

A little later Copernicus exulted in the harmony of his heliocentric system:

> In this arrangement, therefore, we discover a marvelous symmetry of the universe, and an established harmonious linkage between the motion of the spheres and their size, such as can be found in no other way. (9)

His foremost proponent and expositor, Galileo, pointed directly to simplicity as the guide to probability in his *Dialogue Concerning the Two Chief World Systems* ([1632] 1967). Having reviewed the virtues of the Copernican system, Salviati reached a triumphant conclusion:

> See also what great simplicity is to be found in this rough sketch, yielding the reasons for so many weighty phenomena in the heavenly bodies. (327)

Sagredo immediately summarized Salviati's logic:

> I see this very well indeed. But just as *you deduce from this simplicity a large probability of truth* in this system, others may on the contrary make the opposite deduction from it. (327; my emphasis).

Needless to say, Salviati proceeded to offer a devastating criticism bordering on cruelty of those who resisted his deductions.

6. Simplicity[11]

6.1. Principles of Parsimony

Invocations of simplicity are so common that we might barely be aware of how frequently they smooth the passage of our inductive inferences. We ask how a variable *T* is related to a variable *t*. We collect measurements and find that the measured *T* values increase linearly with the *t* values, near enough. We infer without apology to a linear relationship between *T* and *t*. The move is rarely challenged. If it is, then who could resist an impatient retort: "It's the simplest. What else could it be?" This instinctive retreat to simplicity falls short of what is needed if we seek explicit principles that separate the licit from the illicit inductive inferences. Merely being told to choose the simplest is empty without some specification of which is the simpler. And it has no inductive force unless some basis is provided for why that choice does lead to licit inferences.

When explicit statements of a governing principle of parsimony are required, perhaps the most commonly invoked is "Ockham's razor." It is usually reported as

> Entia non sunt multiplicanda praeter necessitatem.
> Entities must not be multiplied beyond necessity.[12]

Edifying as William of Ockham's sentiment is, we might worry that it was merely the abstract speculation of a scholar who himself did not use it in any major scientific discovery. We can have no similar hesitation about a formulation by Isaac Newton, surely one of the most accomplished scientists of all eras. In composing his magisterial *Principia*, he declared a principle of parsimony that would then be used in the development of his "System of the World." Book III of this work introduces "Rules for Reasoning in Philosophy." The first is a principle of parsimony:

> Rule I
>
> *We are to admit no more causes of natural things than such as are both true and sufficient to explain their appearances.*

---

11 The analysis in this section is developed in greater detail in Norton (2021, Chapter 6).

12 William of Ockham's original wording differed but conveyed essentially the same sentiment.

> To this purpose the philosophers say that Nature does nothing in vain, and more is in vain when less will serve; for Nature is pleased with simplicity, and affects not the pomp of superfluous causes. (Newton [1726] 1962, 398)

What are we to make of principles such as these? We cannot find much fault in them as pieces of homely advice. We might lighten the work of our inferential quests if we check the easy options first. However, that practicality falls short of what is needed if the principle is to be a guide to the truth. The facts of the world feel no obligation to conform themselves to what is pragmatically convenient for us. To serve as this guide, the principle must express some fundamental fact about the world: the simpler is more likely true since nature is simple. And it must do it in an unambiguous manner so that it can be applied unambiguously.

These principles fail to meet both requirements. First, as a factual matter, often nature is not pleased with simplicity and can employ a multiplicity of entities or causes. For millennia, traditional matter theories favored less to their detriment. The ancient Greeks presumed four elements: earth, air, fire, and water. The later alchemists presumed fewer still: mercury, sulphur, and salt. As long as the element count was this small, there was little possibility of a serviceable chemistry. Matters were rectified only when Antoine Lavoisier proposed thirty-three elements in his *Elements of Chemistry* ([1789] 1965, 175–76, "Table of Simple Substances"). That set us toward the modern count that exceeds ninety elements. Even with this count secured, there are further multiplicities. All instances of each element are alike chemically. Thus, parsimony would tell us that carbon is made of entities all of the same type. However, not all carbon is the same. It manifests in physically distinct but chemically identical isotopes: ^{12}C, ^{13}C, and ^{14}C.

Second, these proclamations are too ambiguous to be serviceable since they provide no definite means of counting causes and entities. Is the gravitational force of the Sun one cause because it is the force exerted by one large object? Or are there many causes, one for each gravitational force exerted by each atom of the Sun? Is the designer god, against whom Darwin railed, one cause of the many adaptations of living things? Or do we count each individual design decision as a separate cause? Do we understand the electric force of attraction between bodies as an action at a distance effect? Or is it an interaction mediated by an electric field? In one way of counting, the action at a distance theory posits fewer entities. It posits electric charges only. The field view posits these charges and adds the mediating field. In another way of counting, the numbers reverse. If we consider the electric force on some particular body, then the field view attributes it to one thing, the surrounding electric field. The action at a distance account, however, presents the force as the sum of all forces exerted by all of the many charges in the universe.

There is a further ambiguity. We should not multiply entities "beyond necessity." We should admit no more causes than "are both true and sufficient to explain the . . . appearances [of natural things]." Although we might have some intuitive notions of the key terms "necessity" and "explain," the principles are not objective rules until these terms are given unambiguous meanings. Until then, one person's necessity might be another's superfluity.

6.2. Simplicity as a Surrogate

We face a familiar problem. Common inductive practice routinely employs appeals to simplicity. Yet we cannot articulate an explicit principle on which this practice can rely. From the perspective of the material theory of induction, this failure is inevitable. It asserts that there can be no such universally applicable principle of inductive inference.

Understood materially, inductively efficacious appeals to simplicity are always indirect appeals to further inductive inferences. Sometimes these further inductive inferences are sufficiently convoluted that a proclamation of simplicity is a convenient way of avoiding a convoluted narrative or of summarizing one just given. We shall see below that this is the real basis of the Copernicans' appeal to simplicity. In the most straightforward cases, appeals to simplicity are merely veiled appeals to specific background facts that provide the warrants for the inductive inferences at issue. We shall also see below that this is the basis of appeals to simplicity in curve fitting.

Simplicity is a surrogate for further inductive inferences. Appeals to simplicity are otherwise so varied in their details that material analysis cannot supply a more specific characterization.

The material approach resolves some of the ambiguity in Ockham's razor. His "necessity" makes sense as a veiled reference to something inductive: we should not infer to more entities than those to which we are authorized inductively by the evidence. Similarly, Newton's rule limits causes to those sufficient to explain the appearances. If we understand explanation in the abductive tradition, the minimal causes sufficient to explain the appearances are just those to which we should infer inductively as the best explanation. In both cases, the principles of parsimony amount to a simple assertion: infer only to what the evidence permits. Do not go beyond it. This assertion is merely a truism of inductive inference. It is a good practice to follow. The truism replaces and can contradict an independent principle of parsimony. The evidence might well require us to adopt something far from simple. Our best model of particle physics, the standard model, has nineteen independent constants.

6.3. Curves, Tides, and Comets

The most straightforward and most familiar appeal to simplicity arises in curve fitting. We plot measured data points for two variables x and y and then seek the curve that fits them best. Routinely, the curves explored are given by polynomial functions y of x:

linear, quadratic, cubic, quartic, . . . ,

where the functions become less simple as we proceed up the list, in the sense that their definitions require more independent parameters.

The familiar difficulty is that we can always secure a better fit to the data by employing functions further up this list. At some point, inevitably, our curve fit is merely accommodating noise in the data. We are overfitting. The familiar solution is that we forgo some accuracy of fit by choosing a function earlier in the list, usually guided by some explicit statistical criterion. This decision is conceived as balancing accuracy against simplicity.

This description of a familiar inductive practice makes no explicit reference to any particular case. It appears to implement some sort of universal inductive rule grounded in simplicity. This appearance is an illusion, however. Without a context, the above prescription gives incoherent results. We can represent the same data by transformed variables such that the results of the analysis of the transformed problem contradict those of the original problem. For example,[13] we can replace x in the data set with another variable, $z = \sin^{-1}(x)$, and then proceed as before. If we found in the first problem that the simple linear function $y = x$ is the curve of best fit, then that function in the second problem is $y = \sin(z)$. It is not to be found anywhere among the

13 I give a quantitative illustration of this example in Norton (2021, Chapter 6).

finite order polynomial functions y of z since it corresponds to an infinite order polynomial:

$$y = \sin(z) = z - (1/3!)z^3 + (1/5!)z^5 - (1/7!)z^7 + \ldots$$

The standard procedures will never find this infinite order polynomial, for inevitably a procedure will halt at some finite polynomial.

The material theory of induction offers a straightforward escape. The decision on which is the right variable — x or z — is determined by the particular facts of the case at hand. Indeed, the entirety of the analysis is governed by these facts, and they make the determination without resorting to an independent principle of parsimony. These facts control even the most basic supposition of whether it makes sense to seek a curve of best fit at all. Take the example of the variables T and t mentioned above. Suppose that T is the air temperature taken at times t that happen to coincide with midday over a week or two in the spring. This T might increase linearly with t. A curve of best fit would interpolate linearly between the successive temperature measurements and give us incorrect results for times t corresponding to the intervening midnights.

Along with the choice of variables, these facts must specify the list of functions to be used in the curve-fitting procedure. The family chosen must be such that we should expect the true curve to lie earlier in the list. These curve-fitting procedures also depend on a statistical model of the errors confounding the data. A common model assumes independent, normally distributed errors. Any such model is applicable only insofar as it reflects the conditions prevailing factually in the case at hand.

Comet hunting, at least as practiced in the nineteenth century, gives a simple example of how the background facts provide the list of functions to be used in curve fitting.[14] Newtonian mechanics tells us that the trajectory of a comet is a conic section: an ellipse, a hyperbola, or the intermediate parabola. Since the background facts tell us that comets tend to have highly eccentric trajectories, it is hard to distinguish whether they are ellipses or hyperbolas. So the first curve fitted is the intermediate parabola. Then, if the fit is poor, the next curve fitted is an ellipse. It is chosen since an ellipse is the trajectory of comets gravitationally bound to our Sun. Such comets return regularly and

14 I develop in greater detail this example and the example of tidal prediction in Norton (2021, Chapter 6).

are more likely to be encountered by us. Should the ellipse not fit, the comet hunter finally reverts to a hyperbola, the trajectory of a comet that will visit us just once.

Although polynomials are familiar in curve fitting, they are inappropriate for systems with periodic behaviors, such as tides at various coastal locations. Since these tides are periodic, one might expect that the appropriate functions of time *t* are just sin(*t*) and cos(*t*) and their harmonics, sin(2*t*), sin(3*t*), . . . , cos(2*t*), cos(3*t*), . . . We know from the theory of Fourier analysis that linear combinations of these harmonics will return even the most complicated of the possible periodic tidal motions. This expectation underestimates how strongly background facts control the choice of functions fitted to tidal data in the actual practice of tide prediction. The functions routinely fitted to tidal data consist of a sum of harmonics, each with an identifiable physical basis in the background facts. The most important harmonic constituent is the "principal lunar semidiurnal M2" that arises from the tidal bulge raised by the Moon. The next most important is the "principal solar semidiurnal S2" that arises from the lesser tidal bulge raised by the Sun. These two harmonic constituents are just the first of many. In the nineteenth century, William Thomson, who initiated this form of analysis, employed twenty-three constituents, each with a physical basis. For tidal predictions in US coastal regions, the United States National Oceanic and Atmospheric Administration expanded this set to a standard set of thirty-seven constituents. Difficult locations might require over a hundred constituents.

6.4. Ptolemy and Copernicus Understood Materially

The Copernican heliocentric system is favored inductively over the Ptolemaic geocentric system. That favoring is not secured, however, by a factual simplicity of the world. Whatever the simple merits of the geometry of Copernican astronomy, they must be balanced against something far from simple. It requires a sixteenth-century natural philosopher to accept that, contrary to all appearances, the Earth spins on its axis and careens through space around the Sun. Making sense of that — dare I say — is no simple matter.

Providing a proper foundation for the invisibility of this compound motion required the creation of a new science of dynamics in over a century of work by Galileo, Newton, and others. Until this dynamical problem was solved, Tycho Brahe's astronomical system was momentarily a credible compromise. In it, the planets orbit the Sun, and the Sun orbits the Earth, carrying the planets with it. This compromise keeps all of the geometric advantages of the Copernican system while avoiding its dynamical drawbacks. Although in some informal sense Brahe traded simplicity and complexity, there was no formal scheme balancing them, and there was no appeal to a fact of simplicity of the world whose import was unambiguous. Other natural philosophers such as Galileo found a different balance. Brahe was merely seeking an account that fit best with his background facts: the appearance of the motions and the appearance of a resting Earth.

Nonetheless, the Copernicans indicated correctly an evidential superiority of the Copernican heliocentric system over the Ptolemaic geocentric system as far as purely astronomical considerations were concerned. If we view the comparison materially, then we find that the individual elements of the Copernican system were better supported evidentially than those of the Ptolemaic system. The background assumption that warrants inferences in the Ptolemaic system is that, qualitatively, the retrograde motion of the planets is explained in each case by an epicycle-deferent construction. The corresponding inferences in the Copernican system are warranted by the assumption that the planets maintain roughly circular obits but that the retrograde motion of the planets arises from an imposition of the motion of the Earth on them.

In the Copernican system, the appearances of planetary motions then fix many of the details. Corresponding details must be set by independent stipulation in the Ptolemaic system. The relative sizes of the planetary orbits are fixed in the Copernican system, but these sizes must be set by independent stipulation in the Ptolemaic system.[15] In the Copernican system, there are only two possibilities for planets: either their mean positions align with the Sun and their retrograde motions carry them to and fro across the Sun, or they exhibit retrograde motions only when in opposition to the Sun. This conforms to the appearances. The Ptolemaic system can make no corresponding assurance. This conformity must be built in by independent supposition for each planet. These and more differences give the Copernican system a strong evidential advantage.

These last remarks are merely a sketch of a lengthy and complicated collection of inferences demonstrating the evidential superiority of the

15 For an extended account, see Chapter 12 of this volume, "The Use of Hypotheses in Determining Distances in Our Planetary System."

Copernican system. Laying it out in detail is challenging, especially if one is engaged in polemics. There the rhetoric calls for a compelling synopsis. How better to convey the Copernican advantage than by pointing to its simplicity and harmony compared with the Ptolemaic system? Yet it is simpler only in requiring fewer independent posits and more harmonious in that the determination of some features necessitates others. There is no manifestation of a deeper principle of parsimony in nature.

7. Bayes

7.1. The Problem

The forms of inductive inference examined so far have been qualitative. If the Copernican system is better supported by the astronomical evidence than the Ptolemaic system because it requires fewer independent assumptions, then just how much better is that support? Merely "much better" might be all that we can say. To many, that will fall short of what is wanted. Can we not measure support quantitatively? And, if we can, then might questions of strength of support be reduced to objective computations?

This is the promise of Bayesian analysis. The founding tenet of objective Bayesianism is that degrees of inductive support are measured by conditional probabilities. A typical analysis begins with some prior probability distribution, which represents the support accrued by some hypothesis prior to inclusion of the evidence at issue. The import of the evidence on the inductive support of the hypothesis is found by conditionalizing on the evidence, usually through Bayes' theorem, to form the posterior probability. There is no need, I hope, to elaborate since, of all schemes in the modern literature, this one is now best known.

The difficulty with the Bayesian system is that it is too precise and irremediably so. There will be cases in which degrees of support can be represented responsibly by probabilities. They arise in narrowly prescribed problems. For example, since we can recover population frequencies for various genes, we can ask what is the probability that this sample of DNA was drawn from some donor randomly selected from the population? However, evidential questions of a more foundational character are rarely given to us in a context rich in probabilities. Insisting on a Bayesian analysis can be satisfying, then, in the sense that we replace vague notions of strength of support with precise, numerical probabilities. However, the impression of progress is an illusion. The prized numerical precision has been introduced by our own assumptions that do not reflect a corresponding precision in the system investigated. We risk mistaking our manufactured precision for that of the world.

The standard view of a Bayesian account is that probabilities are supplied by default and in abundance. The material approach reverses this. According to it, we are not authorized to supply any probabilities by default. Probabilities can be introduced only when the background facts warrant them, and a thorough analysis displays the pertinent warrants. Adopting that new default protects us from the spurious precision that troubles so much of Bayesian analysis. We can introduce precise probabilities only if the precision of the facts of the context allows them. To do otherwise is to risk asserting results that are merely artifacts of applying an inductive logic ill suited to the problem at hand.[16]

7.2. Sunrises and Laplace's Rule of Succession

The problem of spurious precision has been with Bayesian analysis from the outset. It can already be seen in one of the earliest Bayesian analyses. Laplace asked about the probability that the Sun will rise tomorrow morning given the past history of sunrises. This was already an established question. Before him, Hume had urged that our past history of sunrises gives no assurance of future sunrises. Richard Price, the author of an appendix to Bayes' posthumously published paper, used Bayes' inverse method to compute the odds of a future sunrise.[17] Laplace would now give his application of the probability calculus to the problem. His analysis in 1814 (1902, 19) is a celebrated application of his "rule of succession." To put some formulae on his nonsymbolic narrative, the analysis depended on several assumptions. We assign a probability q to the rising of the Sun.

$$P(\text{rising}) = q \tag{1}$$

16 In Norton (2021, Chapter 10, Section 4), I give examples of such spurious results in the form of the inductive disjunctive fallacy ("Why is there something rather than nothing?") and the lamentable doomsday argument.

17 For more on Hume and Price, see Chapter 6 of this volume; also see Zabell (1989) for more on the history of the rule of succession.

Antecedent to all evidence of any sunrises, we allow that q can have any value from 0 to 1. We represent that latitude by assigning a uniform probability density p to the interval. That is,[18]

$$p(q) = 1 \quad \text{for } 0 \leq q \leq 1 \tag{2}$$

Next Laplace assumed that the individual occurrences or otherwise of a sunrise are probabilistically independent events. These assumptions were sufficient to enable him to compute the probability of a sunrise on the $(n+1)$th occasion given a history of s risings on n past occasions:[19]

$$P((n + 1)\text{th rising} \mid s \text{ risings on } n \text{ past occasions}) = (s + 1)/(n + 2) \tag{3}$$

If the Sun rose on all past n occasions, then the rule of succession gives us

$$P((n + 1)\text{th rising} \mid n \text{ risings on } n \text{ past occasions}) = (n + 1)/(n + 2) \tag{4}$$

The more sunrises we see, the better supported evidentially is the next sunrise. Its probability approaches one arbitrarily closely with enough risings. Laplace immediately translated this probability into a wager: "Placing the most ancient epoch of history at five thousand years ago, or at 182623 days, and the Sun having risen constantly in the interval at each revolution of twenty-four hours, it is a bet of 1826214 to one that it will rise again to-morrow" (1902, 19).[20]

7.3. What Is Wrong with Laplace's Analysis?

This precise quantitative result and its operationalization in a bet are momentarily satisfying and perhaps even thrilling, if numerical precision is the goal. Yet another moment of reflection reveals that the precision attained is fabricated and fanciful. There are two problems (which I will address in the next two sections).

- First, the impression of recovery of a result of some generality is illusory.

18 Lest it pass unnoticed, the probability P and probability density p are distinct and should not be conflated.

19 See the appendix to this chapter for a summary of the computation.

20 The computation of the number of days in 5,000 years as 182,623 is an obvious error, too low by a factor of ten, for 5,000 x 365 = 1,825,000 days or 5,000 x 365.2422 = 1,826,211 days, depending on how one counts days in the year. The odds reported by Laplace of 1,826,214 to 1 indicate that his real estimate of the number of days in 5,000 years is 1,826,213. The erroneous number 182,623 results from dropping the tens digit one.

- Second, a probabilistic analysis is the wrong analysis for the problem as actually posed by Laplace.

I have chosen his analysis for scrutiny here since its simplicity enables us to see both problems quickly. We might imagine the development of the Bayesian approach after Laplace has addressed and resolved these problems. To some extent, this has happened. Where these problems persist most notably, however, is in Bayesian analyses in the philosophy of science. There these methods are routinely applied to problems with vague specifications. The goal is to supplant their vagueness with mathematical precision. This laudable goal, however, can be achieved only by imposing assumptions whose precision is unwarranted by the problems posed. As with Laplace's sunrises, the precision of the ensuing analysis is an illusion of our own manufacture.

7.4. The Failure of Generality

Laplace's "rule of succession" is presented with a suggestion of some sort of general applicability. Perhaps it is a general demonstration that probabilistic analysis defeats Hume's skeptical challenge to inductive inference. Although the application to sunrises specifically is far-fetched, perhaps it shows that probabilistic analysis can solve the sort of inductive problems that Hume identified as insoluble. Or perhaps more modestly it is, at least in simple cases, a convenient starting point for how we are to think of projecting a record of successes and failures inductively into the future.

From the perspective of the material theory of induction, it does none of these. It is a theorem in probability theory, untroubling merely as a piece of mathematics. However, as an instance of inductive inference, it is untethered from real problems in the world. Any inductive rule, such as the rule of succession, can be applied to some particular problem only if the background facts of the domain warrant it. Without that tethering, it is just a piece of mathematics.

To which inductive problems can the rule be tethered? That is, which problems are such that their background facts warrant the rule? We find that there are very few, and they are artificial.[21]

21 We might compare this rule with the ideal gas law in the thermodynamics of gases. It is derived from highly idealized assumptions. Unlike the rule of succession, the ideal gas law applies to a wide range of ordinary gases in ordinary circumstances.

It is no surprise that the rule of succession fails for the real problem of sunrise prediction. The pertinent background facts are rich. Sunrises come about from the rotation of the Earth on its axis, and this rotation can be disrupted only by the most cataclysmic of cosmic events. Absent such a cataclysm, successive sunrises are perfectly correlated; after such a cataclysm, successive failures of the Sun to rise are perfectly correlated. Laplace's assumption of the probabilistic independence of each sunrise fails. If we are serious about predicting such a cataclysm from, say, an errant galactic body, then our analysis must ask about the distribution of such bodies in our neighborhood. What results has to be rich enough to provide a factual basis for any probabilities that might be assigned in predictions of cataclysmic collisions with the Earth.

Laplace had no illusion that his analysis was close to one that accommodated what we know factually of sunrises. He continued the report on the bet quoted above by saying that "this number is incomparably greater for him who, recognizing in the totality of phenomena the principal regulator of days and seasons, sees that nothing at the present moment can arrest the course of it" (1902, 19). This does not appear to be a retraction of his analysis and merely might be a statement that it gives an excessively modest lower bound to the probability appropriate to our real epistemic situation.

If not sunrises, then might Laplace's analysis apply to the expectation of live human awakening? Then biological facts as summarized in mortality tables provide the background facts needed to assess the probability of a human awakening tomorrow given some past history of awakenings. A twenty-year-old male has a twenty-year history of successful awakenings. Mortality tables[22] tell us that a male has a probability of 0.998827 of surviving the next year. Taking the approximation that the probability of a successful awakening each morning in the year is the same, the probability of success the next morning is $0.998827^{1/365} = 0.999996784$. The same computation for a 100-year-old female gives us a smaller probability of awakening the next morning as $0.69845^{1/365} = 0.99901722$. These results differ from what an application of the rule of succession supplies. The rule gives an increase in the probability of awakening with age, not the decrease recovered from mortality tables.

These examples can be multiplied. Laplace's analysis is almost never warranted by background facts. Where does it apply? His own text shows us a

22 Provided by the US Social Security Administration at https://www.ssa.gov/oact/STATS/table4c6.html.

way. The problem of sunrises comes at the end of Chapter 3. Virtually all of the other examples in that chapter are of familiar games of chance and associated randomizers: the tossing of coins, the throwing of dice, and the drawing of black or white balls randomly from an urn. Consider this problem, which Laplace does *not* pose but is sufficiently rich in factual assumptions to warrant his rule:

> An urn contains a very large number of coins, which are biased in all possible ways. The biases are uniformly distributed over all possible values: coins with a chance of heads *q* appear in the urn with the same frequency for all *q* in the entire range from 0 to 1. We select a coin at random from the urn.[23] We toss it 1,826,213 times and find heads on every toss. What is the probability that the next toss is a heads?

It requires only a little reflection to see that all of the conditions for Laplace's rule of succession are satisfied. The background facts warrant the application of the rule. It assures us that the odds of a head on the next toss are 1,826,214 to 1.

Laplace's analysis illustrates a common problem with Bayesian analysis. It has a small repertoire of tractable templates. They include sampling problems, such as drawings from urns, and problems in games of chance, based on physical randomizers, such as thrown dice, shuffled cards, and tossed coins. The supposition is that these templates can be applied to problems that bear only superficial resemblance to the original problems of sampling or games of chance. This supposition mostly fails. Inductive problems in the real world — especially the more interesting ones — rarely are structurally like simple problems of sampling or games of chance.

7.5. Probabilities Are Inapplicable

Laplace's mention of his analysis as applying to sunrises can and, indeed, should be taken only as a colorful embellishment intended to make an arid technical problem appear less dry, for the problem is posed *by assumption* in a factually barren landscape. The problem's formulation fails to provide

---

23 I follow Laplace in overlooking the practical and principled difficulties of selecting randomly from an urn with an infinity (here uncountable) of balls or coins. A safer system spins a pointer on a dial to select a number randomly between 0 and 1. We then construct a coin with that number as its bias.

the background facts required to warrant an inductive inference. To describe the problem as inferring from the evidence of 182,623 sunrises is misleading if taken seriously. Calling them "sunrises" triggers the sorts of background knowledge mentioned above that we are supposed to discount. Successive sunrises are strongly correlated, yet Laplace's analysis makes them probabilistically independent. A better description might be a vaguer evidence statement:

> We have 1,826,213 successes. Will the next occasion be a success?

The only answer is that we cannot say. The evidence is given in a vacuity of background facts. It supports no inductive inference. We need background facts on the nature of the occurrences to warrant an inductive inference. When they are supplied, we can determine just which inductive inferences are warranted. Which they are will vary from circumstance to circumstance. Laplace's analysis will almost never apply.

If we persist in applying a Bayesian analysis and recover results of any strength, where none is warranted, then all we can conclude is that these results are artifacts of a misapplied inductive logic. Once we are alerted to the danger, it is easy to see how Bayesian analysis introduces factual presumptions under the guise of benign analytical machinery. The idea that the unspecified occurrence can be represented by a probability distribution at all is an example. It commits us to factual restrictions that go beyond the factual barrenness presumed. To assign a middling value to the probability, $P(\text{rising}) = q = 0.5$, is not to be neutral. It is to say that, loosely speaking, in situations similar to that of the analysis, we should expect an occurrence in roughly half of them.

Then there is the attempt to represent the complete openness in which the value of q applies. Laplace did his best here by assuming a uniform probability distribution (2) over q. This uniform distribution once again goes beyond the factual barrenness presumed, for that distribution makes many strong claims. It says that a value of q in the interval (0, 0.1) is as probable as a value of q in the interval (0.5, 0.6) but only half as probable as a value of q in the interval (0.5, 0.7). The interval (0, 0.99) is highly probable and its complement (0.99, 1.0) highly improbable. These are strong statements. The absence of background facts means that none of them is authorized.

The difficulty of representing evidential neutrality in a probabilistic analysis is well known. Various techniques known as "imprecise probability" can be used to ameliorate the failure of a uniform probability density to represent adequately a complete indifference over the values of the parameter q.[24] In one approach, we replace the single prior probability density (2) over q by the set of all[25] probability densities over the interval [0, 1]. When we apply the rule of succession, instead of recovering a single probability for the next occurrence, we recover a set of probabilities. In general, there is one for each of the probability densities in the set. That we admit all probability densities gives the appearance of the requisite independence from background facts. That appearance is illusory since we are still assuming that the probability calculus applies at all, even in weakened form. The introduction of this imprecision is fatal, however, to the recovery of a nontrivial result. As we see in the appendix to this chapter, the set of all prior densities includes ones that lead to all possible probabilities from 0 to 1 for the next sunrise. We start assuming that this probability can lie anywhere between 0 and 1 and must end without any restriction on this range. We will have learned nothing from the evidence, no matter how extensive our history of sunrises.

7.6. Bayesian Analysis within the Material Theory of Induction

What are the prospects for Bayesian analysis from the perspective of the material theory of induction? Bayesian analyses can be applied profitably to many specific inductive problems. Given what we know about errant galactic bodies, what should our expectations be for a cataclysmic collision with the Earth that will disrupt our sunrises? Given patients with such and such prognoses, what is their life expectancy? These and many more similar problems are all welcomed by the material theory of induction. In each case, there are identifiable background facts that warrant the application of a probabilistic analysis.

24 Might we escape these problems by adopting subjective Bayesianism? Then the prior probability distribution is merely uninformed opinion and can be chosen freely as long as it preserves compatibility with the probability calculus. This popular approach has had a malign effect if one's interest is inductive support and bearing of evidence. Once one allows opinion free admission into one's system, it becomes very difficult to remove its taint from one's judgments of inductive support. The limit theorems supposed to purge the subjectivity apply in limited, contrived circumstances that do not match the real practice of science.

25 The scope of "all" is vague, but that vagueness is immaterial to the points made here. As a first pass, it designates all integrable functions with a unit norm.

Where Bayesian analysis fails is that it cannot provide an all-embracing framework with formal rules applicable to all problems of inductive inference. It works well on specific problems where the background facts warrant it. But any claim of general applicability, such as is sought in the philosophy of science literature, requires that the framework must be applicable to inductive problems whose background facts fail to authorize a probabilistic analysis. In these cases, persisting in applying a probabilistic analysis risks producing spurious results, artifacts of an inapplicable inductive logic.

8. Conclusion

In this chapter's review of the material theory of induction, only particular instances of inductive inference have been discussed. In each case, the warrant for the inference is found in background facts. For the inference to be licit, these background facts must be truths. Since these facts make claims that commonly extend well beyond direct experience, we must ask what supports the truths of these background facts. The material theory of induction is uncompromising in its answer. The only way that these facts can be supported is by further inductive inferences, which in turn will require warrants in still further inductive inferences. How do all of these inferences fit together? That is the subject of this volume, and I take it up in the next chapter.

Appendix: Laplace's Rule of Succession

Consider $n + 1$ probabilistically independent trials, each with a probability of success q, where q itself is uniformly distributed over the interval [0, 1] according to (2). If there are s successes only among the first n trials, then the probability of success on the $(n + 1)$th trial is given by

$P = P$(success on $(n + 1)$th trial | s successes in first n trials)
$=$ P(success on $(n + 1)$th trial AND s successes in first n trials) /
P(s successes in first n trials)

Since the number of successes s is binomially distributed, we have

$$P = \frac{\int_0^1 q \cdot \frac{n!}{s!(n-s)!} q^s (1-q)^{n-s} p(q)\, dq}{\int_0^1 \frac{n!}{s!(n-s)!} q^s (1-q)^{n-s} p(q)\, dq} = \frac{\int_0^1 q^{s+1} (1-q)^{n-s}\, dq}{\int_0^1 q^s (1-q)^{n-s}\, dq}$$

The integrals can be evaluated using the integral identity

$$\int_0^1 q^A (1-q)^B\, dq = \frac{A!B!}{(A+B+1)!} \tag{A1}$$

for whole numbers A and B. We recover

$$P = \frac{(s+1)!(n-s)!}{(n+2)!} \cdot \frac{(n+1)!}{s!(n-s)!} = \frac{s+1}{n+2} \tag{2}$$

It is the rule of succession (2) of the text.

To show that alternatives to the prior probability distribution (1) can lead to $P = r$ for any r between 0 and 1, consider the family of prior probability distributions:[26]

$$p(q) = \frac{(A+B+1)!}{A!B!} q^A (1-q)^B \quad \text{where } 0 \le q \le 1$$

for A and B whole numbers. Repeating the above calculation for P, we find

$$P = \frac{\int_0^1 q^{A+s+1} (1-q)^{B+n-s}\, dq}{\int_0^1 q^{A+s} (1-q)^{B+n-s}\, dq} = \frac{(A+s+1)!(B+n-s)!}{(A+B+n+2)!} \cdot \frac{(A+B+n+1)!}{(A+s)!(B+n-s)!} = \frac{A+s+1}{A+B+n+2}$$

Rewriting P as

$$P = \frac{A}{A+B} \cdot \frac{1+(s+1)/A}{1+(n+2)/(A+B)}$$

it follows that $P \rightarrow r$ in the limit of $A, B \rightarrow \infty$ such that $A/(A+B) \rightarrow r$. That is, we can bring P arbitrarily close to any nominated $0 \le r \le 1$, merely by selecting A and B large enough in this limiting process. The prior probability $p(q)$ masses all the probability arbitrarily closely to $A/(A+B)$ in the process of

26 Identity (A1) assures normalization to unity.

taking the limit. The limit itself is no longer a function but a distribution, the Dirac delta "function." That is

$$\underset{\substack{A,B\to\infty \\ A/(A+B)\to r}}{Lim}\; p(q) = \delta(q-r)$$

Selection of this distribution as a prior would force P to the value of r exactly, since all intervals of values not containing r would be assigned a zero prior probability.

REFERENCES

Bacon, Francis. (1620) 1952. *Advancement of Learning: Novum Organum: New Atlantis.* Vol. 30 of Great Books of the Western World. Chicago: University of Chicago Press.

Bartha, Paul. 2010. *By Parallel Reasoning: The Construction and Evaluation of Analogical Arguments.* Oxford: Oxford University Press.

Copernicus, Nicholas. (1543) 1992. *On the Revolutions of the Heavenly Spheres.* Translated by Edward Rosen. Baltimore: Johns Hopkins University Press.

Curie, Marie. 1904. *Radio-Active Substances.* 2nd ed. London: Chemical News Office.

Einstein, Albert. 1915. "Erklärung der Perihelbewegung des Merkur aus der allgemeinen Relativitätstheorie." *Königlich Preussische Akademie der Wissenschaften* (Berlin): 831–39.

Galilei, Galileo. (1610) 1957. "The Starry Messenger." In *Discoveries and Opinions of Galileo,* translated by Stillman Drake, 27–58. Garden City, NY: Doubleday Anchor.

———. (1632) 1967. *Dialogue Concerning the Two Chief World Systems.* Translated by Stillman Drake. Berkeley: University of California Press.

Hesse, Mary B. 1966. *Models and Analogies in Science.* Notre Dame, IN: University of Notre Dame Press.

Joyce, George Hayward. 1936. *Principles of Logic.* 3rd ed. London: Longmans, Green.

Keynes, John M. 1921. *A Treatise on Probability.* London: Macmillan.

Laplace, Pierre Simon. 1902. *A Philosophical Essay on Probabilities.* 6th ed. Translated by Frederick Wilson Truscott and Frederick Lincoln Emory. New York: Wiley.

Lavoisier, Antoine. (1789) 1965. *Elements of Chemistry, in a New Systematic Order, Containing All the Modern Discoveries.* Translated by Robert Kerr. New York: Dover.

Miers, Henry A. 1902. *Mineralogy: An Introduction to the Scientific Study of Minerals.* London: Macmillan.

Mill, John Stuart.1882. *A System of Logic: Ratiocinative and Inductive*. 8th ed. New York: Harper and Brothers.

Newton, Isaac. (1726) 1962. *Mathematical Principles of Natural Philosophy*. 3rd ed. Translated by Andrew Motte. Revised by Florian Cajori. Berkeley: University of California Press.

Norton, John D. 2005. "A Little Survey of Induction." In *Scientific Evidence: Philosophical Theories and Applications*, edited by Peter Achinstein, 9–34. Baltimore: Johns Hopkins University Press.

———. 2008. "Ignorance and Indifference." *Philosophy of Science* 75: 45–68.

———. 2010. "Cosmic Confusions: Not Supporting versus Supporting Not-." *Philosophy of Science* 77: 501–23.

———. 2021. *The Material Theory of Induction*. BSPS Open Series. Calgary: University of Calgary Press.

Reich, Eugenie S. 2012. "Flaws Found in Faster-than-Light Neutrino Measurement." *Nature*, February 22. https://doi.org/10.1038/nature.2012.1009.

Rutherford, Ernest. 1913. *Radioactive Substances and Their Radiations*. Cambridge, UK: Cambridge University Press.

Salmon, Wesley C. 1953, "The Uniformity of Nature." *Philosophy and Phenomenological Research* 14: 39–48.

Thouless, Robert H. 1953. *Straight and Crooked Thinking*. London: Pan.

Zabell, Sandy. 1989. "The Rule of Succession." *Erkenntnis* 31: 283–321.

PART I
General Claims and Arguments

2

Large-Scale Structure: Four Claims

1. Introduction

In the previous chapter, I recounted how the material theory of induction treats relations of inductive support individually. That is, to what extent does this specific item of evidence support that proposition? If we think of inductive inference formally, this purely local examination might be sufficient. All that we need for a valid inference, according to a formal theory, is that the evidence and the supported proposition fit appropriately into the empty slots of some licit schema. This local appraisal is incomplete, however, when inductive inference is understood materially. In this approach, there is no fixed repertoire of warranted schemas applicable in all domains. In their place, (true) background facts in each domain warrant the inductive inferences supported in that domain. It follows that the affirmation that some inductive inference is licit requires a further affirmation of the truth of the background fact or facts that warrant the inference. These last facts themselves are contingent and, in the fullest account, must also be secured inductively with appropriate evidence.

Thus, when understood materially, the cogency of inductive inferences and relations of inductive support cannot be appraised fully in isolation. They must be appraised within the context of a larger ecology of relations of inductive support. In this book, I investigate how that larger ecology is configured. In this chapter, I lay the foundation of the material analysis of this large-scale structure. It consists of the following four claims, which I will introduce and defend in this chapter.

1. Relations of inductive support have a nonhierarchical structure.

2. Hypotheses, initially without known support, are used to erect nonhierarchical structures.
3. Locally deductive relations of support can be combined to produce an inductive totality.
4. There are self-supporting inductive structures.

In my defense of these four claims, I will employ extended examples drawn from the history of science. Providing a sufficiently detailed account of these examples within the confines of this chapter is impractical. My approach is to give these accounts in later chapters in Part II, with chapters devoted to each of the case studies. I will recall their results in this chapter briefly only insofar as they are needed.

In Section 2 of this chapter, I argue for the first and most important of the foundational claims listed above, the nonhierarchical structure of relations of inductive support. I address a supposition that relations of inductive support in science or in individual sciences are unidirectional, always proceeding from the less general to the more general. Under this supposition, these relations of support are akin to the relations of support among the successive courses of stones in a tower. Each course is supported only by those beneath it. In its place is a conception of greatly tangled relations of support that cross over one another, failing to respect any orderly hierarchy. They are akin to the relations of support in an arch or vaulted ceiling. Each stone is supported by those beneath it and many others above it and elsewhere distributed over the whole structure. That relations of inductive support form such a massively entangled system is the most prominent feature of the large-scale structure of relations of inductive support according to the material theory. Many further features will depend on it.

In Section 3, I ask how these entangled structures can be discovered. A central result of the material theory is that we first need to know something before we can infer inductively. Otherwise, we have no secure warranting facts for inductive inferences. If initially we know nothing in some domain, then how can we ever learn inductively generalities of infinite scope in the domain? An examination of episodes of scientific discovery gives the answer of the second claim: we proceed by hypothesis. That is, we introduce as hypotheses the facts that would be needed to warrant suitable inductive inferences, and then we make the inferences. In proceeding this way, however, we take on the obligation eventually to return to the hypotheses and provide independent support for them. Only then are our inductive inferences properly secured. The arches or vaulted ceilings of the analogy cannot be constructed simply by piling one stone upon another. To build them, we prop up some stones provisionally by scaffolding and complete the construction. Only then can the scaffolding be removed. The result is a structure each of whose stones, examined individually, is properly supported by masonry. This use of hypotheses is distinct from their use in hypothetico-deductive confirmation. There they are introduced in order to be confirmed themselves. Here they are introduced to mediate in the confirmation of other propositions.

In Section 4, I analyze the intriguing possibility asserted in the third claim found repeatedly realized in cases of inductive support in science. In many cases, the component relations among propositions are individually deductive, even though their combined import is inductive. In this section, I will recall some examples that show how combinations of deductive relations among propositions can, overall, have inductive import.

As a prelude to discussion of the fourth claim, in Section 5 I characterize a mature science as inductively rigid. That means that each proposition of the mature science enjoys strong inductive support from the evidence and that the evidence admits no alternatives. Such a system is intolerant of challenges and generally repels them. If they are successful, then they have a destructive, revolutionary effect. A cascade of strong relations of evidential support propagating through the science will have to be undone.

In Section 6, I develop the fourth claim of the possibility of a self-supporting inductive structure. It is a closed structure in which each proposition is well supported inductively by evidence in the structure through warranting propositions also in the structure. A mature science forms such a structure if we expand its compass to include all of the propositions warranting its inductive inferences, and the evidence and warrants for them, and so on to closure. To see the self-supporting inductive structure, pick any proposition in the science. All of the evidence and warranting propositions needed for its inductive support will be in the structure. That is just the condition that it is inductively self-supporting.

In Section 7, I consider the possibility of nonempirical conditions that might be a necessary supplement to a complete account of the large-scale structure of inductive inference. One might look to a priori principles such as a principle of causality or to the remarkable success of mathematics in formulating physical theories. Such added components, it is argued, fail insofar as they have no empirical foundation; if they do have an empirical foundation, then they lie within the material theory.

In Section 8, I provide a brief preview of what is to come.

2. Nonhierarchical Relations of Inductive Support

Relations of inductive support have a nonhierarchical structure.

2.1. The Hierarchical Conception: The Tower

The original and simplest notion of inductive inference is the notion of generalization from instances. It is codified in the schema of enumerative induction and employed in embellished form by time-honored procedures such as Bacon's tables and Mill's methods. It promotes an oversimplified image of science as an accumulation of generalizations of successively broader scope.

Here is how it looks. In biology, we might start with the particular observations of the flora and fauna of Europe and form generalizations from them. We might then expand our inductive base with particular observations of the flora and fauna of the Middle East, Africa, and Asia. Generalizations concerning them are combined with the earlier generalizations concerning European flora and fauna. We then expand our inductive base even further by introducing knowledge of biological species in the Americas and then the Antipodes. New generalizations concerning them are combined with those achieved earlier to yield generalizations of still greater scope.

We can find similar structures in other sciences. In physical astronomy, we note with Newton that all bodies on Earth gravitate and that all celestial bodies gravitate. We combine the two generalizations to arrive at the greater generalization that all matter gravitates. We note that our Moon and the moons visible to us are nearly spherical, so we infer that all moons are nearly spherical. We infer the same for planets and then eventually for suns and stars.

The result is a stratification of the propositions of a science according to their generality. At the bottom are the least general, the particular facts, commonly conceived as facts of experience or possible experience. As we ascend the hierarchy, we pass to generalizations from them, and then generalizations from them, and so on. The generalizations of the higher layers are supported inductively by those of the lower layers. We descend in the hierarchy by making deductive inferences. They take us from generalizations higher in the hierarchy to those lower in it.

This hierarchy is analogous to the structural support relations among stones in a tower, shown in Figure 2.1. The first course of stones sits on firm ground. It supports the next course of stones, which supports the one above it, and so on to the top of the tower. The firm ground is analogous to experience. It supports the simplest propositions of experience, commonly conceived as propositions about particulars. Each course of stones structurally supports those above it, just as generalizations lower in the hierarchy inductively support those higher up in it.

Figure 2.1. A tower

Although a hierarchical structure of this sort sometimes appears in science, overall it is a poor representation of the organization of propositions in science and the inductive relations among them. It fails for at least two reasons. First (to be developed in Section 2.2), contrary to the tacit supposition, relations of inductive support do not respect the hierarchy of generality. Second (to be developed in Section 2.3), the propositions of science are sufficiently varied in content that their strict partitioning and ordering by generality are unsustainable.

2.2. Relations of Inductive Support Do not Respect the Hierarchy

The hierarchical presumption is that relations of inductive support are unidirectional: they proceed from the less general to the more general. A closer examination of the relations of inductive support within a science shows that this unidirectionality is not respected. Relations of support typically cross over one another. Speaking now only loosely of comparisons of greater and lesser generality, propositions at one level of generality can be supported by a combination of propositions of lesser, equal, or greater generality. The relations commonly are so tangled that no simple ordering of their direction by generality among the propositions of a science is possible.

We shall see more examples below of this lack of respect. It is worth pausing here to visit an especially striking example. It is provided in Chapter 7, "The Recession of the Nebulae." In 1929, Edwin Hubble announced the result that would become the observational foundation of modern cosmological models. Nebulae[1] recede from us with velocities linearly proportional to their distances. Superficially, his analysis looks like the simplest of generalizations. Hubble reported as data the velocities of recession of individual nebulae, as inferred from red shifts in their light, and the distances to these nebulae. This is the level of lesser generality in the hierarchy. He then formed a generalization about all nebulae: their velocities of recession vary linearly with their distances. This generalization resides at a higher level of greater generality in the hierarchy.

Hubble's generalization, it seems, proceeded as we might naively expect, unidirectionally up the hierarchy. As the later chapter shows, his actual inferences were far more complicated and quite unconstrained by this hierarchy. Most troublesome of several problems was that Hubble lacked almost half of the requisite independent distance measurements. His data set reported velocities for forty-six nebulae but included independently derived distance estimates for only twenty-four of them. Hubble was determined, however, to include all forty-six nebulae in his analysis and employed inductive stratagems of some ingenuity and complexity to proceed. In one prominent case, he *assumed* the generality of a linear relationship between the velocities and distances and used it to infer the unknown distances. This inference mixed

1 Hubble's "extragalactic nebulae" or just "nebulae" are, of course, now called "galaxies."

elements from the less general and more general levels to infer propositions in the less general level. Hubble could then test that the inference was successful by using the inferred distances to recover the absolute magnitudes of the nebulae concerned. He checked that these inferred absolute magnitudes conformed to other nebulae of independently known absolute magnitudes.

2.3. The Hierarchy of Generalizations Is Unsustainable

The second false presumption in the hierarchical conception is that it is possible everywhere to partition and order the propositions of a science by generality. Although something like this might be possible in simpler contexts, the presumed partitioning and ordering become impossible to maintain as the propositions of science become more abstract and remote from the specific propositions of observation and experiment. No simple sequence of successive generalizations takes us from the chemical reactions observed in a laboratory to the bonding theory of the complex molecules of organic chemistry, or from the observed emission spectra of gases to the quantum mechanics of the electrons of atoms, or from the motions of the planets to the curved spacetime geometry of general relativity. The inductive pathways from simpler observations and experimental results to the completed theories are sufficiently convoluted that there is no evident basis for comparisons of generality among the intermediate propositions.

For example, ordinary Newtonian mechanics in its various parts treats the distribution of stresses in bodies, the motions of terrestrial projectiles, the flows of fluids, the motions of planets, and much more. How do we rank their many propositions according to their generalities? Is the theory of the distribution of the many stress forces in a complicated architectural structure more general than the analysis of the few gravitational forces acting in a simple problem in orbital mechanics? Or is the latter more general since it treats not just forces but also the motions that they produce? In chemistry, the energy states of a single hydrogen atom are treated by quantum mechanics. Prior to its quantum treatment, the chemistry of hydrogen is treated by a simple phenomenological theory telling us that gaseous hydrogen consists of molecules in which two hydrogen atoms bond. Is the phenomenological theory of the hydrogen molecule more general because it treats bonded hydrogen, whereas the quantum theory of individual atoms does not? Or is the quantum treatment of the hydrogen atom more general since it is part of the more advanced quantum treatment of chemical bonding in which the energy levels of the hydrogen atom play a central role? These questions, and many more like them across the sciences, admit no well-founded answers.

2.4. The Arch

There is no overall partitioning and ordering of the propositions of science by generality. Even when such local orderings appear, relations of inductive support do not respect them. Instead, relations of inductive support are distributed over the propositions of science in a massively entangled network. The simplest instances of this entangled network arise in a crossing over of relations of support whenever we have highly correlated properties. Then a proposition concerning one property can provide support for others at what we might loosely judge to be a comparable level of generality, and those others can provide support in reverse for the original proposition. These relations of support are warranted in turn by the more general proposition of the correlation itself.

For example, stars can vary in many properties, including their effective temperatures, masses, sizes, and elements. A class O star in the Harvard spectral classification system is rare, characterized by a very high effective temperature of the order of 30,000K or greater. Many other properties of stars are strongly correlated with this temperature. A class O star will also have a huge mass and a tremendous luminosity.

Exactly because all of these properties are highly correlated and otherwise unusual, finding one of them in some new star is strong evidence for each of the others. For example, finding that a newly observed star has a very high effective temperature greater than 30,000K is strong evidence that the star is massive. The converse also holds: finding that the star is massive is strong evidence that it has a very high effective temperature. This crossing over of evidential support can be continued for other pairings of properties of class O stars.

There is an architectural analogy to this pair of propositions, each of which provides an inductive warrant for the other. It replaces the analogy to the tower. It is an arch, shown in Figure 2.2. Each side of the arch rests on the firm ground of experience. However, none of the stones higher in the arch is merely supported by the stones beneath it. The stones are also supported by those still higher in the arch and ultimately by those of the other side. One side of the arch, if built without the other, would simple fall down. The two sides mutually support one another.

Figure 2.2. An arch

2.5. Arches Illustrated

In later chapters, I provide more examples of this arch-like crossing over of relations of inductive support. In Chapter 8, "Newton on Universal Gravitation," I describe two cases of pairs of propositions that mutually support each other. The first arises in his "Moon test," in which he argues for the identity of the force of gravity and the celestial force that holds the Moon in its orbit around the Earth. The evidence is the observed accelerations of the Moon toward the Earth and falling bodies at the surface of the Earth. Newton computes the acceleration that the celestial force would yield if it acted at the Earth's surface while strengthening according to an inverse square law. He finds that acceleration to match the observed acceleration of bodies falling at the Earth's surface.

Consider the proposition that the celestial force on the Moon strengthens according to an inverse square law with distance. In this first inference, it is used as an inferential warrant in arriving at the identity of the celestial and terrestrial forces. This usage can be reversed. The proposition of the identity of the celestial and terrestrial forces can also be used as a warrant. Then one can infer from the observed motions that the celestial-gravitational force acting on the Moon strengthens with distance according to an inverse square law. That is, the proposition of the identity of celestial and gravitational force and the proposition of the inverse square law mutually support one another.

In a second example in his account, Newton fits elliptical orbits to the observed positions of the planets. The inference from these positions to their specific elliptical orbits is warranted by the proposition that the planets are acted on by an inverse square law of gravity. Excluding perturbations, that law entails that planets move in conic sections: ellipses, hyperbolas, or parabolas. However, a second argument reverses the proposition that warrants the proposition supported. The key warranting fact is that the elliptical orbits are reentrant. Each planetary year a planet follows the same elliptical orbit. This reentrance, Newton shows, can arise only with an inverse square law of gravity. Taking them together, we find that the specific elliptical orbits of the planets support the inverse square law and that the inverse square law supports the specific elliptical orbits of the planets.

Radiocarbon dating of artifacts provides another illustration of this crossing over of relations of support. It is described in Chapter 10, "Mutually Supporting Evidence in Radiocarbon Dating." In the simplest description, there are two sorts of propositions concerning the dating of artifacts. The H propositions date them by the traditional methods of historical analysis and archaeology. The R propositions date them by estimating how long it took for their content of the radioactively unstable isotope of ^{14}C to decay to the measured levels. The R propositions depend on an accurate knowledge of the original content of ^{14}C captured in artifacts at their formation in different epochs. This knowledge is provided by H propositions: the dating of artifacts by traditional methods. Here H propositions provide evidential support for R propositions. However, the reverse can also happen. Are we sure that no error has crept into the historical methods used to arrive at a traditionally established dating? Then radiocarbon dating can reassure us or correct us. Now R propositions are providing evidential support for H propositions.

I provide more examples of mutually supporting pairs of hypotheses in other chapters. In Chapter 11, "The Determination of Atomic Weights," we see how Avogadro's hypothesis and the law of Dulong and Petit supported each other in chemical investigations of the early nineteenth century. The same relation of mutual support later arose among the chemists' version of Avogadro's hypothesis and the physicists' version of the hypothesis within the kinetic theory of gases. In Chapter 9, "Mutually Supporting Evidence in Atomic Spectra," we find the Ritz combination principle providing support for the quantum theory. Then later the quantum theory provides support for a corrected version of the Ritz combination principle.

2.6. The Vaulted Ceiling

The examples above of pairs of mutually supporting propositions are exceptional for their simplicity. It is far more common for these relations of mutual support to be embedded within a much larger network of inductive relations of support in a science. The Newtonian example is not of an *isolated* structure since the various hypotheses in it figure in relations of support for other propositions in science.[2] In general, relations of support cross over one another in many different ways and at many different levels. One then finds that even a small part of science can be part of a prodigious array of relations of support connecting it with neighboring sciences and then beyond them to the farthest reaches of science.

The analogy to a single arch does not capture this richness. An analogy to a dome is a little better. Stones in each part of the dome depend for their support on stones in many other parts. A still better analogy is to a massively complicated vaulted ceiling, as shown in Figure 2.3. It consists of many interconnected domes and arches. The integrity of the entire structure depends on the mutual support of all of its parts.

2 For example, an inverse square law is presumed in the computations associated with Cavendish-type experiments that determine the magnitude of the gravitational constant G. The law is also used to infer that spherical planets act gravitationally, as if their masses were concentrated at their centers; to infer that certain comets move on hyperbolas; and to compute the behavior of terrestrial tides.

Figure 2.3. A vaulted ceiling in the Commons Room, Cathedral of Learning, University of Pittsburgh; image by John D. Norton

This interconnectedness of relations of inductive support provides mature science with its monolithic structure. One cannot reverse one part without destabilizing the remainder of the structure. A vivid example of an effort to reverse one part comes with the persistent creationist efforts to remove evolutionary theory from biology. The problem that creationists face is that evolutionary theory is inductively entangled with the other sciences. In their challenge to that theory, creationists find that they need to impugn the great age of the Earth in favor of a much younger Earth, whose age is determined from biblical scholarship. Hence, they must impugn modern uniformitarian geology. It is based on an old Earth whose major geological features were formed slowly over eons. They must impugn the radiological methods used to date both organic artifacts and rocks, which ultimately will lead to conflicts with radiochemistry. They must also dispute standard cosmology since it also calls for an ancient Earth. This forces them, then, to question observational and theoretical astronomy and the physics on which it depends.

The size of the network of support relations in mature sciences leads to a combinatorial explosion in the number of support relations that directly or indirectly bear on the propositions of the component sciences. This effect gives depth to the inductive security of each part. A fully worked-out example would help us to see this security more clearly. Unfortunately, displaying the complexity of such a network in all of its detail is a task too large for this chapter or this book. However, we can get a good sense of the density and richness of these structures by visiting just small pieces of them in the examples developed in the chapters that follow.

2.7. Vaulted Ceilings Illustrated

In Chapter 11, "The Determination of Atomic Weights," I recount the immense difficulties faced by chemists in the early nineteenth century in determining relative weights of atoms. The problem arose in Dalton's *New System of Chemical Philosophy* of 1808 and 1810. Dalton knew, for example, that 8g of oxygen combines with 1g of hydrogen to make water. To infer from this that the molecular formula of water is H_2O, he needed to know that an oxygen atom is sixteen times as massive as a hydrogen atom. He had no table of atomic weights to consult and no way to determine them, so he just assumed that the ratio was 8-1. The result was that he arrived, famously, at the molecular formula for water of HO. Dalton was trapped in a circularity: to know the correct molecular formulae, he needed to know the relative weights of atoms, but he could learn the relative weights of atoms only from the molecular formulae.

One might imagine that this circularity was easily broken. It was not. The task required the efforts of chemists over roughly half a century. Chapter 11 recounts Cannizzaro's celebrated solution circulated at the Karlsruhe conference of chemists in 1860. Cannizzaro relied on Avogadro's hypothesis, the law of Dulong and Petit, and an extensive set of measurements of the physical properties of a wide range of substances to determine their molecular formulae. The determinations were complicated, and I have done my best to present them in Chapter 11. For my purposes here, the key fact is that the molecular formulae were not just determined but also overdetermined. That means that some subset of them could be used to provide inductive support from some other part and vice versa.

For example, once Cannizzaro had determined that hydrogen and oxygen gases are diatomic, H_2 and O_2, his gas density data enabled him to fix the molecular formula for water as H_2O. Or he could start with this molecular formula for water and find that oxygen and hydrogen are diatomic. This is just a glimpse of a massive tangle of relations of inductive support in Cannizzaro's analysis. For example, that hydrogen gas is diatomic entered into similar overdetermined relations of support concerning compounds of the halogens: chlorine, bromine, and iodine.

In Chapter 9, "Mutually Supporting Evidence in Atomic Spectra," I provide another illustration of this sort of tangle of relations of inductive support. Energetically excited hydrogen gas emits light. It emits only specific frequencies of light whose measurement became an important project for spectroscopists in the late nineteenth century and early twentieth century. Those frequencies divided into well-structured sets of lines, found in different parts of the electromagnetic spectrum: the infrared, the visible, and the ultraviolet. These sets or "series" were named after the spectroscopists who measured them: the Lyman, Balmer, Paschen, Brackett, and Pfund series.

The series were connected by a simple arithmetic relationship first noted by Rydberg but exploited by Ritz in 1908 as his "principle of combination." The key fact was that the lines of all of the series could be generated by taking the arithmetic differences of a set of terms. For Ritz, this fact provided a useful heuristic. He could apply his combination principle to the lines of a known series and predict a new, hitherto unobserved, series. The approach proved to be successful, and immediately Ritz could report a new line conforming with his prediction.

For my purposes, what is important is that the full set of lines in all of these series is overdetermined once one adopts Ritz's principle. That means that one can take the lines of one series and infer from them to the existence of another series. What results is a tangle of relations of inductive support. This structure, fortunately, is much easier to comprehend, as Chapter 9 shows, since it is recoverable by simple arithmetic additions and subtractions.

2.8. The Firm Ground of Experience

In the arch and vaulted ceiling analogies, the ground that supports the masonry corresponds to the empirical basis of the science. This basis does not depend on any simple-minded or strict distinction between observational and theoretical propositions, for I follow the now common view that a clear distinction between them cannot be made. Rather, I mean by it what is commonly taken in a present science as its supporting empirical facts. They can be far removed from direct human observations.

For example, one of the most stable and most important observational facts supporting modern cosmology is that space is filled with a 2.7K background of thermal radiation. This simple-sounding fact was secured over decades only after extraordinary efforts, some of which are recounted in Chapter 9, "Inference to the Best Explanation: Examples," of *The Material Theory of Induction* (Norton 2021). Among the difficulties faced, to establish a thermal character in a radiation field, one must have measurements made at many different frequencies. Only then can the energy distribution characteristic of thermal radiation be established.

A related observational fact of modern cosmology is that galaxies are observed to recede from us with a velocity that increases linearly with distance. Although the observation is now routinely reported without much hesitation in modern treatments, it was subject to a searching critique in the later twentieth century by Halton Arp. He argued that the red shift in light from the galaxies could not be interpreted as resulting from a velocity of recession since objects with very different red shifts appeared to be connected spatially. An extensive debate was needed to refute his hesitations (for details, see Norton 2023).

The analysis of just what might be meant by the empirical facts of a science is a project that goes beyond my concerns here. My view is that Nora Boyd's (2018a, 2018b) analysis provides the best modern treatment. Boyd allows that all such empirical facts are entangled with theory. However, she argues, these facts can still be used to decide among competing theories through a process of winding back to the provenance of the facts. When we seek to use some empirical fact to decide between two theories, we wind back through the various stages of the formation of the fact. If sufficient data have been preserved, then eventually we come to a point at which enough of the theoretical encumbrance has been removed for the fact to provide a neutral basis of comparison for the two theories.

3. The Role of Hypotheses in the Discovery of Inductive Relations of Support

Hypotheses, initially without known support, are used to erect nonhierarchical structures.

3.1. The Discovery Problem

The discussion in the previous section concerns relations of inductive support, independent of human knowledge of them. A further question of great importance is how we can learn these relations. Only then do they assist us in our inductive exploration of the world. If the totality of facts connected by relations of inductive support were delivered to us as a completed whole, then it would be a straightforward matter to check that all of the requisite relations of inductive support obtain. This is a science fiction scenario. It is what would happen were we to stumble onto a copy of the fictional *Encyclopedia Galactica* of some advanced alien civilization. In it, entire sciences hitherto unknown to us would be delivered to us in their totality.

In real life, our explorations proceed more haltingly. The guiding rule of the material theory of induction is that "you must already know something to be able to infer inductively." We cannot know that some inductive inference is licit unless we are assured of the truth of the warranting fact. Yet, if we are in the early stages of investigation in some new field, then commonly we know rather little, and it is likely too little to proceed with assured inductive inferences of any great reach.

This is a problem faced by all new sciences. The strategy used almost universally is to proceed provisionally. We might not know which are the general facts of some domain, but sometimes we can determine which propositions are plausible candidates for the facts that would warrant the inductive inferences sought. To use a familiar term, these plausible propositions are "hypotheses." We can then proceed provisionally under the supposition that our hypothesis is a fact and infer to the propositions that it would warrant were it a fact. The key element is that the supposition is provisional. Conclusions drawn or inductively supported using the hypothesis themselves have only provisional status. They remain so until we find inductive support for the warranting hypothesis. We have incurred an inductive debt in proceeding to the conclusions, and they are properly secured only when that inductive debt is discharged by finding support for the warranting hypothesis.

Hypotheses have a natural analogue in the procedures for building arches, domes, and vaulted ceilings. A masonry arch, dome, or vaulted ceiling cannot be built simply by piling stones one upon another. As soon as a few stones have been placed, the highest ones would be without adequate support and would fall. The standard procedure is to use scaffolding, known technically as "centering." As shown in Figure 2.4, traditionally it consists of a wooden framework. The stones are set on top of the framework. Prior to the completion of an arch, these stones are not properly supported by its other stones. Their support is only provisional since the wooden centering will be removed eventually. Here they are analogous to hypotheses whose support is also only provisional. When all of the stones of the arch have been placed, the centering can be removed. Now the remaining stones of the arch fully support each other. This final stage of construction is analogous to discharging the evidential debt taken by introducing the hypothesis. As the full investigation is completed, further inductive support, anchored eventually in experience, is provided for it.

Figure 2.4. Wooden centering used in the construction of the Waterloo Bridge

3.2. Hypotheses Illustrated

The chapters that follow provide illustrations of this use of hypotheses. In several of them, the use of hypotheses is invited by a specific problem. Scientists find themselves trapped in an evidential circle. Commonly, there are two related quantities to be determined. To find the first, the scientists need to know the second, but initially it seems that they cannot know the second unless they already know the first. They are trapped. A suitably chosen hypothesis is used routinely to break the circle.

Chapter 12, "The Use of Hypotheses in Determining Distances in Our Planetary System," is an extended study of this use of hypotheses. Consider the earliest efforts to determine distances to celestial bodies. The Moon subtends an angle of about half a degree in our visual field. If we knew the diameter of the Moon, then simple geometry would let us compute the distance to the Moon. However, we do not know its diameter precisely because we do not know how far it is from us. Determining its distance and diameter forms the troublesome evidential circle. The Sun also subtends an angle of about half a degree in our visual field. Determining its distance from us is blocked by the same evidential circle. Determining distances to the planets is even harder since naked eye astronomy cannot resolve their disks. They are just points of light in the sky.

Chapter 12 recounts how ancient and later astronomers sought to break out of this evidential circle by ingenious geometrical triangulations or, as it is known in the astronomical context, measuring parallax. These efforts met with limited success. Ancient astronomers were unable to measure the tiny parallactic angles accurately enough. In the seventeenth century, using telescopic aids, a fairly good parallactic measurement of the distance to Mars was achieved. However, even with telescopic aids, *direct* parallactic measurements of the key Earth-Sun distance were not achieved as late as the nineteenth century.

From the outset, to fill the gaps, hypotheses were called into service. They were not used to fix the distances directly, only to provide hypothetical estimates of the ratios of the distances. Then all that was needed was a single distance determination, such as the distance to the Moon or to Mars, and the remaining distances could be computed from the ratios. What makes this case study revealing is that, in addition to a success story, it recounts failures. They arose when independent evidential support could not be secured for the hypotheses, and eventually they were rejected. The chapter recounts three attempts.

The earliest were Pythagorean/Platonic proposals that recovered the ratios from musical harmonies and simple arithmetic relations. A later proposal was incorporated into Ptolemy's geocentric cosmology. Ptolemy proposed a plausible distance ordering for the celestial bodies. He recovered the ratios of their distances from the further hypothesis that their orbits are packed together as closely as the geometry of the compounded circles of his system allowed if intersections of the circles are precluded. Neither Pythagorean nor Ptolemaic proposals were able to secure independent evidence. Their inductive debt was not discharged, and they were abandoned.

They were replaced by Copernicus' , heliocentric hypothesis. Through it, the ratios of the planetary orbital distances were readily recoverable from terrestrial measurements. Unlike the earlier systems, the Copernican hypotheses gained evidential support from both within and without. Most important was its conformity with Newton's mechanics. Newton had used the more fully developed heliocentric astronomy of his time as an essential premise of his argument for universal gravitation. In another example of the crossing over of relations of inductive support, the direction of inductive support was reversed. Newton's mechanics soon became strong evidence for the details of Copernican astronomy.[3]

The dependence of solar system distance measurements on the heliocentric theory persisted. The most accurate estimates of the key Earth-Sun distance in the eighteenth and nineteenth centuries came from careful measurements of the transits of Venus across the face of the Sun. The Earth-Sun distance could then be recovered from them by geometric triangulations. These calculations still relied on the heliocentric theory's determination of the ratios of the orbits of the Earth and Venus.

Further illustrations of the use of hypotheses to break evidential impasses have already appeared earlier in this chapter. We saw how Dalton was trapped in an evidential circle concerning atomic weights and molecular formulae. He sought to break the circularity by hypothesizing that the correct molecular formulae used the simplest ratios available. The hypothesis failed to secure independent support and was abandoned. The circularity was broken later through two hypotheses: Avogadro's hypothesis and the law of Dulong and Petit. The evidential debt incurred in supposing them was discharged eventually through the mutual support of these two hypotheses and the support provided for them from the emergence of the statistical mechanical treatment of gases in physics.

We also saw that Hubble was stymied in his efforts to use the data from all forty-six nebulae for which he had measurements by a lack of independent distance measurements for twenty-two of them. Chapter 7, "The Recession of

3 The inversion in this relationship is seen most clearly in the ability of the Newtonian system to provide corrections to the heliocentric astronomy of Newton's time. The planets orbit not in ellipses but in precessing ellipses. What came to be known as Kepler's third harmonic law was corrected to accommodate the finite mass of the Sun. The importance of successive approximations in Newton's and later work has been explored by Smith (2002, 2014).

the Nebulae," recounts how Hubble was still able to incorporate these twenty-two nebulae into his analysis by means of hypotheses that gave him indirect indications of their distances. At various stages of his analysis, he hypothesized that the linear relationship among the other twenty-four nebulae also held for these twenty-two, that the absolute magnitude of the brightest star in each nebula is the same, and that the absolute magnitudes of nebulae in a cluster are confined to a small range common to all nebulae.

In the early-twentieth-century analysis of atomic spectra, we saw how the discovery of new series was advanced by the Ritz combination principle. It was introduced as a hypothesis. It gained the requisite independent evidential support with the emergence of modern quantum theory, in which it was recovered as a consequence of Bohr's atomic theory.

These last illustrations have been mostly of successes secured at least eventually. This happy outcome is not assured. A prominent example of a failure is provided by the steady state cosmology of the mid-twentieth century. It was based on the hypothesis of the "perfect cosmological principle," first advanced by Bondi and Gold (1948). According to it, the universe is homogeneous on the large scale, not just spatially but also over time. The way in which we see the universe now, on the large scale, is the way in which it has always been and will always be. A definite cosmology now follows. Its most striking feature is the continuous creation of matter. Unless matter is continually created throughout space, expansion of the galaxies would lead to a dilution of its average matter density and violate the perfect cosmological principle. The steady state cosmologists took on a massive evidential debt in hypothesizing the perfect cosmological principle. They were never able to establish independent evidence for the hypothesis, and they were never able to repay the debt. Most notable was the failure of the steady state theorists to accommodate Penzias and Wilson's discovery in 1965 of the cosmic background radiation, and the competing "big bang" or "primeval fireball" hypothesis eventually proved to accommodate it handily.[4]

3.3. This is *not* Hypothetico-Deductive Confirmation

This use of hypotheses might appear to be similar to the hypothetico-deductive approach to confirmation. These are accounts of confirmation based on the principle that a hypothesis is inductively supported when it

4 For a brief account of this last competition, see Norton (2021, Chapter 9).

successfully entails true evidence deductively.[5] The essential difference lies in the goal of introducing the hypotheses in an evidential analysis. In hypothetico-deductive confirmation, hypotheses are introduced so that the evidence can confirm them according to the hypothetico-deductive principle. In the applications within the material theory, hypotheses are introduced to mediate in the confirmation of *other* propositions. The confirmation of the hypothesis is a task reserved for later investigations. The hypothesis is expected to be confirmed not hypothetico-deductively but by other inductive inferences with their own material warranting facts.

4. Deductive Inferences in Inductive Structures

Locally deductive relations of support can be combined to produce an inductive totality.

4.1. Inferences that Are or Are Nearly Deductive

There is a striking feature of many of the inferences in this text and in my earlier text, *The Material Theory of Induction* (Norton 2021). Although the inferences contribute to relations of inductive support, many of them are close to being deductive inferences or might actually be deductive inferences. That is, when combined with the warranting fact, the inference *from* the evidence *to* the conclusion to be supported is often deductive. The direction of the inference here is important. It is not merely the deductive inferences of hypothetico-deductive support. In the latter, the deduction passes from the hypothesis or theory to the evidence. That direction has now been reversed.

Here are some examples. Chapter 1 of *The Material Theory of Induction* (Norton 2021) recalled Curie's inference from the crystallographic properties of the few samples of radium chloride at her disposal. Curie inferred to the generality of these crystallographic properties. I identified the warrant for her inference as

> (Weakened Haüy's Principle) *Generally*, each crystalline substance has a single characteristic crystallographic form.

5 For an elaboration of this principle and the extensive problems associated with it, see Norton (2005).

When this weakened principle is used to warrant Curie's inference, it is the qualification "generally" that makes the inference inductive. It accommodates the possibility of polymorphism, that one crystalline substance might manifest in more than one crystallographic form. The inductive risk taken by Curie is small, especially if we assume that her generalization was tacitly limited to crystals of radium chloride prepared under conditions comparable to those in her laboratory.[6] If we drop this qualification and revert to Haüy's original conception, the warranting fact would be

> (Haüy's Principle) Each crystalline substance has a single characteristic crystallographic form.

Under this warrant, Curie's inference would be a deduction.

Chapter 2 of *The Material Theory of Induction* (Norton 2021) recounted Galileo's inference concerning his law of falling bodies. Galileo had found that, in equal time intervals, a body in free fall successively covers distances in the ratios of 1-3 to 5-7. He generalized this sequence of ratios to the sequence of odd numbers. In this inference, I argued that Galileo had used the warranting fact that the ratios of 1-3 to 5-7 were present no matter the time interval used in the measurement. It then followed, deductively, that the only possible general law was of the sequence of odd numbers. Indeed, the deductive inference needs as a premise only the ratio of 1-3 and its invariance under a change of the unit of time.

There are, it turns out, other well-recognized, historically important examples in which the inference from evidence to our theories is deductive. These cases have been codified as "demonstrative inductions." Their inferences are demonstrative in the sense that they are deductions. However, they are called "inductions" to reflect an older usage of the term as referring to inferences from particularities to generalities. My contribution to this literature in Norton (1993) was to trace how quantum discontinuity was established in the early decades of the twentieth century. The essential datum was Planck's formula in 1900 for the distribution of energy over the different frequencies of black body radiation. In the early analysis, it was shown that assuming discontinuities in energies enabled one to deduce the Planck formula. Poincaré and Ehrenfest soon showed that the direction of deduction could be reversed.

6 I thank Pat Corvini for emphasizing this point to me.

With suitable background facts, it was possible to deduce quantum discontinuity from the evidence of the Planck formula.

4.2. Support that Is Locally Deductive but Globally Inductive

In deductive inferences, the conclusions are at best logically equivalent deductively to the premises or logically weaker than them. So it appears that deductive or near-deductive inferences to our conclusions cannot give what we seek from inductive investigations. We seek an expansion of our knowledge. These deductive inferences are merely rearranging and returning to us all or part of what we have already supposed.

This pessimistic expectation is not realized, however, once we recall that relations of support within inductive structures are not hierarchical but massively entangled. That enables the entangled relations of deductive support to combine to provide inductive support in the overall structure. This circumstance arises when we have sets of propositions that mutually support each other deductively. Nonetheless, to accept the totality is to accept propositions logically stronger than the evidence.

Striking examples of this combination of deductions arise in Newton's arguments for universal gravitation and his inverse square law of gravity. I have already sketched them above and provide a more detailed exposition in Chapter 8, "Newton on Universal Gravitation." To recall, the first example arises in his "Moon test." In it, he showed that terrestrial gravity is the same force as the celestial force holding the Moon in its orbit around the Earth. To show it, Newton reckoned that, if the force acting on the Moon strengthens with the inverse square of distance as the Earth is approached, then it would accelerate terrestrial bodies with just the accelerations actually found at the Earth's surface. The logic of the Moon test involves two hypotheses:

> $H_{\text{inv. square}}$: The celestial force acting on the Moon is strengthened by an inverse square law with distance at the Earth's surface.
>
> H_{identity}: Terrestrial gravitation and the lunar celestial force are the same.

In the context of Newton's Moon test, drawing from the evidence of the accelerations of the Moon and terrestrial bodies in free fall toward the Earth, each of these hypotheses can be deduced from the other. That is, each hypothesis provides a warrant for a deductive inference from the evidence to the other hypothesis. The two hypotheses combined are the result of the Moon test analysis. Their conjunction is inductively supported by the evidence of lunar and terrestrial accelerations.

The second example has a similar structure. The most basic results of Newton's celestial mechanics reside in two hypotheses:

$H_{ellipses}$: The planets move in their specific elliptical orbits.

$H_{inv.\ square}$: The planets are attracted to the Sun by a force that varies with the inverse square of distance.

Against the background of the observed positions of the planets and the laws of Newton's mechanics, each hypothesis could be deduced from the other. Indeed, Newton employed a subtle variant of the usual way of inferring between these two hypotheses. In the case of the near-circular orbits of the planets, he needed only the datum that the planetary orbits are reentrant. That is, in a planetary year, each planet returns to its starting point. He could then show that this reentrance was a sensitive test for deviations from the inverse square law. The observed exactness of the reentrance entailed the exactness of the inverse square law. Once again the overall inductive import of the analysis was that the evidence of the observed positions of the planets supported inductively the conjunction of the two hypotheses.

Chapter 9, "Mutually Supporting Evidence in Atomic Spectra," provides another example with a similar structure. I noted above that the Ritz combination principle enables inferences of support among the different series of the hydrogen spectrum. As detailed in the chapter, these inferences are deductive. Using the Ritz combination principle as a premise, from the Balmer series, we can deduce the Paschen, Bracket, and Pfund series. These deductions can be reversed as well. Adding the premise of only a single line from the Balmer series, we can deduce the entire, infinite Balmer series from the Paschen series. There are infinitely many series in the hydrogen spectrum, although only finitely many have been observed. The series are closely connected by further deductive relations such that we can infer deductively from any series to any other series by means of the Ritz combination principle and, if needed, the additional premise of a finite set of suitably selected lines. Although these interrelations are deductive, the final import is inductive. The Ritz combination principle and the finitely many spectral lines observed provide inductive support for the entire system of infinitely many series, each with infinitely many lines.

There might be, for some, an air of paradox in the idea that we can combine deductive relations to yield a structure with inductive import. That impression is mistaken. These cases are actually more secure inductively than many of those considered in earlier sections. In those earlier cases, inductive relations of support are combined to produce structures with overall inductive import. Inductive risk is introduced both in the component relations of inductive support and in the combined structure. If those component relations of support are deductive, then this first source of inductive risk is eliminated.

5. The Maturity of a Science

5.1. Inductive Rigidity

A preparation for the discussion of the fourth and final claim is the characterization of what constitutes mature sciences. They are characterized by inductive rigidity. That is, each proposition of the science is well supported evidentially so that a change in the proposition is not allowed by the evidence for the science. There is no assurance that a science can achieve maturity. In the early stages of the development of a science, important propositions are entertained hypothetically. They are not fixed rigidly. As the development continues, further relations of inductive support are found, the hypotheses gain evidential support, and their provisional status is discharged. If this process is completed, then the science achieves maturity such that each of its propositions is well supported.

Once this maturity is achieved, the inductive rigidity of a mature science is widely recognized among its practitioners. Challenges to the science are treated as tiresome, moribund exercises. A skeptic might doubt some proposition in a mature science. In response, someone competent in the science would be able to display the evidence that supports the proposition. In the case of special relativity, this is a dialogue with which I have some personal experience. The theory has been challenged routinely by critics since its inception over a century ago. Many of its foundational propositions have been disputed, at one time or another, unsuccessfully. The light postulate of the theory asserts that all inertially moving observers find the same speed *c* for light in vacuo. It is initially a puzzling postulate. Imagine an inertially moving observer chasing at high speed after a light signal that moves at *c*. That observer will not find the light signal slowed from *c*, even in the slightest.

This perplexing result makes the postulate a favored target. However, that postulate has direct support from de Sitter's analysis in 1913 of light emitted from distant double stars. Its deeper support derives from the Lorentz covariance recoverable from Maxwell's electrodynamics, in turn supported by a plethora of individual experiments in electricity and magnetism.[7]

This maturity is a goal that proponents of a theory strive to achieve, and standard textbook sciences commonly come close to achieving it. It is not uncommon, however, for the full achievement of the goal to be incomplete in parts of the theory. There propositions might achieve general acceptance while lacking proper support. The falsification of such a proposition is usually associated with great excitement and even a momentary sense of crisis. However, precisely because the falsified propositions never were strongly supported, their failure can be absorbed into theory.

On September 19, 1957, Francis Crick announced what came to be called the "central dogma" of molecular biology. It speaks, in various forms, of a unidirectional pathway of synthesis within cells from DNA to RNA to proteins. The reverse pathway is prohibited. Although the dogma was widely adopted, there was little real evidence for it. It was a simple and comfortable idea that fit with a denial of the Lamarckian inheritance of acquired characteristics.[8] When it was discovered that certain viruses could implement the reversed pathway from RNA to DNA, the result was readily incorporated into molecular biology. *Nature* published an excited editorial, "Central Dogma Reversed," in 1970.

In the twentieth century, many new particles were discovered. It was assumed routinely that the laws governing them would respect parity. That is, they would not distinguish left from right. In retrospect, there was no good evidence for this assumption other than that it had become routine in the physical laws discovered earlier. Then, in 1964, Cronin and Fitch discovered experimentally that the weak interaction in particle physics can violate charge-parity conservation. In another example, the hard-to-detect neutrinos had long been attributed to a zero rest mass. This had seemed to be a reasonable assumption. The early determinations of the neutrino rest mass pointed to a quantity in the neighborhood of zero. However, as neutrino physics developed, it became clear that a tiny mass had to be attributed to neutrinos.

7 For historical details, see Norton (2014).

8 Here I rely on Cobb (2017).

That would enable the process of neutrino oscillation in which neutrinos migrate over the three different flavors in which neutrinos manifest. This oscillation explained experimental and observational anomalies, most notably a dearth of measured electron neutrinos emitted by the Sun.[9]

In these last cases, anomalous evidence could be absorbed into the existing theories since the propositions that they contradicted lacked the strength of evidential support of other parts of the theory. Had these other better-supported parts been contradicted, the outcome would have been more troublesome. A well-supported proposition is tightly bound with so much more of the theory. Should it fail, it would bring down much more of the theory with it.

Although particle physics could absorb nonzero neutrino masses, matters would have been quite different had the OPERA Collaboration (2011) measurement been correct. Its measurements, it announced, appeared to show that neutrinos were propagating faster than light. If correct, then this would have destabilized particle physics. It would have contradicted a fundamental posit of the governing quantum field theory, the locality of quantum field operators. Particle physics was saved for the time being.

The inductive rigidity of a mature science does not make the science incorrigible. It is simply a statement of the best that can be gleaned from the evidence. No matter how strong the inductive support of a science, some inductive risk is associated with it. When incontrovertible evidence does emerge that contradicts a well-supported proposition within a mature theory, the result can be and usually is a breakdown of the theory. Rigid steel cables have some elasticity, but they will snap if overextended. What ensues is a revolution in science, a popular topic of investigation in the history of science.

These revolutions commonly occur when the science is extended beyond domains in which it was first developed and in which its evidential base is found. Newton's seventeenth-century mechanics was developed on an evidential base of slow-moving objects, such as falling stones and orbiting planets. Special relativity emerged when developments in nineteenth-century electrodynamics gave reliable results concerning much faster propagations at the speed of light. Special relativity, in turn, fails when we move to domains of intense gravitation, as Einstein found through his general theory of relativity.

---

9 For a review, see Gonzalez-Garcia (2003).

All of these superseded theories, however, remain evidentially well supported as long as we consider only the evidence of the domains for which they were devised. Although general relativity and relativistic cosmology now tell us that Euclidean geometry can fail when applied to spaces of cosmic extent, Pythagoras' ancient theorem remains as reliable as it ever was for the builders of houses, castles, and skyscrapers.

5.2. A Distributed Vindication

Although the inductive rigidity of a mature science is a commonplace for its practitioners, its demonstration would be a massive task. The network of interrelated propositions is enormous for any real science. A full display of the evidence and inductive relations supporting each goes well beyond what is possible in a book chapter. Indeed, for a well-developed science of great scope, displaying this rigidity in detail likely would be beyond the capacities of any single author. Rather, the requisite knowledge, though likely not fully known to any one scientist, is distributed over the full community.

This distribution is illustrated by our proper confidence in the laws of conservation of energy and momentum and our expectation that no proposal for a perpetual motion machine can succeed. Given the variety of types of proposals advanced over the centuries, a full inventory of the evidence against them would be prohibitively long. In each case, it is not enough merely to assert generically that the conservation of energy and momentum prohibits the operation of the machine. A full analysis requires us to display where the details of the mechanism proposed conflicts with other propositions in established science.[10] Different proposals will call on different expertise in the different sciences in which the proposals are formulated. We can be confident, however, that for each new proposal there is an expert in the community familiar with the pertinent science and able to respond.

A recent illustration is the "EmDrive" proposal for spaceship propulsion brought to the attention of a larger scientific community by a *New Scientist* article (Mullins 2006). It consists of microwaves in a chamber such that, it is proposed, the forces exerted by the microwaves in many directions on the chamber walls do not entirely cancel out. They leave a small net force that can propel the chamber. In this, it is unlike any other scheme of propulsion known. All known schemes produce propulsion by driving some form of

10 For a history of these proposals, see Ord-Hume (1977).

matter in the opposite direction to the thrust sought. A rocket expels hot gases. An airplane projects a current of air or hot gases behind it. A ship's propeller projects a stream of water behind it. The forward force on the rocket, airplane, or ship is balanced by an equal and opposite (reaction) force on the driven matter, as required by Newton's third law of motion. This driven matter carries rearward momentum. The conservation of momentum then assures us that the rocket, airplane, or ship gains forward momentum in the opposite direction. That is what accelerates it.

The EmDrive violates the conservation of momentum. It is a closed device supposed to set itself into motion without any ejected matter or a reactive force. Although the proposal is prima facie extremely implausible, interest in it has been remarkably stable and is matched only by the tenacity of skeptical critics. Part of the positive interest lies in wishful thinking. If it works, then it is a device that could power starships! Another reason for its endurance lies in the small magnitude of the force predicted. Detecting it requires the most delicate experiments. As critics have pointed out, such experiments can easily produce spurious results if all of the confounding effects[11] are not properly controlled.

The resulting literature is too extensive to survey here. Recounting one exchange, however, is sufficient to illustrate how the distribution of expertise works. Harold White and his collaborators at the NASA Johnson Space Center are proponents of these microwave propulsion systems. In a technical paper, White and March (2012) proposed that the reactionless thrust might arise through the Casimir force of the quantum vacuum. This is specialized physics. As they acknowledge in their introductory paragraph, classical electrodynamics precludes a reactionless force. Indeed, that classical electrodynamics conserves momentum is a result readily accessible to anyone with a serious, college-level course in electrodynamics. The Casimir effect, however, is more arcane. It is a force produced by quantum fields in a vacuum. Its basic mechanism is not so obscure. However, it is more demanding to develop a theoretical analysis of it that would securely preclude the reactionless force proposed by White and March. Such an analysis is within the expertise of Trevor Lafleur (2014), a physicist specializing in plasma physics. His analysis finds no basis for the reactionless force in the quantum vacuum.

11 Such confounders can be subtle. For example, Tajmar et al. (2018) report such a confounder in the coupling between electrical cables in the experimental setup and the Earth's magnetic field.

6. Inductively Self-Supporting Structures

There are self-supporting inductive structures.

6.1. Inductive Closure: That Is All There Is

A self-supporting inductive structure is a set of propositions such that each one in the set is well supported evidentially; the evidence supporting it is in the set of propositions; and the propositions that warrant the relations of inductive support are propositions within the set. This set is inductively closed.

We have already seen such self-supporting inductive systems in the small. If we take the background propositions from among which they proceed as fixed, then they are found in the examples above of pairs of mutually supporting hypotheses and of networks of inductive support such that the relations of support cross over one another in a bewildering tangle. The more difficult and interesting problem is whether such systems arise on the large scale and whether they are embodied by our mature sciences. I will argue in the subsection below that, if a mature science is properly characterized by the rigidity described in the previous section, then the material theory entails that it is a self-supporting inductive structure.

Before proceeding, it will be helpful to address directly the sense that such structures are paradoxical. They might sound akin to lifting oneself into the air by pulling on one's own bootstraps. However, there is no paradox. If one can affirm that each proposition in the set is well supported individually in virtue of other propositions in the set, then there is nothing more that can be asked. The analogy to pulling oneself up by one's own bootstraps fails.[12] A better architectural analogy is to some elaborate sculpture whose total stability appears to be impossible, yet it still stands. A simple example is the tensegrity icosahedron of Figure 2.5.

12 In the imagined scenario, we hover in midair by pulling on our bootstraps. The pulling force is supposed to counter the force of gravity. This analysis neglects another force. The upward force from the bootstraps in tension is balanced by the downward force from the corresponding compression in our legs. The force of gravity remains unbalanced, and the eager bootstrap puller falls to the Earth.

Figure 2.5. Plan and elevation of a tensegrity icosahedron; model and photographs by John D. Norton

In a superficial description, it seems to be impossible that such a tensegrity structure can stand. There are six rods connected only by cords in tension. One end of each of three rods rests on the table surface. All of the remaining rods and their parts are held suspended above the table surface. No rod directly touches any other rod. Their *sole* connections are through cords in tension. Such a structure, it seems, would collapse into a pile of rods and cords. Must not a rod, supported only by cords in tension, anchor those cords on another rod still higher in the structure? And must not that rod be held by cords tied to another still higher rod? And so on in an infinite regress? Yet there are only six rods, and it stands.

On closer examination, we can inspect any rod individually and affirm that it is supported securely by cords attached to both ends. That is true for any rod that we examine. That is all that is needed for the structure to stand. We need no additional, holistic condition beyond the condition that each rod individually is supported.

It is the same with self-supporting inductive structures. If we can affirm that each proposition individually is well supported inductively, then nothing further need be demanded. Of course, if we were tacitly to assume a hierarchical structure for relations of inductive support, then these self-supporting inductive structures are impossible. Then at least some of the propositions needed to warrant all of the inductive inferences in the structure could not be inductively supported themselves within a finite structure. An infinite regress would ensue. However, as argued in detail above, this hierarchical assumption is incorrect.

One might still harbor reservations. These self-supporting inductive structures necessarily harbor circularities in the relations of support. That these circularities are benign I argue at length in Chapter 3, "Circularity." Or one might accept that such structures exist but that they make the import of evidence equivocal since our evidence might support many such systems. In Chapter 4, "The Uniqueness of Domain-Specific Inductive Logics," I argue that a mechanism, native to the material theory of induction, precludes this danger.

6.2. Mature Sciences Are Self-Supporting Inductive Structures

We can now see that mature sciences are inductively self-supporting. This conclusion requires that the compass of a mature science is expanded enough, possibly even to embrace neighboring sciences, so that inductive closure is secured. That means that we can select any proposition in the mature science and will find, within that compass, the evidence that inductively supports the proposition along with the propositions that warrant the inductive support.

This assertion of self-support supposes that we can expand the compass of a mature science sufficiently to secure closure. We can imagine that sequences of inductive inferences and warranting propositions form an infinite chain that outstrips finite description so that no finite expansion is adequate. I do not see how, as a matter of inductive logic, such a chain can be dismissed without further examination of its details. Perhaps it is possible. However, I do not see that it arises in actual practice in our mature sciences. If that were the case, then the inductive rigidity of a mature science would not be accessible to us. Yet our repeated experience in the history of science is that we do have mature sciences that display just the inductive rigidity described here.

7. Nonempirical Components of the Large-Scale Structure of Inductive Support

This chapter provides an account of the large-scale structure of inductive support that uses only materially warranted inductive inferences or relations of inductive support. One might accept that much of this large-scale structure is captured by the material theory. However, it might be tempting to imagine that the material account still needs to be supplemented by deeper, nonempirical truths if the account of the large-scale structure is to be complete. Such deeper truths would be beyond normal evidential scrutiny and thus outside the reach of the material theory.

To make it plausible that no such added components are viable, in this section I consider and reject some candidates.

7.1. The Universal Logic of Induction

The least adventurous proposal for the added component is that the large-scale structure requires at least some universal rules of inductive inference or some general calculus of induction. Perhaps we do need to assume the universal applicability of the probability calculus to all relations of inductive support, as Bayesians seem to hold. The failure of all such universal rules has been argued at length in the *Material Theory of Induction* (Norton 2021) and reviewed in Chapter 1. There is no need for these arguments to be repeated here.

7.2. Kantian Synthetic, A Priori Propositions

Might the very viability of induction at all depend on a Kantian synthetic a priori proposition? Such a proposition would be factual, but it would require no evidence since its truth — supposedly — can be established a priori: that is, independently of experience. Since the literature on this one idea could occupy many lifetimes, I dare express only my view that this literature has failed to provide viable examples of synthetic a priori propositions that could serve this function. Kant's original proposals did not fare well. It might have been appealing to imagine in the eighteenth century that, as an a priori certainty, space could never manifest to us other than as Euclidean. However, those who have absorbed the variant spatial geometries brought by general relativity find it otherwise. The geometry of space is not something determinable a priori but a subject for empirical investigation.

The mode of failure of this one proposal for a synthetic a priori proposition afflicts all of the proposals. If they make a definite, factual assertion, then they end up failing empirically. If they escape empirical refutation by vagueness, then they make no factual assertion and are empty.

7.3. Causality

Might we seek such a condition in a principle of causality? It is a Kantian principle and has an enduring popularity outside Kantian circles. The principle asserts that every effect is brought about in a regular manner by some cause. Might such a supposition be a precondition for science and thus for inductive inferences in science? I have criticized this conception at length elsewhere.[13] In short, the problem is that the terms "cause" and "effect" are so poorly specified that the principle is factually vacuous. We can always implement the principle in any scenario simply by artful choices for what the terms designate. Things in the world do connect in a myriad of interesting ways. What those ways are cannot be stipulated a priori but must be discovered empirically.

7.4. Mathematics

It is often found remarkable that mathematical descriptions of the world are so fertile and powerful. Might the supposition of a mathematical structure

---

13 See, for example, Norton (2003, 2016, 2024).

of the world be a prior condition necessary at least for the physical sciences? There is much to say on this supposition. The main point of relevance is that the supposition itself is open to empirical testing. We have tested it and found that it applies to a surprisingly large range of phenomena. This means that, in the absence of any deeper, a priori vindication, it is a contingent fact to be learned inductively. In this regard, it is no different from the other warranting facts of the physical sciences. It is not an obstacle to the material warranting of inferences but a part of it.

An illustration shows how the proposition is not an a priori principle but open to the possibility of failure empirically. Much of modern physics presumes that its basic laws are to be written as differential equations. That fundamental presumption has been challenged by Stephen Wolfram (2002). His "new kind of science" seeks to replace these differential equations in physics by discrete algorithms and cellular automata. It is a most radical proposal. Wolfram has continued to press his approach, but its reception among physicists remains poor. Their skepticism is not based on an assertion that, as an a priori matter, the physical world must be governed by differential equations. Rather, as Becker (2020) reports briefly, the doubt is that Wolfram's methods can recover the present results of physics with the same scope and accuracy. The concern is empirical. The proposal lacks powerful enough inductive support to supplant existing methods.

Nonetheless, we can still ask what the prospects are for an a priori justification for the mathematical character of nature. These prospects are poor, in my view, since it is doubtful that there is a deep truth in the supposed mathematical character of nature. Rather, I harbor an enduring concern that our deference to the power of mathematical descriptions is excessive. The supposed truth is empty unless the specific mathematics favored by nature is specified. Yet the only way that we know to identify the right mathematics among many choices is empirical. Thus, I find it hard to be moved by a celebrated and poetic confession attributed to Heinrich Hertz: "One cannot escape the feeling that these mathematical formulas have an independent existence and an intelligence of their own, that they are wiser than we are, wiser even than their discoverers, that we get more out of them than was originally put into them."[14] On the contrary, I am in awe not of the formulae but of the creativity

14 As quoted in Bell ([1937] 1953, 16). The quotation is unsourced and seems to be the origin of later repetitions. Recently, Shour (2021) has tracked down the origin of the remark in

of mathematicians who made them. New physical theories commonly come in clumsy mathematical clothing. Each new physical theory is taken as a challenge by mathematicians to find formulations in which the new theory looks mathematically simple and natural. The ensuing mathematics fits the world not through some preordained harmony but merely retrospectively through our ingenious and artful contrivances.[15]

To see the process, one need only recall the inadequacies of geometry as Euclid formulated it for the celestial mechanics of the seventeenth century. Kepler sought to use the Platonic solids in a nestled geometric structure to explain the relative orbital sizes of the planets. We now regard the whole project not as reflecting the inner mathematical constitution of the world but as dependent on barren mathematical coincidences. One can only wonder at Newton's labors in his *Principia* to develop his celestial mechanics using simple Euclidean geometry so poorly suited to the task. The theory becomes so much more elegant and transparent when re-expressed in the later methods of vector calculus, contrived in part precisely for this purpose.

7.5. The Ineffable

Finally, when explicit attempts to identify these nonempirical conditions fail, one might be tempted by the idea that these conditions are present but ineffable. They are so deeply enmeshed in our ways of thinking that, it is speculated, we cannot discern them. This appears to me to be the last defense of a failing program. These conditions have powerful consequences in connecting facts, and these connections are fully accessible to us. Yet the conditions that underwrite these connections are supposed to be opaque to us. The supposition of their invisibility makes them irrelevant. What matters are the contingent connections that they supposedly induce among the facts of the science, and we can be secure in accepting these connections only if we can affirm or support them through methods accessible to us.

Hertz's published writings. (I thank Marc Lange for letting me know of Shour's paper.)

15 For another expression of this view in counterpoint to Einstein's later Platonism, see Norton (2000, Appendix D).

8. Conclusion

The four claims defended in this chapter form the basis of the material understanding of the large-scale structure of relations of inductive support. These claims by no means exhaust the questions that one might raise about this large-scale structure and the accompanying skeptical challenges to the material understanding. Some of these questions and challenges will be raised in the chapters to come in Part I, and the claims defended in this chapter will be used to answer them. I will ask in Chapter 3, if the structure is nonhierarchical, does it harbor circularities? (Yes.) Are they benign? (Yes.) What of uniqueness? I will ask in Chapter 4. That is, can a finite body of empirical evidence, even if extensive, yield a unique, self-supporting structure? (Yes.) Or must we forever contend with multiple, competing, self-supporting structures? (No.) Relations of inductive support are nonhierarchical and circular. Does this mean, I will ask in Chapter 5, that the material theory of induction is just a coherentist epistemology? (No.) And finally, in Chapter 6, what of *the* problem of induction? Is the material theory prone to the traditional problem? (No.) Is there an analogous problem residing in a fatal regress of warrants? (No.)

These are all good questions and worthy challenges. I will show that the material approach to inductive inference has ample resources for answering them.

REFERENCES

Anonymous. 1970. "Central Dogma Reversed" (editorial). *Nature* 226: 1198–99.

Becker, Adam. 2020. "Physicists Criticize Stephen Wolfram's Theory of Everything." *Scientific American*, May 6, 2020. https://www.scientificamerican.com/article/physicists-criticize-stephen-wolframs-theory-of-everything/.

Bell, Eric Temple. (1937) 1953. *Men of Mathematics*. Vol. 1. London: Penguin.

Bondi, Hermann, and Thomas Gold. 1948. "The Steady-State Theory of the Expanding Universe." *Monthly Notices of the Royal Astronomical Society* 108: 252–70.

Boyd, Nora. 2018a. "Scientific Progress at the Boundaries of Experience." PhD diss., University of Pittsburgh.

———. 2018b. "Evidence Enriched." *Philosophy of Science* 85: 403–21.

Cobb, Matthew. 2017. "60 Years Ago, Francis Crick Changed the Logic of Biology." *PLoS Biology* 15, no. 9: e2003243.

Dalton, John. 1808. *New System of Chemical Philosophy. Part I.* Manchester: R. Bickerstaff.

Gonzalez-Garcia, M.C. 2003. "Neutrino Masses and Mixing: Evidence and Implications." *Reviews of Modern Physics* 75: 345–402.

Lafleur, Trevor. 2014. "Can the Quantum Vacuum Be Used as a Reaction Medium to Generate Thrust?" https://arxiv.org/pdf/1411.5359.pdf.

Mullins, Justin. 2006. "Relativity Drive: The End of Wings and Wheels." *New Scientist* 2568. https://www.newscientist.com/article/mg19125681-400-relativity-drive-the-end-of-wings-and-wheels/.

Norton, John D. 1993. "The Determination of Theory by Evidence: The Case for Quantum Discontinuity 1900–1915." *Synthese* 97: 1–31.

———. 2000. "'Nature in the Realization of the Simplest Conceivable Mathematical Ideas': Einstein and the Canon of Mathematical Simplicity." *Studies in the History and Philosophy of Modern Physics* 31: 135–70.

———. 2003. "Causation as Folk Science." In *Causation, Physics and the Constitution of Reality*, edited by Huw Price and Richard Corry, 11–44. Oxford: Oxford University Press.

———. 2005. "A Little Survey of Induction." In *Scientific Evidence: Philosophical Theories and Applications*, edited by Peter Achinstein, 9–34. Baltimore: Johns Hopkins University Press.

———. 2014. "Einstein's Special Theory of Relativity and the Problems in the Electrodynamics of Moving Bodies that Led Him to It." In *Cambridge Companion to Einstein*, edited by Michel Janssen and Christoph Lehner, 72–102. Cambridge, UK: Cambridge University Press.

———. 2016. "Curie's Truism." *Philosophy of Science* 83: 1014–26.

———. 2021. *The Material Theory of Induction*. BSPS Open Series. Calgary: University of Calgary Press.

———. 2023. "Inductive Inferences on Galactic Redshift, Understood Materially." In *Current Debates in Philosophy of Science*, edited by C. Soto, Synthese Library 477, 227–46. Cham, Switzerland: Springer.

———. 2024. "The Metaphysics of Causation: An Empiricist Critique." In *Alternative Approaches to Causation*, edited by Yafeng Shan, Chapter 4. Oxford: Oxford University Press.

OPERA Collaboration. 2011. "Measurement of the Neutrino Velocity with the OPERA Detector in the CNGS Beam." https://arxiv.org/abs/1109.4897.

Ord-Hume, Arthur W.J.G. 1977. *Perpetual Motion: The History of an Obsession*. London: George Allen and Unwin.

Shour, Robert. 2021. "Heinrich Hertz on the Wisdom of Equations." Preprint at https://www.researchgate.net/publication/351564092_Heinrich_Hertz_on_the_wisdom_of_equations.

Smith, George E. 2002. "The Methodology of the Principia." In *The Cambridge Companion to Newton*, edited by I. Bernard Cohen and George E. Smith, 138–73. Cambridge, UK: Cambridge University Press.

———. 2014. "Closing the Loop: Testing Newtonian Gravity, Then and Now." In *Newton and Empiricism*, edited by Zvi Biener and Eric Schliesser, 262–350. Oxford: Oxford University Press.

Tajmar, Martin, Matthias Kößling, Marcel Weikert, and Maxime Monette. 2018. "The SpaceDrive Project — First Results on EMDrive and Mach-Effect Thrusters." https://tu-dresden.de/ing/maschineanwesen/ilr/rfs/ressourcen/dateien/forschung/folder-2007-08-21-5231434330/ag_raumfahrtantriebe/SPC-The-SpaceDrive-Project-First-Results-on-EMDrive-and-Mach-Effect-Thrusters.pdf?lang=en.

White, Harold, and Paul March. 2012. "Advanced Propulsion Physics: Harnessing the Quantum Vacuum." https://www.lpi.usra.edu/meetings/nets2012/pdf/3082.pdf.

Wolfram, Stephen. 2002. *A New Kind of Science*. Champaign, IL: Wolfram Media.

3

Circularity

1. Fear of Circles

The nonhierarchical structure of relations of inductive support admits circularities. They are inevitable once we examine a large enough set of these relations. The circles can be small, as when two propositions mutually support each other. The circles can be large, as when extended chains of relations of support eventually connect back to their starting points. Some will find the mere presence of these circles disturbing, apparently with good reason. In debates, philosophical or not, defeat is assured if your opponent can expose your reasoning as circular. In formal structures, circularities are vicious and must be eliminated, often by the most elaborate of novel theorizing. The damning verdict is automatic and unanswerable. You have found a circularity? There is no need to waste any more thought on the enterprise. It is fatally flawed. The perpetrator of a circularity might be expected to resort to all manner of sophistry. But escape is impossible, and the ultimate collapse of the enterprise is inevitable.

Such is the fear of circles, *horror circulorum*. I have written this chapter for those in its grip. My goal is to provide them with therapy, for the horror is based on an oversimplified view of circularities. It neglects the many forms that circularities can take. Some are as fatal as this dark view fears. Many are benign, and, as we shall see, others are even essential to a theoretical structure. To ban them entirely would restrict unnecessarily the scope of our theorizing. To show this, I provide a small classification of circularities, according to how they affect the logic of the structure in which they appear. It will show that the circularities of relations of inductive support are benign and even essential.

There are three categories. The first is composed of the "vicious" circularities explored in Section 2. They lead to logical inconsistencies and underwrite the dark view of circularities as fatal defects. When such circularities arise in inductive structures, they are transient and eliminated by suitable adjustments to the propositions in the structures. The second, explored in Section 3, includes circularities in structures whose content is left indeterminate. They might merely be failed arguments or intermediate stages of development on the way to the third type. Or, if they are ineliminable, they might be the basis of a convention. The third, described in Section 4, encompasses the circularities that are part of a well-behaved structure whose content is uniquely defined, without contradiction. This is the case of the relations of inductive support of a mature science. In Section 5, I summarize how the circularities in relations of inductive support appear in the taxonomy. The mechanism identified in Chapter 4, "The Uniqueness of Domain-Specific Inductive Logics," leads to a convergence toward inductive structures with univocal import.

That benign circularities are possible is the tonic that can cure *horror circulorum*. It tells us that mere identification of a circularity in some system is a starting point, not an end point. If we want to take the next step and damn the system for the circularity, then there is a positive obligation to establish that the specific form of circularity present is harmful. This cannot be done, I believe, for the circularities in a mature science. They are benign.

2. Vicious Circularity

A "vicious circularity," as I use the term here, is a set of circular relations in some formal structure that leads to a contradiction.

2.1. The Idea

The term "vicious circle" has long been familiar in treatises on logic. Kirwan (1807) already found its use established. Curiously, the formal definition that he gave was merely of question begging, "*petitio principii*," which I describe in more detail in Section 3. Kirwan wrote of

> . . . that mode of argumentation called the *vitious circle* [*sic*], in which one point is proved by another, and this other is proved *solely* by the first; so that the proofs are mutual and under the same point of view. (441–42; Kirwan's emphasis)

That what is described is really question begging is made clear by Munro's (1850) treatise, whose exposition follows Kirwan's closely. Munro illustrated the circle in the following way:

> The whole of Dr. Brown's elaborate lectures on the nature of virtue amounts to nothing more than a vicious circle. We approve of actions, because they are right; and they are right, because we approve of them. (231)

More curiously, Kirwan's own example was of a circle that produced a contradiction. His definition of a vicious circle was immediately illustrated by the self-refutation of skeptics:

> Thus the sceptics argue that we ought to doubt of every thing, because human reason is fallible, and may deceive us. And since reason may deceive us, we should doubt of the validity of the reasons that induce us to doubt. (Kirwan 1807, 442)

The idea of a vicious circle as essentially leading to a contradiction was cemented by Bertrand Russell's work in mathematical logic. In reflecting on Cantor's proof that there can be no greatest cardinal number, Russell arrived at what came to be known as Russell's paradox. It was given an early elaboration in "The Contradiction" (Russell 1903, Chapter X). The paradox concerns sets and their members. Some sets can have other sets as their members. Naively, we easily accept that some can even be members of themselves. A set of sets can be a member of itself, for example. But what of those sets that are not members of themselves? What of the set of all such sets? The supposition that there is such a set immediately produces a contradiction. If it is a member of itself, then it is not a member of itself. But if it is not a member of itself, then it is a member of itself. The contradiction arises essentially through the circular relationship between the set and its members.

Although the paradox looks at first like a minor annoyance easy to circumvent, it was immediately recognized as a deep problem for set theory and the foundations of mathematics. It shows that sets cannot be defined merely as an extension of any property. That is, we cannot say "consider the set of all things that have property P," where that property, expressed as some formula, can be freely chosen.[1] The most searching and elaborate investigations were needed to give set theory a noncontradictory foundation. One avenue was the development of the axioms of Zermelo-Fraenkel set theory. Russell's path led to the theory of types, found in his joint work with Alfred North Whitehead, *Principia Mathematica*. There Russell and Whitehead reinforced the odious character of vicious circles. The first named section in the first volume, Chapter 2, is "The Vicious-Circle Principle," and one of its formulations was that "whatever involves *all* of a collection must not be one of the collection" (their emphasis). Breaches of this principle, they announced, were to be called "vicious-circle fallacies" (Russell and Whitehead 1910, 40).

The vicious circle of Russell's paradox derived from its imprudent use of self-reference. Such imprudence proved to be a fertile source of analogous paradoxes. Russell (1908) provided a convenient compendium. It began with the now classic Epimenides:

> Epimenides the Cretan said that all Cretans were liars, and all other statements made by Cretans were certainly lies. Was this a lie? The simplest form of this contradiction is afforded by the man who says "I am lying"; if he is lying, he is speaking the truth, and vice versa. (222)

Its structure matches that of Kirwan's (1807) example of the self-refuting skeptics. The inventory continued with Russell's set paradox and a list of other related paradoxes familiar to readers of the literature, including Berry's paradox, Richard's paradox, and the Burali-Forti contradiction. These paradoxes provided the impetus for a century of philosophical work on truth in the foundations of formal logic. It was designed to find ways of precluding paradoxical sentences such as "this sentence is false" or finding unparadoxical ways of including them. The contradictions that follow from self-reference became one of the most powerful tools of formal logic. They are the basic device used in Gödel's famous demonstration of the incompleteness of arithmetic.

To philosophers who have any interest in formal matters, all of this is so elementary as to have become part of "what everyone knows." At the same

1 This troublesome principle has been called "the intuitive principle of abstraction" (Stoll [1963] 1979, 6).

time, these foundational investigations have forged an automatic and enduring link between circularity and contradiction. So the *horror circulorum* is established.

2.2. Vicious Circles in the Material Theory of Induction?

Are the circularities of inductive support vicious? Nothing compels it. As we shall see below, one can have circularities that are not vicious: that is, they produce no contradictions. Such are the circles arising among the relations of inductive support for mature sciences. The inductive support of these mature sciences is secure and even unassailable, and it would not be so if contradictions could be found within it.

This is the situation with the evidential support of a mature science. However, prior to this mature stabilization, contradictions can and do arise among the relations of support. Developing sciences are commonly built upon hypotheses whose evidential grounding has not been secured. Sometimes these hypotheses fail, and that failure manifests in contradictions. In 1917, Einstein presented the first relativistic cosmology using the assumption that the universe is static. His hypothesis was soon contradicted by Hubble's discovery of the recession of the galaxies.

These contradictions are not manifestations of an ineliminable, foundational flaw in the very idea of inductive support. They are unlike the vicious circularities of naive set theory, whose circularities forced us to reconceive the very idea of a set and of the truth of propositions. Rather, they are a natural part of the work of fallible investigators. The structures that they produce are fallible but malleable, and it is a routine part of investigations to reform the structures to eliminate them. Einstein discarded his assumption of a static universe, and other theorists began to explore the notion of a dynamic, expanding universe compatible with general relativity. These contradictions and adaptations are of no more concern than an accounting error in a budget. Perhaps a receipt was mistyped or an expense neglected. It is a simple but tedious exercise to find the error and correct it. There has been no fundamental breach of a principle of arithmetic that would forever preclude the use of budgets.

The radiocarbon dating of historical artifacts, described in Chapter 10, "Mutually Supporting Evidence in Radiocarbon Dating," shows how these contradictions arise within a circle and are remedied. Artifacts are dated by two means. The first is traditional historical analysis. The second is the measurement of the radioactive ^{14}C ("carbon 14") content of the artifact. What results are two sets of propositions, one historical and the other radiocarbon. Each should support the other. When radiocarbon dating methods were first explored, it soon became apparent that there were recalcitrant discrepancies in the dating provided by the two means. That is, there were contradictions within the circular relations of mutual support among the radiocarbon dates and historical dates.

The elimination of these contradictions became a major focus of research on radiocarbon dating methods. Radiocarbon dating depends essentially on knowing the original ^{14}C content of the artifact. That content is halved for each ^{14}C half-life of 5,730 years. It was natural to suppose that these original levels match those of artifacts formed today. It soon became apparent, though, that this assumption was the source of the contradictions. These levels have varied over historical times. Theoretically grounded reconstruction of these original levels proved to be unworkable. Instead, these levels were reconstructed by means of the historically known ages of artifacts. The corrections needed were collected in a calibration curve, as is shown in the later chapter. Using such curves, the radiocarbon and historical datings of artifacts were adapted to one another in the precise manner needed to eliminate the contradictions. After that adaptation, each set of datings could be used to check and affirm the other set. The circularity among the two sets of propositions remained but without contradiction.

In a similar vein, the structures of inductive support for a mature science can be disrupted by new, empirical discoveries. The disruptions manifest as contradictions that can be removed by adjustments to the inductive structures. Because of the rigidity of relations of inductive support in a mature science discussed in Chapter 2, the adjustments likely will propagate through the entire structure. They will have revolutionary import.

Newton's seventeenth-century mechanics prevailed for over two centuries. Its inductive support apparently was unassailable. One of its basic results was that the velocity of a uniform observer was to be added to or subtracted from that of any propagation to recover the velocity that the observer would find for it. The new evidence of Maxwell and Lorentz's nineteenth-century electrodynamics destabilized Newton's mechanics. Under Einstein's careful scrutiny, the electrodynamics revealed that light propagation violated this simple Newtonian result. The speed of propagating light was always the same, no matter the uniform motion of the observer.

The contradiction was resolved when Einstein realized that space and time themselves, at high speeds, do not behave as Newton had concluded. The evidence and relations of evidential support leading to Newton's theory were not discarded. Rather, their limited scope was now recognized. They could be applied only to systems moving at much less than the speed of light. This restriction was readily implemented. Newton drew the evidence for his mechanics from the motions of ordinary falling bodies, moons, and planets. These are all bodies whose speeds are much less than that of light. The evidential base of Einstein's special relativity embraced that of Newton's mechanics for lower speeds and that of electrodynamics for higher speeds. Einstein's new physics required alterations to every physical theory in which space and time played roles. The alterations propagated through physics with revolutionary import.

3. Indeterminate Circularities

A less troublesome form of circularity arises when the circles produce no contradictions but leave the structure indeterminate. The indeterminacy might not be obvious since the analysis might be offered as determinate.

3.1. Begging the Question

A familiar example, known since Aristotle, is circular reasoning, "begging the question"[2] or *petitio principii.* It is a form of reasoning that pretends to establish a conclusion but only gives the illusion of doing so. Richard Whately's (1856) *Elements of Logic* gives what seems to be a standard definition for nineteenth-century work.[3] Alerting us in a preface ("advertisement") that he uses brackets to indicate equivalent meanings, Whately tells us that

> . . . "*petitio principii*" ["begging the question"] takes place when a premiss [*sic*], whether true or false, is either plainly equivalent to the conclusion, or depends on it for its own reception. (184)

2 This is the original sense, to which I adhere. A recent usage gives the expression the meaning of "inviting the question."

3 Mill (1882, 571) reports Whately's treatment extensively.

He continues to note the delicacy of the identification, for unobjectionable deductive inferences will have this character in case a premise entails the conclusion and vice versa. Such is the case for inferences that demonstrate the equivalence of two physical conditions, such as the equivalence of the "Thomson" and "Clausius" forms of the second law of thermodynamics. To be worthy of the label *petitio principii*, there must be some sense that the inference is used deceptively, to pretend that more is gained than really is, "obliquity and disguise being of course of most importance to the success of *petitio principii* . . ." (222).

Examples are easy to find. One is a religious figure or tract for which infallibility is to be concluded since the figure or tract itself declares its infallibility. The more interesting cases of begging the question arise when the circularity is sufficiently hidden that its presence is easily overlooked. Mill provides an example:

> Plato, in the *Sophistes*, attempts to prove that things may exist which are incorporeal, by the argument that justice and wisdom are incorporeal, and justice and wisdom must be something. Here, if by *something* [Mill's emphasis] be meant, as Plato did in fact mean, a thing capable of existing in and by itself, and not as a quality of some other thing, he begs the question in asserting that justice and wisdom must be something; if he means any thing else, his conclusion is not proved. (1882, 574)

A more extended example, I contend, arises in the many demonstrations of probabilism: Dutch book arguments, decision theoretic representations, the accuracy-based scoring rule argument, and so on. In *The Material Theory of Induction* (Norton 2021, Chapters 10 and 11), I argue that all of these proofs proceed by employing premises in which the basic assumptions of probabilism are already present in disguised form. When their presence is identified, the demonstration collapses. It then becomes easy to see that arbitrary adjustments to the rules of the betting scenario, to the properties of the preferences assumed, or to the scoring rule used can lead to variant, nonprobabilistic calculi.

For my purposes here, the essential fact is that this sort of circular reasoning fails to determine the conclusion sought. Nonetheless, the conclusion sought might be true, or it might be false.

3.2. Circularities that Produce Conventions

Similar indeterminacy-producing circularities arise among magnitudes in science. The indeterminacy is then often taken as evidence that the magnitude of some quantity can be set as a convention. Perhaps the best-known examples arise in relativity theory and in geometry. In his special relativity paper in 1905, Einstein argued that we cannot affirm the simultaneity of spatially separated events, factually, by light signals (or, analogously, by any other means). Any scheme that uses light signals to ascertain the relative timing of such events requires that we know how fast light propagates in one direction. A natural scheme requires, for example, that we know that light propagates at the same speed from a place A to a place B as it does in the reverse direction. Yet to know this we must be able to determine how quickly light propagates from one place to another. This determination requires that we can already compare the timing of events at these two places.

Einstein (1920, 22–23) summarized our predicament: "It would thus appear as though we were moving here in a logical circle."[4] The significance of this circle is that there are no independent facts separately for the simultaneity of spatially separated events and the speed of light propagating between them. Rather, we can choose freely as a convention either the simultaneity relation or this speed. Then the other is determined. Here is how Einstein put it:

> That light requires the same time to traverse [the forward path] as for [the reverse path] is in reality neither *a supposition nor a hypothesis* about the physical nature of light, but a *stipulation* which I can make of my own free will in order to arrive at a definition of simultaneity. (23; Einstein's emphasis)

His foremost expositor in this matter, Hans Reichenbach (1958, 126–27), summarized a more extensive analysis of the same circularity:

> Thus we are faced with a circular argument. To determine the simultaneity of distant events we need to know a velocity, and to measure a velocity we require knowledge of the simultaneity of distant events. The occurrence of this circularity proves that simultaneity is not a matter of knowledge, but of

4 "Man scheint sich also hier in einem logischen Zirkel zu bewegen."

a coordinative definition, since the logical circle shows that a knowledge of simultaneity is impossible in principle.

Under Poincaré's and Einstein's inspiration, Reichenbach argued for a structurally analogous convention that he called the "relativity of geometry" (Section 30). It depends on a similar logical circle. One can determine that the geometry of a space is Euclidean or otherwise by the expedient of surveying it with measuring rods. The essential condition is that the rods are rigid ones that measure distances truly. The complication, Reichenbach urged, is that rods can be acted on by what he called "universal forces" that equally distort all bodies. This complication creates the circle. We cannot know which universal forces, if any, are acting on a rod unless we already know the true geometry of the space. The circle is resolved by declaring that we may select the geometry of space conventionally. We merely posit the universal forces needed so that our rod measurements give us that geometry.

For completeness, I should mention that both conventionality theses were hotly debated in the later twentieth century without any clear resolution. Those opposed to the conventionality claims urged that there were other nonconventional means to break the circles. We do not need to take sides in this debate for my concerns here.[5] All that we need to see is that these circular dependencies among physical quantities can leave the quantities indeterminate. Even a fairly modest empiricism must be troubled by the idea of quantities whose values cannot be determined by any physical measurement or observation. If the indeterminacy is sustained, then the comfortable resolution is to assert that there is no physical fact for these values. They may be chosen arbitrarily: that is, as a convention.

3.3. Indeterminate Circularities in Relations of Inductive Support

It is possible for this sort of circularity to arise among relations of inductive support. If they prove to be ineliminable, then we might expect an empirically minded scientist to proceed as above. If we are sure that no evidence can

5 However, I incline toward the anticonventionalist view. For an elaboration, see the chapters "The Conventionality of Simultaneity" and "Geometric Morals" in Norton (2007). Reichenbach's supposition of universal forces is troublesome since, if the mode of analysis is accepted, then analogous suppositions can be used to establish the conventionality of any physical magnitude measured by some instrument.

break the circle, then we have concluded that these are propositions whose truth is immune to evidential scrutiny. Such propositions are leading candidates for conventional stipulation. Indeed, conventional stipulation, by the supposition of the case, will make no difference empirically.

The more common situation is the one that arises in the examples of circular dependencies recounted in the earlier chapters and explored in greater detail in the later chapters. The circularities might be such initially as to leave the quantities of interest indeterminate. However, further investigation brings new facts to bear that break the circularity. A focus on exactly such investigations can become a major stimulus for further research.

Dalton's original atomic theory was trapped in a circle, as detailed in Chapter 11. To know the correct molecular formulae of substances, Dalton needed to know the relative weights of the atoms combined in them. But he could know those relative weights only if he already knew the molecular formulae. This meant that his theory was compatible with water having a huge array of different molecular formulae: H_2O, HO, HO_2, and many more. He was free to stipulate any of them without fear that the meager evidence at his disposal would contradict his choice. Dalton chose HO. Had the circularity been unbreakable, we might have settled eventually on a curious sort of atomic theory in which the relative masses of the atoms could be set arbitrarily, much as we arbitrarily set the zero point for the potential of a Newtonian gravitational field. As we now know, this freedom was transient. It still took over half a century of further work to bring enough additional facts to bear to break the circle and recover H_2O.

The determination of celestial distances involved similar indeterminacy-producing circularities. Our earliest efforts to determine the distances to the Moon and the Sun were troubled by one. We could measure the angular sizes of these bodies so that, if we knew their diameters, then we could infer the distances to them. However, we needed to know just these distances to determine their diameters. In Chapter 12, "The Use of Hypotheses in Determining Distances in Our Planetary System," I describe how diligent analysis by ancient astronomers was able to break the circularity and produce estimates of the diameters and distances.

A similar circularity appeared in Hubble's classic paper in 1929 on the recession of the nebulae. Hubble had apparent brightness measurements for forty-six nebulae. To convert those measurements to distances, he needed to know the absolute brightnesses of these nebulae. Then, using the fact that brightness diminishes with the inverse square of distance, he could recover the distances to the nebulae by comparing how bright they seem to be with how bright they really are. However, for twenty-two of them, Hubble lacked absolute brightness determinations. Absent other information, to know their absolute brightnesses, he first needed to know their distances. This closes the circle, leaving the distances to these twenty-two nebulae indeterminate. As recounted in Chapter 7, "The Recession of the Nebulae," Hubble brought further statistical considerations to bear to break the circularity and recover determinate, if fallible, distances for these twenty-two nebulae.

In sum, this sort of indeterminacy-producing circularity can arise among relations of inductive support. It presents no foundational challenge to the very notion of inductive support. There are several possibilities, none foundationally troublesome. The circularities can be broken by further scientific investigations. If ineliminable, then they might prove to arise from conventions. Or they might be ineliminable simply because of a paucity of evidence. Certain sorts of historical facts are obvious candidates. We might like to know many details of some ancient civilization. However, if sufficient archaeological evidence has not been preserved, then we have no choice but to settle for indeterminacy, not of the facts but of what the evidence can determine about them. That is just how it should be.

4. Determinate Circularities

The most benign circularities are those that arise in determinate structures. Then none of the issues of contradiction or indeterminacy arises. This sort of circularity is so widespread and familiar that it rarely arouses complaints.

4.1. Elementary Examples

Simple computations of determinate magnitudes often involve circularities. An easy example is the computation of the black area B and the white area W of the yin-yang symbol of Figure 3.1.

Figure 3.1. The yin-yang symbol

From the symmetry of the figure, we have

$$B = W$$

It is half of a circular dependency. Assuming that the total figure has unit area, we also have

$$W = 1 - B$$

This is the second half of the circular dependency. There is nothing troublesome in the circularity. The two equations are solved uniquely to give

$$B = W = 1/2$$

A slightly fancier computation is the standard way that the following infinite sum is evaluated:

$$S = 1/2 + 1/4 + 1/8 + 1/16 + \ldots$$

In a familiar manipulation, the sum is doubled to yield

$$2S = 1 + 1/2 + 1/4 + 1/8 + \ldots = 1 + S$$

This last equation expresses a circular dependence but is readily solved to give us the sum $S = 1$.

The only danger in this otherwise benign computation is that we must be assured antecedently that the infinite sum does have a definite, finite value. Even if we are not assured that the sum is finite, the circularity can still give us a determinate result. Consider

$$S = 1 + 2 + 4 + 8 + \ldots$$

It is doubled to yield

$$2S = 2 + 4 + 8 + 16 + \ldots = S - 1$$

This circular equation in S has two solutions. $S = -1$ can be discarded if we preclude a negative sum. The applicable solution is S = infinity.

Finally, we might ask whether these circularities could mislead us when the sum sought is badly behaved. Such is the case with Grandi's series, whose sum we might try to write as

$$S = 1 - 1 + 1 - 1 + \ldots$$

Of course, there is no such sum. The partial sums oscillate indefinitely between 0 and 1. If we proceed formally, we might write

$$S = 1 - 1 + 1 - 1 + \ldots = 1 - (1 - 1 + 1 - \ldots) = 1 - S$$

This circular equation in S has a unique, finite solution, $S = 1/2$. This value cannot be the ordinary arithmetic sum of Grandi's series, for there is no such sum. However, if we consider generalized notions of summation that might be applied here, then we can take this circular dependency as part of the conditions of adequacy of the generalized notion. An example of such a generalized notion is the Cesàro sum. It proceeds by taking the *arithmetic average* of the first n terms in the series. The sum of the entire series is just the limit of this average as n goes to infinity. The Cesàro sum for Grandi's series is 1/2.[6]

4.2. An Extreme Example

These examples of benign circularity have been elementary. They serve to show, however, that circularities within well-defined structures are common and unremarkable. At the other extreme, we can have similarly benign circularities in exotic structures. Most striking of them is one that directly challenges the historical stimulus of *horror circulorum*, Russell's Vicious-Circle Principle. The principle prohibits circularities, such as sets that are members of themselves. In response, the edifice of modern set theory, as exemplified in the Zermelo-Fraenkel system, was built precisely to preclude such circularities.

6 In another approach, we consider $S(a) = 1 - a + a^2 - a^3 + a^4 - \ldots = 1/(1+a)$ for $0 < a < 1$. We *define* $S(1) = \mathrm{Lim}_{a \to 1} S(a)$, and it does have the value 1/2.

All of this changed with the appearance of Peter Aczel's non-well-founded set theory or hyperset theory. It provides an account of sets that allows for just the sort of circularities prohibited by Russell's principle but without inducing his paradoxes. The details of the theory go well beyond what I can review here. Most briefly, the approach drops the Foundation Axiom of the Zermelo-Fraenkel system and replaces it with the Anti-Foundation Axiom. The import of the transition can be seen in the case of the simplest circularity in set membership. Following Barwise and Etchemendy (1987, 37–41), it is the set Ω, defined circularly by the fact of its self-membership:

$$\Omega = \{\Omega\}$$

That is, the set Ω is defined as that set that has itself as its sole member. If we substitute for Ω, then we can rewrite the set as $\Omega = \{\{\Omega\}\}$. Continuing, we have $\Omega = \{\{\Omega\}\} = \{\{\{\Omega\}\}\} = \{\{\{\{\Omega\}\}\}\}$. A full substitution leads to an infinite nestling of set memberships:

$$\Omega = \{\{\{\{\{\{\{ \ldots \}\}\}\}\}\}\}$$

Precisely this infinite nestling of set memberships is prohibited by the Zermelo-Fraenkel Axiom of Foundation. All such nestlings, according to it, must terminate finitely. Aczel's Anti-Foundation Axiom allows it because it can be given a definite graph theoretic representation,[7] and moreover the axiom asserts its uniqueness.

In this set Ω, we have just the sort of circularity that should trigger *horror circulorum*, a set that is its own member. However, that very circularity defines a determinate, unique structure in non-well-founded set theory.

4.3. Intermediate Examples

Between these elementary and exotic instances of benign circularities, there are many more instances, all of them part of unremarkable, routine science. A great achievement of nineteenth-century physics was Maxwell's electrodynamics. Its basis, in a modern formulation, are the four vector differential equations known as "Maxwell's equations." In the simplest case of electric and magnetic fields in vacuo, these equations fix the electric field strength vector **E** and the magnetic field strength vector **H**. Using the older Gaussian

7 It is just $\Omega \rightarrow \Omega \rightarrow \Omega \rightarrow \Omega \rightarrow \ldots$.

system of units (in which the equations are simpler) and standard notational conventions, the first two equations are just

$$\nabla.\mathbf{E} = 0 \text{ and } \nabla.\mathbf{H} = 0$$

These equations do not govern how the fields evolve over time, such as when electromagnetic waves propagate. Their time evolution is recovered from the next two equations, which exhibit a tight circular dependence. The third is

$$\nabla \mathrm{x} \mathbf{H} = (1/c)\ \partial \mathbf{E}/\partial t$$

It asserts that a time-varying electric field produces a rotational magnetic field, whose lines of force form circles around those of the electric field. The fourth equation is

$$\nabla \mathrm{x} \mathbf{E} = -(1/c)\ \partial \mathbf{H}/\partial t$$

It asserts an analogous process: a time-varying magnetic field produces a rotational electric field, whose lines of force form circles around those of the magnetic field.

This circular dependence among quantities like **E** and **H** is common. A second and much more elaborate set of circularities arises in Einstein's gravitational field equations for his general theory of relativity. They are used to determine the basic quantity of the theory, the metric tensor. It is, expressed in coordinate-based components, a matrix of ten quantities: $g_{ik} = (g_{00}, g_{01} = g_{10}, g_{02} = g_{20}, \ldots, g_{33})$. These ten quantities are fixed by Einstein's ten second-order, coupled, nonlinear, partial differential equations. Through their coupling, they harbor an elaborate set of circular interdependencies among the components g_{ik}.

Although circularity is inherent in both Maxwell's equations and Einstein's equations, they produce determinate structures: that is, allowing for standard-gauge freedoms, both admit well-posed initial value problems. Loosely speaking, this means that, if we determine the configuration of fields for the present moment, then their evolution into the future is uniquely determined. We have no trouble using these equations to determine precisely how radio waves propagate and how black holes form.

4.4. Determinate Circularities among Relations of Inductive Support

Circularities among physical quantities arise routinely as a benign feature of determinate structures in physical theories. Similarly, it is routine for the structures of inductive relations in a science to harbor circularities, even as the bearing of those relations is univocal. Such is the most common case among the examples of circularities in inductive structures seen in this chapter and elsewhere in this book. For example, Dalton's original atomic theory was beset with a circularity. Subsequent research removed it and gave us determinate molecular formulae and atomic weights. The ancient circularities that troubled the determination of distances to celestial bodies were resolved so that we now have precise determinations of them. Hubble's analysis in 1929 was hampered by a circularity that precluded the direct determination of distances to twenty-two of the forty-six nebulae in his data set. After further investigations, the distances to these closer nebulae are no longer in doubt. Although radiocarbon and historical dating of artifacts involves circular dependencies, we now have sufficient cross-checking of the methods that the original uncertainties have been eliminated.

5. Conclusion

Circularities arise routinely among rich structures of evidential support. They are no mere accidents. Rather, they are part of what enables a mature science to establish the familiar solidity of its evidential support. For those in the grip of *horror circulorum*, their presence is a source of concern and doubt. In this chapter, I have sought to demonstrate that this fear is unfounded.

Some circularities are worrisome. Such are the vicious circularities whose contradictions forced us to abandon the naive notion of a set and to develop elaborate theories of truth. We saw in Section 2 that there can be circularities that produce contradictions in relations of inductive support. However, they are not of the same type that would force us to abandon the very idea of inductive support. They are transient difficulties resolved by further investigations.

Other circularities do not produce contradictions but leave their structures underdetermined. That is troublesome only if it is pretended otherwise. It is this deception that renders begging the question objectionable. Otherwise, these circularities can be employed usefully to establish the conventionality of a physical magnitude. In the case of relations of inductive support, these indeterminacies can arise in intermediate stages of investigation. If they prove to be ineliminable, then perhaps we have found a hidden convention, or perhaps insufficient evidence exists for us to learn definitively about the target system.

Most commonly, the indeterminacies are eliminated by further investigations. They lead to an inductive structure with univocal import characteristic of a mature science. As Section 4 recounts, in this they are like many of the circularities among physical quantities in science untroubled by indeterminacies. That they arise commonly in mature sciences is not happenstance. In the next chapter, "The Uniqueness of Domain-Specific Inductive Logics," I will argue that this uniqueness results from a definite mechanism. If there are competing systems, then the competition is unstable. If one system gains an advantage by learning facts favorable to it but weakening its competitor, then it follows from the material conception of inductive inference that this strengthens the inductive reach of the first while diminishing that of the second. As long as further evidence is available and investigators pursue it, this instability is self-reinforcing and leads to the unique admissibility of the first system.

REFERENCES

Barwise, Jon, and John Etchemendy. 1987. *The Liar: An Essay on Truth and Circularity.* New York: Oxford University Press.

Einstein, Albert. 1905. "Zur Elektrodynamik bewegter Körper." *Annalen der Physik* 17: 891–921.

———. 1920. *Relativity: The Special and the General Theory.* 3rd ed. Translated by R.W. Lawson. London: Methuen.

Hubble, Edwin. 1929. "A Relation between Distance and Radial Velocity among Extra-Galactic Nebulae." *Proceedings of the National Academy of Sciences* 15: 168–73.

Kirwan, Richard. 1807. *Logick: Or an Essay on the Elements, Principles and Different Modes of Reasoning.* 2 vols. London: Payne and Mackinlay.

Mill, John Stuart. 1882. *A System of Logic, Ratiocinative and Inductive.* 8th ed. New York: Harper and Brothers.

Munro, H.H. 1850. *A Manual of Logic, Deductive and Inductive.* Glasgow: Maurice Ogle and Son.

Norton, John D. 2007 (last revised 2024). *Einstein for Everyone.* https://sites.pitt.edu/~jdnorton/teaching/HPS_0410/index.html.

———. 2021. *The Material Theory of Induction.* BSPS Open Series. Calgary: University of Calgary Press.

Reichenbach, Hans. 1958. *The Philosophy of Space and Time*. Translated by Maria Reichenbach and John Freund. New York: Dover.

Russell, Bertrand. 1903. *The Principles of Mathematics*. Vol. 1. Cambridge, UK: Cambridge University Press.

———. 1908. "Mathematical Logic as Based on the Theory of Types." *American Journal of Mathematics* 30: 222–62.

Russell, Bertrand, and Alfred North Whitehead. 1910. *Principia Mathematica*. Cambridge, UK: Cambridge University Press.

Stoll, Robert R. (1963) 1979. *Set Theory and Logic*. New York: Dover.

Whately, Richard. 1856. *Elements of Logic*. 8th ed. New York: Harper and Brothers.

4

The Uniqueness of Domain-Specific Inductive Logics

1. The Challenge Posed

According to the material theory of induction, the inductive relations within a mature science form a self-supporting structure.[1] That is, the propositions of the science derive their inductive support entirely from an extensive body of empirical evidence, such that each proposition in a theory is supported individually by this body of evidence through the mediation of other propositions. Those other propositions are supported in turn in the same way.

This raises a challenge: what assurance do we have of the uniqueness of the resulting relations of inductive support? We should not expect such an assurance for a developing science sustained only by a fragmentary body of evidence. In such cases, the evidence is too weak to determine unique relations. But what of the case of a mature science in which the body of evidence is sufficiently expansive to provide strong evidential support for all of the propositions of the science? Is such a science uniquely supported? Might there be a second science whose propositions contradict the first science but is as strongly supported in all of its parts by the same body of evidence?

Were there such cases, the result would be inductive anarchy, and it would be of an especially troublesome kind within the context of the material theory of induction. Each set of facts proposed by each science would support its own inductive logic. Since the facts disagree, the resulting logics would not agree on the bearing of evidence. One could find propositions in a science

1 My thanks to James Woodward for helpful comments on an earlier draft of this chapter.

supported inductively or not according to which of the inductive logics is employed.

Perhaps we can find reasons to expect such multiple systems. If we think of relations of support as analogous to relations of structural support in a building, then we can erect very different self-supporting systems of masonry on the same foundations. So why do we not have multiple systems of inductive logic?[2] The underdetermination thesis in its strongest form is the grim speculation that no body of evidence, no matter how extensive, can determine the content of a theory. Inductive pessimists who find this speculation appealing will expect multiple systems as a matter of course. My goal in this chapter is to refute this inductive pessimism by means of three arguments.

First, if the underdetermination thesis were true, then all sciences, even the most mature, would be awash in incompatible competitor sciences that enjoy comparable inductive support by the evidence. As a matter of history, this is not the case. Rather, as I will review briefly in Section 2, once a science achieves maturity, its competitors are discarded, and a single science prevails and endures. Since the underdetermination thesis is accepted in some literatures as a truism of evidence, in Section 7 I review briefly why it is really a poorly grounded speculation, better called the underdetermination conjecture.

Second, competing relations of support derive from competing theories that make incompatible factual assertions. As I will argue in Section 3, the empirical character of science requires that such factual differences be reflected in differences of empirical evidence, or they lie outside the scope of empirical sciences. It follows that empirical evidence can always decide for some and against others of the competing theories. (I will develop this concern in Sections 7 and 8 in further discussion of the underdetermination thesis.)

---

2 Here the analogy to buildings is weak and misleading, for we can imagine a flat terrain on which we can erect many great cathedrals of differing design, selected according to our whims. However, a body of evidence analogous to this featureless terrain is bereft of evidential value. It can sustain only the thinnest of inductive logics such as the relations of "completely neutral support" described in *The Material Theory of Induction* (Norton 2021). The analogy improves somewhat if we imagine building on a complicated and richly structured terrain that admits only specific modes of construction. The empirical foundation of our science should be structured richly enough to direct us to a fuller content of the science itself.

Third, as I will relate in Section 4, there is a natural mechanism peculiar to the material theory of induction that favors the emergence of uniqueness. Understood materially, the competition between scientific theories is dynamically unstable as long as continuing attention is given to the full exploration of the evidence. If one theory gains an evidential advantage over another, then that theory's inferential powers are enhanced. According to the material theory of induction, facts warrant inductive inferences. Thus, the evidentially strengthened theory has secured more facts and with them a strengthened warrant to infer inductively to still more facts. The competing theory is correspondingly weakened. If this process continues, then it amplifies the advantage in a positive feedback loop and leads one theory to dominate and eventually eliminate its competitors.

These instabilities are illustrated in Section 5 with several examples. Two later chapters provide more extended examples. In Chapter 14, "Stock Market Prediction: When Inductive Logics Compete," we see that there are multiple systems currently in use for predicting price movements in the stock market. The chapter shows that they are in unstable competition and that a proper pursuit and weighing of the evidence would lead to one dominating the others. In Chapter 13, "Dowsing: The Instabilities of Evidential Competition," I recount how the practice of dowsing emerged in the sixteenth century. Even then it was a controversial practice. Two views competed: the proponents of dowsing and skeptics who argued that the practice was ineffective. Over the ensuing centuries, the evidential case for the skeptics made self-reinforcing advances that successively undermined the scientific credibility of dowsing until it collapsed.

In concluding sections, I consider standard challenges in the literature to the uniqueness claimed in this chapter. What of challenges to any theory by unconceived alternatives? Does not the already mentioned underdetermination thesis preclude uniqueness? What of observationally equivalent theories? In Sections 6, 7, and 8, I discuss each question and argue that none supports a cogent challenge. In Section 9, I argue that a material approach to inductive inference fares better at accommodating the uniqueness of inductive support of mature science than do formal accounts. In Section 10, I provide a brief summary and conclusion.

2. The Uniqueness of Mature Sciences

Once a science reaches maturity in its domain of application, it stabilizes and remains fixed. The effect is so familiar that we need to recall only a few instances. At the level of precision required for virtually all applications, Euclid's ancient geometry has sufficed up to the present day. Deviations from it arise, according to general relativity, only when we venture well beyond the realm in which Euclid's geometry found its evidential support: that is, we explore systems with intense gravity or those on cosmological scales. At the level of precision for even the most exacting dynamic systems, Newton's seventeenth-century mechanics has sufficed up to the present day. Deviations appear only in domains remote from those in which Newton's mechanics is well supported evidentially. Examples of these remote domains are systems moving close to the speed of light or those at atomic scales, where quantum effects are important. The chemistry of common materials is based on a system of elements secured in the nineteenth century, deriving from the work of Lavoisier and its codification in the periodic table of Mendeleev. The diversity of geological structures derives from Lyell's early-nineteenth-century uniformitarianism, and the variety of life forms derives from Darwin's mid-nineteenth-century theory of evolution. The examples can be multiplied. The uniqueness of mature sciences contradicts the proliferation predicted by the underdetermination conjecture.

It can be tempting to imagine that the dominance of one mature science does not derive from the weight of evidence. It is, we might speculate darkly, merely a reflection of local conditions such as external social factors or political pressures or even the concerted fraud of scientists. Of course, aberrations are possible when local conditions eclipse the proper weighing of evidence. When they arise, such aberrations do not survive changes of location and time. Lysenko's mid-twentieth-century corruption of biology in Soviet Russia depended on his political power and support. Lysenkoism failed when that support was lost. It was bad science, unsupported by evidence. What is distinctive about mature sciences is their uniformity across culture and time. The geometry of Euclid might have been codified in fourth-century BCE Alexandria, yet it long escaped its Alexandrian roots to become the geometry used internationally and for millennia, without serious challenge, until tiny corrections were required by general relativity in the twentieth century.

3. Competition Is Empirically Decidable

Competing systems of relations of evidential support derive from competing theories. They compete in the sense that they make incompatible factual claims about the world. Since science is empirical, such competition cannot be sustained indefinitely. The empirical character of science requires that the factual claims of a theory must be supported inductively by the evidence of observation and experiment.[3] To respect this empirical character, the competition among incompatible factual claims of competing theories must be resolvable by observation and experiment. If their factual differences are beyond observation or experiment, then whatever constitutes these differences lies outside empirical science.[4] It follows that there must be some possible observation or experiment capable of deciding among competing theories. The competition will be resolved as long as scientists are diligent and inventive enough in their pursuit of empirical evidence.

A radical, skeptical view holds that there are limited prospects for this sort of comparison. The worry is that observations are so theory laden that they are useless for comparisons of theories. Theories become, to use Kuhn's (1996, Chapters X, XII) expression for paradigms, "incommensurable" or, more simply, beyond cogent comparison. I do not share this skepticism. Theories can be compared on their adequacy to the empirical evidence and are routinely thus compared. The best account of this comparison is provided by Nora Boyd's (2018a, 2018b) empiricism, already mentioned in Chapter 2. Boyd shows that, if we are to decide between two theories on the basis of some item of evidence, the procedure is to wind back toward the provenance of the evidence. We continue until we have stripped away enough of the theoretical encumbrances to have freed the statements of evidence of entanglement with the theoretical presumptions of either theory.

---

3 To preclude confusion, the empiricism advocated here is what I call "small e" empiricism. It is the widely held view that we can learn our sciences only from our experiences. It is distinct from antirealist versions of "big E" Empiricism, such as van Fraassens' (1980) constructive empiricism in which *all* that we know of the world is *only* what we can or could experience directly.

4 Further analysis might be needed, however. The two theories might appear to be different only since they merely represent the same facts in different guises. Perhaps one or both theories contain content superfluous to the empirical successes of the theories.

These decisions need not be immediate. However, when empirical evidence favors one theory over another, it introduces an instability that must be resolved. Competing theories are responsible to all of the empirical evidence in their domains of application. A faltering theory can choose to ignore or discount unfavorable empirical evidence only temporarily while awaiting rescue from further evidence. Alternatively, the faltering theory can make internal adjustments to accommodate the unfavorable evidence. Such adjustments weaken the theory and make it more prone to further weakening.

These considerations would not apply to pairs of theories in one domain whose empirical content is so disjointed that they never disagree on what is observable while still retaining their identities as distinct theories. Although I grant this possibility in principle, I have had trouble finding real examples. Candidates can be sought in theories that treat some domain at very different scales both in size and in time. Perhaps neuroscience and psychology are cases in which both theories treat what is essentially just brain activities. They use different theoretical devices without intersecting or intersecting much empirically. Although this disjointed character is possible, neuroscientists in particular work energetically to breach it. I discuss another candidate briefly in Chapter 14, "Stock Market Prediction: When Inductive Logics Compete." There are different systems for predictions of moves in stock prices. Insofar as one system might make predictions only in the shorter term and another might make them over the longer term, it might be possible for them to proceed from disjointed factual bases. Although this is a possibility in principle, it does not seem to have been realized.

4. Inductive Competition Is Unstable

When one theory, in competition with another, gains a slight evidential advantage, it follows from the material nature of inductive inference that this advantage will be amplified. Facts warrant inductive inference, and the more facts a theory has secured the more it can infer inductively.

The role of hypotheses in a developing science can make this process of amplification potent. As we have seen, when the body of evidence supporting a science is meager, or the import of the existing evidence has not yet been fully explored, the scientists proceed in their investigations by positing hypotheses of suitable strength to warrant their inferences. These hypotheses must eventually be given suitably strong evidential support. During the preliminary period, it is possible to sustain multiple systems of facts and the inductive logics that they induce. Systems in competition will be distinguished by their employment of incompatible hypotheses. The viability of these multiple systems is fragile and unstable. If one system gains a small advantage through the import of novel evidence or a novel interpretation of existing evidence, then that small gain strengthens the system, in particular lending more support to its founding hypotheses. The competing systems are correspondingly weakened. This momentary advantage can persist and be amplified, or a weakened system itself can find new evidence that restores its support. However the competition might play out, its dynamics is unstable and overall tends to favor further strengthening of the system that has gained a small inductive advantage. The tendency then is for the advantaged system to be strengthened still further, whereas those in competition find it harder to recover. The dynamics drives toward the dominance of one system and the elimination of others.

5. Illustrations of Instability

A detailed examination of the competition described in Section 4 in particular cases would be lengthy. In later chapters, I provide such examinations in the cases of competing systems of stock market prediction and the historical competition between proponents and skeptics of dowsing. Here I can describe other cases only briefly. To do so, I draw from the convenience provided by Chapter 9 of *The Material Theory of Induction* (Norton 2021). As part of its analysis of the argument form "inference to the best explanation," the chapter reviews pairs or sets of theories in competition. We can see in these examples how each theory gains an evidential advantage while disadvantaging its competitors. Here I will not recount the details of the competing theories but only the dynamics of the competition. I refer readers to this chapter in *The Material Theory of Induction* for further details and citations of the pertinent literature.

5.1. Darwin's *Origin of Species*

In his *Origin of Species*, Darwin developed his theory of the origin of diverse biological forms through natural selection. It is portrayed throughout as in competition with the proposal that this diversity arises from the independent creation of each of these forms. Darwin argued that advantageous features of organisms arise through one process, their selection by nature. However, independent creation must attribute each new feature to a new decision by a Designer to create each organism just as it is. More telling are examples of organisms with features that have no apparent advantage. Why do terrestrial geese, for example, have webbed feet when webbing is useful only in water? Darwin gives an evolutionary account: terrestrial geese evolved from aquatic geese. Independent creation can attribute the webbed feet only to a capricious decision by the Designer.

With each successful accounting of advantageous and otherwise anomalous attributes, Darwin's original hypothesis of natural selection gains evidential support. Each of these successes weakens the competing hypothesis of independent design, which accumulates a growing burden of independent and capricious design decisions. The accumulation of these successes amplifies the evidential advantage of natural selection. It is moved from plausible speculation to a well-supported proposition while its competitor, independent creation, languishes.

5.2. Lyell's *Principles of Geology*

Uniformitarian geology asserts that present-day geological features were produced slowly by processes still acting in the present. Lyell's *Principles of Geology* made the case for it. Lyell was in a polemical dispute with competing catastrophist theories. They accounted for the same features by processes not currently acting and often of great violence. The initial advantage of the catastrophists was that it is natural to imagine great mountains and deep valleys as created by sudden, momentous events. Lyell chipped away at this advantage by showing how one geological feature after another can arise from currently acting processes. To use an example that he promoted, a competing account of fossils is that they arise in stone from a "plastic virtue, or some other mysterious agency." Lyell, however, accounted for them in terms of the fossilization of ordinary living things.

The evidential dynamic is similar to that of Darwin's case for natural selection.[5] With each uniformitarian success, Lyell's uniformitarian hypothesis is strengthened and its evidential advantage amplified, whereas support for special and even mysterious catastrophist processes is weakened.

5 That is not surprising since Lyell's work was an inspiration for Darwin.

5.3. Thomson's Cathode Rays

J.J. Thomson's paper in 1896 on "Cathode Rays" is celebrated as establishing that cathode rays consist of negatively charged particles, soon to be known as "electrons." Thomson, at the time, was embroiled in a debate with Philipp Lenard over the nature of these cathode rays. Thomson advocated for a particle account. Lenard defended the competing view that they are radiative, which then meant that they were waves propagating in the ether. Lenard had argued against a matter theory of cathode rays akin to Thomson's by noting that the rays persist even when the cathode ray tubes are completely evacuated. That is, there is no matter in the tubes to comprise the rays. Only ether remains. The rays, he concluded, had to be processes in the ether. Thomson's analysis depended on his experimental results that cathode rays are deflected by magnetic and electric fields exactly as if they are charged particles in rapid motion. Lenard struggled to accommodate these items of evidence in his account of ether. He could only speculate that Thomson's magnetic field had somehow disturbed the ether so that the rays would bend. This vagueness further weakened his retreating theory.

Thomson pressed his advantage with a coup de grâce. Waves in the ether bend because their velocity varies from place to place. This is how light is refracted by media of differing optical densities. A uniform magnetic field would disturb the ether in the same way in every place. Thus, elementary wave optics precludes it from bending cathode rays. However, uniform magnetic fields do bend the rays. Thus, the evidence that gave strong support to Thomson's particle theory is the same evidence that undid Lenard's ether wave theory.

The evidential advantage of Thomson's hypothesis is amplified by its accommodation of further evidence. For example, a metal vessel catching cathode rays becomes negatively charged, as one would expect if the rays are streams of negatively charged particles. An ether wave theorist might seek to dismiss this as an accidental artifact of the experimental arrangement. That escape ceases to be plausible once the charged particle hypothesis has an evidential advantage.

5.4. Einstein and the Anomalous Motion of Mercury

In November 1915, an exhausted Einstein was putting the finishing touches to his general theory of relativity. In that month, he found to his great joy that his new theory accounted exactly for a long-standing anomaly in the orbit of Mercury that so far had resisted explanation. His theory's success with Mercury was immediately recognized as an evidential triumph. The history does not follow the pattern of one theory gaining a slight evidential edge, which is then amplified. The accounts competing with Einstein's theory had all been discredited by the time of his completion of general relativity. However, if we consider the logical relations among the competing theories, independently of their order of emergence historically, then we see the same pattern of competition and amplification of slight evidential advantages.

The natural competitor to Einstein's theory is that the anomalous motion of Mercury arises from gravitational effects fully within Newtonian theory. It results from the perturbative effects of further, unrecognized matter. The "further matter" hypothesis has an initial advantage. It had become routine for astronomical anomalies to be resolved by the identification of further matter. For example, irregularities in the orbit of Uranus could be accounted for by the mass of a more distant, unrecognized planet. That led to the discovery of the planet Neptune. General relativity, however, is an exotic theory of extraordinary complexity mathematically. That it happens to return precisely the anomalous motion of Mercury is interesting. But it is hardly decisive evidence for the theory when standard Newtonian theory has a proven track record of accommodating just such anomalies by prosaic means.

However, these prosaic means falter. The various formulations of the favored, further matter hypothesis successively fail when evidence capable of separating the competing formulations is accommodated. If the further matter was located in a planet, "Vulcan," then its position was calculable, but no planet was observed there. Further possibilities located the matter in a slightly flattened Sun or in a dispersed cloud of matter surrounding the Sun that produces the zodiacal light. Neither proved to be viable. With each failure of the further matter hypothesis, the fortunes of Einstein's theory rose. Another possibility was an adjustment to the exponent in Newton's inverse square law of gravity. Although that exponent can be adjusted to accommodate the anomalous motion of Mercury, it fails to fit well with the motions of the remaining planets. Einstein's theory, however, has no adjustable parameters. It cannot accommodate any other motion of Mercury. Seen against this accumulation of failures of competitors, Einstein's theory rises as the only viable alternative.

5.5. Big Bang and Steady State Cosmology

In the mid-twentieth century, the prominent decision for cosmology was between the big bang and steady state theories. Later textbook accounts point to Penzias and Wilson's announcement in 1965 of their discovery of cosmic background radiation. It was, they said, the observational fact that confirmed the big bang theory and refuted the steady state theory. We are led to imagine the competition as ending abruptly.

That is not what happened. There was no immediate decision favoring big bang cosmology. It did gain a small advantage since the big bang cosmologists of the time — notably Dicke's group at Princeton University — had predicted something like it. However, the big bang cosmologists of the 1960s were reluctant to claim a definitive victory in print and with good reason, for the import of the evidence was still equivocal. Rather, it took roughly three decades for the decision between the two to be definitive.

Three developments were needed during these decades. First, considerably more observational work was needed. We now report Penzias and Wilson as observing thermal radiation of a cosmic origin of 2.7K. However, to affirm that a radiation field is thermal requires measurements across the spectrum. Penzias and Wilson had only measured one wavelength, 7.4cm. Many more measurements were needed and in fact were undertaken in the decades following. The incontrovertible evidence of a thermal spectrum was provided by NASA's COBE satellite in 1989.

Second, big bang cosmology needed to establish that it did indeed predict such thermal radiation. This required the development of precise cosmological models. In them, the radiation that we now measure is the remnant of radiation in a hot early universe that decoupled from matter when the cosmic fireball had cooled to 3,000K. That decoupled radiation is cooled to 2.7K by the expansion of the universe. Many components of this big bang account have to work correctly. The most troublesome is establishing that the early cosmic fireball was an equilibrium thermal system to which a temperature can be assigned in the first place. One could simply assume thermal equilibrium from the outset. It would be better, however, if cosmic processes in the early universe would produce this equilibrium. That was precluded in the cosmological models popular in the 1960s and 1970s by the so-called horizon problem. It showed that matter in those models was expanding so fast that it could not interact enough to achieve thermal equilibrium. The standard solution has been to invoke an early inflationary phase in the expansion of the universe.

The ready acceptance of this inflationary account illustrates the amplification of earlier successes. Until a big bang cosmology has some strong support, the inflationary addition is merely a speculative supplement to an already speculative theory. Once the big bang dynamics is supported, however, an inflationary phase is easy to accept as its natural completion.[6]

Third, it needed to be shown that steady state cosmology cannot accommodate the cosmic background radiation. This is by no means obvious, for thermal radiation can be acquired cheaply by theorists. All they need is some system to come to thermal equilibrium. Steady state theorists sought this through various avenues. One was that there is a slight opacity to space itself. Radiation from the continuous process of creation of steady state cosmology would be absorbed and reradiated through this slight opacity, thereby arriving at a thermal equilibrium. This proposal failed since the amount of opacity needed would be too great to allow observation of distant radio sources. Other efforts by steady state theorists, such as iron whiskers to thermalize starlight, also failed. This illustrates how an evidentially disadvantaged theory is further weakened by the need for successively more far-fetched repairs.

These three developments led to the decision in favor of big bang cosmology. That decision came slowly. Big bang cosmology enjoyed only a slight advantage at the outset. It grew steadily as observational results and theoretical developments favored it while efforts by steady state theorists to accommodate the same evidence faltered.

5.6. Arp and Bahcall on the Origin of Galactic Red Shifts

While the publicly more visible debate between big bang and steady state cosmologies proceeded, a narrower, less visible debate unfolded among astrophysicists and astronomers on the observational foundations of these cosmologies. Both big bang and steady state cosmologies assumed an expansion of the universe. Its evidential support lay in the finding by astronomers, starting most prominently with Hubble in 1929, that the galaxies are receding from our galaxy with a velocity that, on average, increases linearly with their

6 However, doubts linger about whether a period of inflation really does solve the horizon problem or whether it merely relocates it into the need to fine-tune initial conditions in a still earlier phase of cosmic expansion.

distances from our galaxy. (Hubble's analysis in 1929 is the subject of Chapter 7, "The Recession of the Nebulae.") An inference to a distance-dependent velocity of recession proceeded from the observation that light from the galaxies is uniformly shifted to the red end of the spectrum, with the shift increasing linearly with distance. This red shift was interpreted as deriving from a velocity of recession.

That the red shift in a galaxy's light resulted from its velocity of recession was disputed energetically by Halton Arp, a well-established astronomer. His case against this association grew in the 1960s and was regarded as sufficiently serious to merit a direct confrontation at the meeting of the American Association for the Advancement of Science (AAAS) on December 30, 1972, in Washington, DC. There Halton Arp faced John Bahcall, an astronomer at the Institute for Advanced Study in Princeton, New Jersey, there to defend the standard view.

I need not here rehearse the details of the debate. I have recounted them elsewhere (Norton, 2023) and refer readers to this source for elaborations. What matters for my purposes here is that the confrontation of Arp and Bahcall provides another illustration of the unstable dynamics of competition among theories. Is the red shift of light from galaxies the result of their motion of recession, as Bahcall affirmed? Or is it the result of some other source, as Arp argued? Each laid out his case.

Bahcall based his case on the evidence, available in multiple forms, that the red shift of light from the galaxies varies roughly linearly with the distances to those galaxies. Establishing that linear dependence was his major concern. The connection to a velocity of recession was provided by the then favored expanding universe cosmologies: they all required a linear relation between the velocity of recession of a galaxy in our vicinity and its distance from us.

Arp's case depended on his own extensive observations of galaxies. Arp had amassed an extensive collection of cases of galaxies that appeared to be physically connected but had very different red shifts. A physical connection would mean that the associated galaxies must be at roughly the same distance from us. Their marked differences in red shift could not then derive from a linear dependence of red shift on distance.

The two views in competition were sufficiently strong to merit serious examination at the AAAS meeting in 1972. However, the competition was unstable. Bahcall's view was already the recognized view. As his position strengthened subsequently, Arp's dissenting view was correspondingly weakened.

We can trace this instability in the competition in three areas. The first area was new astronomical data, which continued to conform to Bahcall's view. Arp's view, however, was weakened by investigations indicating that the physical associations so central to his case were merely fortuitous alignments in our sky of objects separated by great distances.

The second area was the connection to cosmology. Bahcall's view conformed to then standard cosmologies. If one applies general relativity to the sorts of matter distributions observed by the astronomers, then a dynamic cosmology ensues. It might be contracting or expanding. However, a static universe, such as Einstein had originally proposed in 1917 and Bahcall needed, was unstable and thus precluded.

Just as Bahcall's view was supported by then standard cosmology, so too his view of the linear dependence of red shift and distance provided support for the cosmology. It was the observational basis of the expansion of the universe. The outcome was a magnification of his evidential advantage. His evidential success strengthened support for expanding universe cosmologies, and their strengthened support then further enhanced his position.

Arp's view, however, found no support in existing cosmology. If the red shift was not derived from a velocity of recession, then the ensuing cosmology was one of an overall static mass distribution that lay outside standard cosmology. To preserve the viability of his critique, Arp needed to presume a static cosmology for which there was no real independent support. The evidential processes that enhanced support for Bahcall's view simultaneously weakened support for Arp's view.

The third area in which the instability manifested was in the physical basis of the red shift. Bahcall's standard view could employ a simple basis, ready to hand. The velocity of recession of galaxies in an expanding universe cosmology led directly to it. With that source precluded, Arp had no correspondingly established physics from which to derive the red shifts. He resorted briefly to speculation, such as "tired light."

Quasars proved to be a decisive test. They are luminous bodies with great red shifts. In the standard view, they must be very distant from us and thus have enormous intrinsic luminosity. Initially, in this view, it was hard to explain the enormous energies that it supposed for these bodies. Arp's alternative was that they are merely nearby objects, highly red shifted, but not of such great intrinsic luminosity. Quasars were subsequently identified as the enormously energetic nuclei of a galaxy, likely holding a supermassive black hole. Once again the evidential success of Bahcall's standard view was magnified. The view supported the immense energy and distance of quasars, and establishing a physical basis for their immense energy then enhanced support for Bahcall's standard view. Arp, however, was unable to provide a cogent physical basis for the high red shift of quasars if they are assumed to be nearby objects.

As Bahcall's standard view went from strength to strength, Arp's dissident view faltered and was dropped from serious consideration.

5.7. More Illustrations

Chapter 9 of *The Material Theory of Induction* (Norton 2021) recounts two more competitions: oxygen versus phlogiston theories in the late eighteenth century and corpuscular versus wave theories of light in the nineteenth century. The details of their competition are too involved to admit compact summaries. I can extract one result, however.

At the crudest level, oxygen theory prevailed over phlogiston theory when Lavoisier's experiments required that oxygen must be attributed to a conserved weight. Phlogiston theory faltered since the same experiments required that phlogiston be attributed to a dubious negative weight, levity. Similarly, a major factor in the decision on theories of light came with Fizeau and Foucault's measurements of the speed of light in air and water. The corpuscular theory required the speed to increase in a denser medium, whereas the wave theory required it to decrease in a denser medium. The experiments found a decrease in the speed.

We see here that theories in competition are responsible to the same experiments and that careful exploration can find experiments that only one of the theories can accommodate. Although we might doubt that just one experiment can be decisive, that responsibility still plays a major role in the dynamics that leads one theory to prevail over its competitors.

6. Unconceived Alternatives

The instability illustrated in these examples arises in the competition between two theories. Is that enough to make the case? Might we worry that there is a third, fourth, or fifth, as yet unimagined or unarticulated, theory lurking in the wings, such that evidence cannot separate one of them from our favorite theory? The possibility of such further theories has been defended, notably by Stanford (2006), as "unconceived alternatives."

They do not provide the sort of threat that one might imagine. They open the possibility that our current best theory might not be the one best supported by the evidence. That is not the question here. The question is whether the best supported theory is unique. That can be the case even when the theory that we happen to favor most strongly is not the best supported.

For unconceived alternatives to challenge uniqueness, they must provide us with a theory challenging our favored theory that is equally well supported, assuming that our favored theory is the best supported given the evidence, or the challenger theory must provide us with two unconceived alternatives equally well supported and still better supported than our favored theory.

The analysis already given indicates that such an achievement lies beyond what unconceived alternatives can supply. As long as these alternative theories differ in some factual claim, their difference must be open to adjudication by observation and experiment, even if that adjudication might not be practical immediately. Otherwise, their differences lie outside empirical science.[7]

7. The Underdetermination Conjecture

If one seeks literature to contradict this chapter's claim of uniqueness, the natural reference is the "underdetermination thesis."[8] Loosely speaking, it asserts that no body of evidence, no matter how extensive, can pick out a theory uniquely as the one best supported inductively. The thesis is then used to advance the tendentious claim that our commitment to any theory, even those of the most mature sciences, always relies on the addition of other factors, possibly social, psychological, pragmatic, or conspiratorial. The thesis is mislabeled as a "thesis" insofar as theses are commonly taken to be propositions for which we have good evidence. It is, as I will now argue, merely a conjecture that has never secured proper support. It can be stated for my purposes here as

7 The closest that the literature can provide, for these theories, balanced perfectly evidentially, arise as the observationally equivalent theories used to support the underdetermination thesis. In Section 7, I explain why these examples fail in their purposes.

8 For an introduction, see Stanford (2017).

Underdetermination conjecture: any body of empirical evidence, no matter how extensive, will provide inductive support for multiple, mutually exclusive sets of propositions such that no one set is distinguished as enjoying the strongest support.

This conjecture should be distinguished from the weak, de facto claim that at some definite moment the extant evidence for a theory might fail to determine it. This circumstance commonly arises in newly emerging sciences. If the science matures, then it is merely a transient shortcoming. Otherwise, it is not.

The full conjecture is remarkably strong in its pessimism. It applies to all bodies of evidence and theory. Thus, it is astonishing that the conjecture has never advanced beyond what for many is merely a comfortable hunch. For them, the conjecture seems to be plausible and welcome. If one is inclined to it, then easy but inadequate examples might be enough motivation. The evidence can tell us of a correlation between children who watch cartoons and children who behave violently in the playground.[9] That evidence leaves undetermined which causes which, or if there is a common cause of both, or if the correlation itself is mere happenstance. The example merely illustrates de facto underdetermination. Randomized control trials can decide among the possibilities.

Once it has been mentioned enough in the literature, the plausibility of the conjecture for some makes it easy to lose sight of the fact that there is no cogent demonstration of the conjecture. The arguments offered in favor of the underdetermination conjecture have been subjected to repeated analysis and have failed scrutiny. The arguments can be shown to neglect much of the existing work in inductive inference and to make dubious claims concerning observationally equivalent theories. Laudan and Leplin (1991) and Norton (2008) explore these failures, too extensive to be developed in detail here.

The simplest and most common demonstration of the conjecture rests on an inadequate account of inductive inference. A single body of empirical evidence can be entailed by many different sets of hypotheses, with suitable boundary conditions and auxiliary assumptions. With a naive hypothetico-deductive account of confirmation, it would then follow that they are all equally well supported inductively. This naive account has long been

9 This example is from the opening paragraph of Stanford (2017).

subjected to criticism from many perspectives. Consider the standard geological and evolutionary account of the origin of fossils. Compare it with a revisionary theory claiming that the Earth and its rock strata were created five minutes ago, complete with an intact fossil record. Since both entail the same evidence, we would have to say that both are equally well supported. The standard response in the literature is sketched in Section 5, "Hypothetical Induction," in Chapter 1. It is that bare hypothetico-deductive confirmation must be supplemented by further conditions to enable discrimination in such cases. We might be told, for example, to assign greater support to the more explanatory hypothesis or to the simpler one.

Within the material theory of induction, merely entailing the evidence by itself does not confer inductive support for a hypothesis or theory. The entailment must happen in the right way: each of the parts of the propositions in the theory must be supported inductively in accordance with the requirements of the material theory. The supposition that the creation occurred exactly five minutes ago, as opposed to ten or fifteen minutes or a millennium ago, must be supported. The revisionary theory can provide no discriminating evidence. In comparison, standard geology does provide extensive evidence for its chronology of the formation of the Earth.

The transition from hypotheses that merely entail the evidence to an evidentially well-supported body of propositions is difficult and can take a long time. We see in Chapter 12, "The Use of Hypotheses in Determining Distances in Our Planetary System," that, in spite of sustained and ingenious efforts, a system of orbital sizes for the planets of our solar system was not firmly established until the eighteenth and nineteenth centuries. Indeed, at the most general level, the nature of inductive inference is sufficiently irregular, according to the material theory of induction, that there can be no sufficiently expansive framework sufficiently precise to admit a cogent demonstration of the conjecture.

After I drafted this chapter, Sam Mitchell sent me his book published in 2020. It also seeks to undo the skepticism concerning the reach of evidence associated with Duhem and Quine. His concern is specifically to respond to the claim that the import of evidence is always holistic. We cannot be assured that contradicting evidence refutes any specific hypothesis, the inductive pessimists insist. They suppose that any such judgment requires auxiliary hypotheses that can be the real culprits in the contradiction. Mitchell disagrees.

His analysis agrees on many points with the one developed here and is most welcome.

8. Observationally Equivalent Theories

Theories that have the same observable consequences are frequently displayed in the literature on the underdetermination thesis as "observationally equivalent theories" or "empirically equivalent theories."[10] They serve to illustrate the underdetermination thesis since, it is asserted erroneously, no evidence can favor one over the other, and they are used in an attempt to make the case for the underdetermination thesis.

Do these observationally equivalent theories pose a threat to the uniqueness urged in this chapter? Here I will recount briefly why they do not. I will use a simple example of a pair of observationally equivalent theories. For a more expansive inventory of examples and for more detailed, critical analysis of the underdetermination thesis along the lines below, see Norton (2008).

In the early seventeenth century, purely astronomical observations of the relative positions of the Sun, Moon, and planets could not discriminate two systems. The first was the familiar Copernican heliocentric system. The second was the Tychonic geocentric system. The observational equivalence followed assuredly from the simple fact that the Tychonic system could be generated merely by relocating the point of rest in the Copernican system from the Sun to the Earth but otherwise preserving all relative motions.

This example and the others like it fail to sustain any interesting conclusions about the limited reach of evidence for two reasons. First, if the competing theories differ in something factual, then the empirical character of science requires that the difference should manifest in something observable. The Copernican and Tychonic systems differ in whether the Earth or the Sun is at rest. Purely astronomical facts about the relative positions of the Sun, Moon, Earth, and other planets cannot decide, for they provide no notion of rest. They can be separated, however, if we ask after the physical forces acting among the bodies of the solar system. Newton's later physics distinguished bodies moving inertially from those that accelerate. Inertial motion becomes

10 Here I resist this latter expression because of its vagueness. If two theories have identical observational consequences, it does not follow that they are supported equally by observations. That is, one can still be favored empirically over the other, as I argued in the preceding section.

the Newtonian surrogate for rest. At most, either the Earth or the Sun can be in inertial motion. When we seek the gravitational forces acting between the bodies of the solar system, that body must be the Sun and not the Earth. We decide in favor of the Copernican system.[11]

This decision was possible because subsequent investigations in a broader domain, that of gravitational physics, provided the further evidence needed to separate the systems. This possibility remains for every case of observationally equivalent theories. Insofar as they differ on anything factual and they lie within empirical science, we cannot preclude new evidence separating them. Indeed, we should expect determined investigators to find such evidence.[12]

Second, if we set aside the possibility of new evidence, then there is a second failing of all the cases of observationally equivalent theories in the literature. If the case is to be presented in the literature, then it must be possible to demonstrate in the confines of tractable publication that the two theories really are observationally equivalent. For example, there is a simple recipe for converting the Copernican system into the Tychonic system. We take the motions of the Copernican system and simply subtract vectorially from them the motion of the Earth. The result is a system of motions with the Earth at rest but agreeing with the Copernican system in all relative motions.

When such a translation is available, we cannot preclude the possibility that the two theories do not differ in anything factual. Rather, they are merely different presentations of the same theory. If we restrict considerations only to the relative positions of bodies in the solar system, then this is the case for the Copernican and Tychonic systems. They differ only in the designation of which body is at rest. But that designation lies outside the body of facts pertinent to our restricted domain. It is, as far as they are concerned, merely an empty stipulation.

---

11 As an exercise, one might like to contemplate whether some distribution of masses might enable the Tychonic system to conform to Newtonian gravitation theory. One would require, for example, that the Earth must be much more massive than the Sun so that the Sun orbits the Earth and not vice versa. We can then no longer account for the motion of Venus, whose maximum elongation from the Sun is between 45° and 47°. It would be pulled out of its orbit around the Sun by the far greater attraction of the Earth or, failing that, display significant perturbations because of the Earth's attraction.

12 Here the historical sciences might provide an exception. The totality of evidence recoverable from some archaeological site, for example, might leave questions about the site unanswered. The failure is not the result of a lack of power of inductive inference but merely the paucity of evidence.

This possibility threatens all cases of observationally equivalent theories. That they can be interconverted opens the possibility that they are merely the same theory. They differ only in their descriptions and in superfluous posits of no factual import. It is possible and sometimes enticing to mistake these posits as having factual import, even though they manifest in nothing observable. The most familiar example in real science concerns a suitably refined version of Lorentz's ether-based electrodynamics and the relativistic electrodynamics that Einstein introduced in 1905 with his special theory of relativity. The two are observationally equivalent, and, as far as experimentation was concerned in the first decade of the twentieth century, they were treated as the same theory. However, Lorentz insisted that the ether factually has a state of rest, contrary to Einstein's principle of relativity. The difficulty was that nothing observable — no experiment — could determine just which of the infinity of inertial states of motion was that ether state of rest. The mainstream of physics soon came to discount the ether state of rest as fictional.

9. Formal Accounts

Since the material theory of induction can meet the challenge, it is proper to ask whether formal accounts of inductive inference can also meet it. They do not do well with it and for reasons associated directly with their formal character.

First, as I argued at some length in *The Material Theory of Induction* (Norton 2021), the rules of various formal systems are poorly articulated, and an ambiguity in their import is inevitable. Consider, for example, the use of arguments by analogy to infer the properties of light. Light is analogous to sound in that both have a wave character. The pitch of sound is analogous to the color of light. However, sound needs a medium in which to propagate, the air, analogous to the discredited nineteenth-century luminiferous ether. This difficulty does not arise in a different analogy. In it, light is taken as analogous to rapidly moving corpuscles. Then light, like corpuscles, can propagate in vacuo without the support of a medium. Yet the corpuscles of the nineteenth-century and earlier theories have no wave-like properties. Just how are we to weigh the conflicting successes and failures of these different analogies? The general rules in the literature are too vague and hedged to give us a definite answer.[13]

Second, there are multiple formal schemes for inductive inference and no clear guides to which to use in any application. Take, for example, argument by analogy and inference to the best explanation. Neither of the analogies of light to sound and light to rapidly moving corpuscles recovers the phenomenon of light polarization. Sound waves are longitudinal, whereas polarization derives from the transverse character of light waves. That is, neither of the familiar analogies provides an explanation of polarization. Rather, the best explanation of polarization is that light is not analogous to either sound or corpuscles.[14]

Which formal scheme should be applied where? In particular cases, we might use prudence to decide and have things work out tolerably well. However, we do so in the absence of unambiguous metalogical rules.

Third, the Bayesians are confident that they have a solution. Their scheme, they believe, embraces and explains all others and can recover uniqueness through various limit theorems. This confidence can be sustained only as long as they ignore the enduring and insoluble problem of the priors. The Bayesian system is not and cannot be self-contained. The selection of prior probabilities must be made outside the normal processes of conditionalization by Bayes' theorem. Yet these priors can be so selected as to protect almost any bias. The simplest illustration arises when we have two theories T_1 and T_2 that deductively entail the same evidence E. Then we have equal likelihoods: $P(E|T_1) = P(E|T_2) = 1$. An application of Bayes' theorem then tells us that

$$P(T_1|E) \,/\, P(T_2|E) = P(T_1) \,/\, P(T_2)$$

That is, our comparative assessment of the relative support afforded to the two theories by the evidence, the ratio of posterior probabilities $P(T_1|E) / P(T_2|E)$, is determined entirely by whatever external judgments led us to the ratio of prior probabilities $P(T_1) / P(T_2)$. Bayesians face an unwelcome dilemma: either set these priors arbitrarily so that the final judgment is arbitrary, or seek

13 See Chapter 2 of this text and Chapter 4, "Analogy," in *The Material Theory of Induction* (Norton 2021).

14 Might we try the analogy to waves propagating along a flexible rope since they are waves of transverse displacement? This analogy fails to recover the behavior of polarized light in polarizing filters. The best explanation of the behavior is that, when it comes to polarizing filters, light is not analogous to waves on a flexible rope.

guidance from other accounts of inductive inference. This problem troubles all formal calculi of inductive inference, or so I have argued in Chapter 12, "No Place to Stand: The Incompleteness of All Calculi of Inductive Inference," in *The Material Theory of Induction* (Norton 2021). None can be self-contained but can only return nontrivial results insofar as nontrivial inductive content is introduced from outside the scope of the calculus.

It is fortunate that scientists do not try to conform their judgments of inductive support algorithmically to these conflicting and ambiguous formal schemes, for that would induce inductive anarchy.

10. Conclusion

In this chapter, I have sought to establish that the threat of multiple and equally well-supported systems of inductive inference has been parried. The escape derives from the empirical character of science. Competing systems of inductive logic derive their competing factual warrants from different theories within science. When these warranting facts differ, their differences must manifest in something accessible to possible observation, or they lie outside empirical science. When the pertinent observations are secured, they will strengthen one of the theories while weakening its competitors.

This escape is enhanced by the close integration of the facts of a science and its relations of inductive support, asserted by the material theory of induction. The integration promotes a positive feedback dynamic that accelerates the strengthening of one system of relations of support at the expense of its competitors. As more of the factual claims of a science are sustained by the evidence, the growing body of supported fact authorizes stronger inductive inferences within the domain of the science. That in turn leads to inductive support for still further facts. As one theory ascends, even if haltingly, its competitors will fall. When sufficient evidence is available, the accumulation of these processes will lead to the dominance of one science and its associated relations of inductive support while its competitors are eliminated. The uniqueness and inductive solidity of mature sciences in their domains are expected and explained.

REFERENCES

Boyd, Nora. 2018a. "Scientific Progress at the Boundaries of Experience." PhD diss., University of Pittsburgh.

———. 2018b. "Evidence Enriched." *Philosophy of Science* 85: 403–21.

Kuhn, Thomas S. 1996. *The Structure of Scientific Revolutions*. 3rd ed. Chicago: University of Chicago Press.

Laudan, Larry, and Jarrett Leplin. 1991. "Empirical Equivalence and Underdetermination." *Journal of Philosophy* 88: 449–72.

Mitchell, Sam. 2020. *Fault Tracing: Against Quine-Duhem*. Berlin: De Gruyter.

Norton, John D. 2008. "Must Evidence Underdetermine Theory?" In *The Challenge of the Social and the Pressure of Practice: Science and Values Revisited*, edited by Martin Carrier, Don Howard, and Janet Kourany, 17–44. Pittsburgh: University of Pittsburgh Press.

———. 2021. *The Material Theory of Induction*. BSPS Open Series. Calgary: University of Calgary Press.

———. 2023. "Inductive Inferences on Galactic Redshift, Understood Materially." In *Current Debates in Philosophy of Science*, edited by C. Soto, Synthese Library 477, 227–46. Cham, Switzerland: Springer.

Stanford, Kyle. 2006. *Exceeding Our Grasp: Science, History, and the Problem of Unconceived Alternatives*. New York: Oxford University Press.

———. 2017. "Underdetermination of Scientific Theory." In *The Stanford Encyclopedia of Philosophy*, winter 2017 ed., edited by Edward N. Zalta. https://plato.stanford.edu/archives/win2017/entries/scientific-underdetermination/.

van Fraassen, Bas. 1980. *The Scientific Image*. Oxford: Oxford University Press.

5

Coherentism and the Material Theory of Induction

1. Introduction

At the large scale, relations of inductive support within the material theory of induction are nonhierarchical and admit circles. In this aspect, these relations are similar to the relations of justification within coherentist theories of epistemic justification in epistemology. To me, this was a welcome coincidence and offered the possibility that the coherentist literature would be useful in addressing outstanding issues in the large-scale structure of material relations of inductive support. Are circularities in the overall structure troublesome? Are they vicious? Do they leave inductive structures underdetermined? These questions do have satisfactory answers within the material theory. I have developed them in the two preceding chapters. In developing these solutions, it became apparent that the coherentist literature was consistently unhelpful in solving these problems. Further investigation showed that this was no mere oversight by coherentist philosophers. Rather, the framework of coherentism in epistemology is sufficiently different from that of the material theory of induction that its problems are only similar to but not identical with those of the material theory. Moreover, the resources accessible to coherentist analysis prove to be weaker than those accessible in the material theory, and coherentists at best can provide weaker solutions to the problems.

In this chapter, I will review the similarities and dissimilarities between the two approaches with the purpose of substantiating the appraisal just given. In Section 2, I will recall the basic claims of coherentist theories of justification, and in Section 3 I will note the aspects in which they are similar to the large-scale structure of relations of inductive support of the material theory. In Section 4, I will catalog the many dissimilarities. Coherentism is holistic, whereas the material theory is local (Section 4.1). Coherentism takes global coherence as its basic relation, in which the material theory takes a local relation of inductive support as its basic relation (Section 4.2). Coherentism defines itself by its opposition to foundationalism in epistemology. As a result, it faces significant difficulties in accommodating the role of the world in its justifications. The material theory has no corresponding problem (Section 4.3). Where the material theory is a theory of inductive logic independent of human cognition, coherentism takes beliefs as the relata for relations of justification and must give an account of how these relations appear in an agent's cognition (Section 4.4). Finally, of lesser concern, the common exemplars of beliefs for coherentism are prosaic beliefs in ordinary life. The material theory is designed to be an account of evidential relations in the sciences. The result is that coherentists identify different problems from material theorists as pressing and emphasize different aspects of the relations of support (Section 4.5).

In Section 5, I will review the common problems facing both coherentism and the material analysis. Are the circularities in their structures harmful? Do these structures allow multiple, equally admissible systems? How does the world inform the relations of support and justification? Are the structure's justifications indicative of truth? In efforts to answer these questions, it is argued that coherentism has fared poorly, whereas the material theory has not.

In Section 6, I will review the recent Bayesian literature on coherentism. The primary goal of that literature is to vindicate or disprove the idea that overall coherence leads to truth-conducive justification. That means that it proceeds from the holism of coherentism. Since the material analysis of the large-scale structure of inductive support is not holistic, the Bayesian analyses are tangential to it. In Section 7, I will arrive at a negative appraisal of the entire Bayesian project of examining coherentism. It is misplaced since the notion of coherence itself is not a probabilistic notion. The probabilistic formalization is premature since the notion of coherence remains poorly articulated. Finally, it follows from the material theory of induction that a probabilistic framework is not general enough to provide any universally applicable results on coherence. Results derived using the false presumption of universal applicability are ill founded.

2. Coherentist Theories of Epistemic Justification

Coherentist theories of justification came to prominence in the later part of the twentieth century through the work of several philosophers, most notably Keith Lehrer (1974, 1990, 2000) and Laurence BonJour (1985). The literature seems to have lost its momentum in the early 2000s. Presumably, part of the reason was that BonJour (1999) abandoned the approach after recognizing the gravity of the problems facing it.[1] The coherentist approach was subsequently revived by Bayesians, notably Bovens and Hartmann (2003) and Olsson (2005). These Bayesian accounts proceed from the assumption that a coherentist approach is already at hand. They seek to express and assess the core notion of coherence in probabilistic terms. Often their results contradict coherentism. My impression is that BonJour's *The Structure of Empirical Knowledge* (1985) provides the best articulated version of coherentism and in a form that can be connected most readily with the concerns of the material theory of induction. Hence, I will draw from his treatment.

The coherentist theories of justification (henceforth just "coherentism") should be distinguished from a coherentist theory of truth, such as that articulated in Rescher (1973). The latter gives an account of what it is for a proposition to be true,[2] whereas the coherentist theories of justification seek only the grounds under which an agent is justified in holding a belief.

The core claim of coherentism has been stable over the decades. Lehrer (1974, 154) gives it thus:

> Justification is a reciprocal relation of coherence among beliefs belonging to a system. According to a coherence theory, a belief is completely justified if and only if it coheres with a system of beliefs.

More recently, Olsson (2017), in his article in the *Stanford Encyclopedia of Philosophy*, gives it thus:

> According to the coherence theory of justification, also known as coherentism, a belief or set of beliefs is justified, or justifiably

1 "With the exception of work being done by Bayesians, few epistemologists are presently working on coherentism" (Murphy 2020, Section 6); "coherentism is pretty obviously untenable, indeed hopeless" (BonJour 1999, 139).

2 They "hold that truth is to be simply *identified* with coherence" (BonJour 1985, 88).

> held, just in case the belief coheres with a set of beliefs, the set forms a coherent system or some variation on these themes.

The stipulations (Lehrer) "if and only if" and (Olsson) "just in case" are strong and prove to be troublesome. They reflect the conception of coherentism as the alternative to foundationalism.

In the foundationalist conception, beliefs are justified inferentially by other beliefs. A regress ensues as we trace back along the chains of the justificatory inferences. The regress is halted by positing basic beliefs that terminate the chains since they themselves are not justified inferentially. The obvious candidates for these basic beliefs are those somehow given to us directly by experience. This direct connection with experience is what allows us or even compels us to accept them. The existence of some form of basic beliefs is the distinctive thesis of foundationalists.

In opposing foundationalism, coherentists seek to escape the regress argument by denying the linearity of relations of justification. Instead of tracing the chains of justification back to these anchoring beliefs, coherentists urge that tracing the chains merely takes us on a tour of our system of beliefs that eventually will cycle back to our starting point. There is no need for terminal beliefs to anchor our chains of justification. All that is needed is that our full system of beliefs forms a coherent system.

3. Similarities

It is in the articulation of this last conception that coherentist writing comes closest to resembling the description in Chapter 2 of the large-scale relations of inductive support in the material theory of induction. BonJour introduces the coherentist escape from the foundationalist's regress argument as follows:

> According to the envisaged coherence theory, the relation between the various particular beliefs is correctly to be conceived, not as one of linear dependence, but rather as one of mutual or reciprocal support. There is no ultimate relation of epistemic priority among the members of such a system and consequently no basis for a true regress. Rather the component beliefs of such a coherent system will ideally be so related that each can be justified in terms of the others, with the direction of argument on a particular occasion of local justification

> depending on which belief (or set of beliefs) has actually been challenged in that particular situation. And hence, a coherence theory will claim, the apparent circle of justification is not in fact vicious *because it is not genuinely a circle*: the justification of a particular empirical belief finally depends, not on other particular beliefs as the linear conception of justification would have it, but instead on the overall system and its coherence. (1985, 91–92; BonJour's emphasis)

Correspondingly, as we saw in Chapter 2, the material theory of induction introduces a large-scale structure of relations of inductive support that is nonhierarchical and contains circles of dependency of all sizes.

4. Dissimilarities

Although the similarities noted above are striking, there are many dissimilarities between coherentism and the material theory of induction. The theories differ on so much that they are best understood as distinct. In this section, I will review the main differences.

4.1. Holism versus Localism

BonJour (1985, 91) distinguishes "local" and "global" levels of justification. The local level contains justifications for a belief or small set of them that take the larger system of belief for granted. The global level concerns the justification of the belief system in its entirety. It is this global level, BonJour argues, that has been neglected and becomes the basis of coherentism. Earlier he gives priority to the global over the local:

> According to a holistic view [coherence theory], it is such a system of beliefs which is the primary unit of justification; particular beliefs are justified only derivatively, by virtue of membership in such a system. (24)

He then insists on the importance of the global level for *any* admissible view:

> For the sort of coherence theory which will be developed here — and indeed, I would argue, for any comprehensive, nonskeptical epistemology — it is the issue of justification as it

> arises at the latter, global, level which is in the final analysis decisive for the determination of empirical justification in general. (91)

More tersely, BonJour avers that "the basic unit of justification for a coherence theory is an entire system of beliefs" (103).

The conception is holistic. The justification for a belief derives from its relationship to a belief system that, in its totality, is coherent. This large-scale coherence is constitutive for justification.

This holistic conception brings recalcitrant problems for coherentism, for the strength of a belief system is gauged by its coherence at the global level. The obvious and immediate difficulty is that systems can vary in the strength of their coherence in different parts. How are these varying strengths to be combined into a single global measure? There are many ways to provide synoptic measures of varying quantities. When the quantities are numerical, we can choose from among arithmetic means, geometric means, medians, modes, and much more. Since we do not have a precise measure of coherence, we cannot even begin to ask which is the appropriate synoptic assessment. Thus, the strength of the justification of any particular belief in the system depends on a univocal judgment of the strength of its coherence, where no such univocal judgment is available. The problem is compounded when we seek to decide among competing belief systems. We must arrive at judgments of each system's coherence clear enough to sustain univocal comparison.[3] This difficulty will return below when I report that Bayesians have found it impossible to identify a single probabilistic measure of coherence.

The material theory of induction, by its constitution, is a local theory. It inverts BonJour's conception of the global as the primary unit and the local as derivative. That is, materially, the basic relation is the inductive support accrued to each proposition by the evidence and warranting facts. The totality of all of these relations is the large-scale structure. Whatever notion of support applies to this structure as a whole is derivative. It results from the combination of the local relations of inductive support. Similarly, that there

3 BonJour concedes this problem: "But the main work of giving such an account [of coherence], and in particular one which will provide some relatively clear basis for *comparative* assessments of coherence, has scarcely been begun, despite the long history of the concept" (1985, 93–94; BonJour's emphasis).

are circular interdependencies is not constitutive of inductive support. It is a derivative result recovered only after all of the local relations of inductive support are combined.

Since the material theory does not anchor the support of individual propositions to the coherence of the entire system, it escapes the difficulty just sketched for coherentists. The inductive support for some proposition is dependent in turn on the inductive support for its evidence and warranting fact. We can trace this support back through the support for further propositions and might end up touring through much of the pertinent science. The overall strength of support for the original proposition derives from the summation of these relations of support. Such summations can deliver different overall strengths of inductive support for different propositions. They do not depend, as a holist would require, on a single measure of the coherence of the science as a whole.

Thus, the material theory accommodates cases of science in which coherence is strong in some places but not in others. It allows the strength of inductive support for individual propositions to reflect these differences, as it should if coherence matters at all.

Quantum mechanics is one of our most successful scientific theories. It underpins much of modern science, from particle physics to the physics of condensed matter and semiconductors, to modern theories of chemical structures and reactions, and to more. However, in places, it lacks coherence. Most notably, quantum measurement is a recalcitrant problem. There are multiple competing accounts of it. Their persistence is a clear sign that none is correct or, at least, that none is demonstrably so. A second area of difficulty is that quantum field theory breaks down at sufficiently high energies. This is revealed by the appearance of infinite energies whose presence needs to be controlled by computational techniques such as renormalization.

These weaknesses reflect a lack of coherence in those parts of quantum theory. They will affect some results of quantum theory more than others. These differences will then be reflected in differences of inductive support assigned by the material theory. For example, quantum theory has met with extraordinary success in accounting for the emission spectra of the elements. According to the material theory, the support for this account propagates through much of quantum theory. It does so in a way insensitive to the vagaries of quantum measurement and thus can be strong. Matters are different with the familiar claim that quantum measurement as a "collapse of the wave packet" consists of an instantaneous effect that has propagations faster than light. The strength of inductive support for this superluminal propagation depends in turn on the strength of support for this particular approach to quantum measurement. Its status remains unclear, and there are competing accounts of quantum measurement that do not include this collapse as a physical process.

The differences in inductive support for these two propositions are recovered fully from a summation of the iterated supports. Neither is traced back to a univocal measure of the coherence of quantum theory as a whole.

4.2. Coherence versus Inductive Support

The problems just sketched for coherence theories derive from the supposition that coherence is assessed holistically. Those problems are compounded by the lack of a clear articulation of the notion of coherence, whether understood holistically or locally. At an intuitive level, the idea is simple enough. Coherence is a matter of how well "a body of beliefs 'hangs together,'..." to use BonJour's (1985, 93) expression. Giving a clearer account, however, presented BonJour with so many obstacles that he began with a disclaimer that his response was "deliberate — though I think, justified — evasion" (94). What follows (94–100) is a four-component "outline" of the notion of coherence. A system of belief is coherent to the extent that it is

- logically consistent;
- probabilistically consistent;
- explanatory; and
- includes significant conceptual change.

Logical and probabilistic consistency are enhanced by the extent of inferential relations among beliefs. The explanatory strength of the system rises as the extent of the explanatory anomalies falls. Finally, the inclusion of significant conceptual change is justified by noting that such changes commonly come with scientific advances.

The difficulties of this outline are all too clear. Since it depends on four conditions, the possibilities for internal conflict are great. For example, the early forms of quantum theory in the first decades of the twentieth century were extraordinarily explanatory. That was their appeal. However, equally clearly, they were logically inconsistent.

The deeper problem is that articulation of the notion of consistency now depends on further theories, most notably of probability and explanation. The presumption is that, elsewhere, there are cogent accounts of each.[4] That is not so. These are troubled notions. I spent considerable effort in *The Material Theory of Induction* (Norton 2021) showing that these notions fail to function as routinely expected. There is a default presumption that, whenever we have some sort of uncertainty or indefiniteness, probabilities capture it. Chapters 10–16 were devoted to showing that this presumption has no good foundation and leads to mistaken judgments. Regarding "explanation," there is no single, universal understanding of the term. In Chapters 8 and 9, I argued that there is no distinctive notion of explanation able to power inductive support, even in the canonical and celebrated examples of inference to the best explanation.

BonJour has no stomach for any real defense of his account. "A fully adequate explication of coherence," he admits, "is unfortunately not possible within the scope of this book (nor, one may well suspect, within the scope of any work of manageable length)" (1985, 93). Matters did not improve, and BonJour later wrote that "the precise nature of coherence remains a largely unsolved problem" (1999, 124).

Lehrer (1990, 2000) offers a different account of coherence. It is narrower and takes explanation as the core notion within what he calls "The Explanatory Coherence Theory of Truth" (2000, Chapter 5). The basic definition is thus:

> *S* is justified in accepting that *p* if and only if the belief of *S* that *p* is consistent with that system *C* of beliefs having a maximum of explanatory coherence among those systems of beliefs understood by *S*, and the belief that *p* either explains something relative to *C* that is not explained better by anything which contradicts *p* or the belief that *p* is explained by something relative to *C* and nothing which contradicts it is explained better relative to *C*. (2000, 105)

Impressive as this definition appears to be, its content is obscure as long as the core notion of explanation invoked in it remains vague. Subsequent

4 BonJour (1985, 93) writes that the "various detailed investigations by philosophers and logicians of such topics as explanation, confirmation, probability, and so on, may be reasonably taken to provide some of the ingredients for a general account of coherence."

discussion of the then current state of accounts of explanation proves to be of no help in clarifying the notion. Lehrer reviews the "immense literature" on the topic and arrives at the sober conclusion that

> This literature illustrates most clearly the futility of hoping to find an explication of explanation to which we can fruitfully appeal in our articulation of the explanatory coherence theory. (2000, 106)

Lehrer (1999) provides a correction to his earlier accounts.[5] Using formulations similar to his (1990, Chapter 6; 2000, Chapter 6), he derives coherence from the notion of an "acceptance system":

> To summarize, my acceptance of *p* coheres with my evaluation system if and only if all objections to my acceptance of *p* are beaten or neutralized on the basis of my evaluation system. (1999, 247)

There is considerable discussion of how this coherence with the evaluation system is to be understood. The overall import is not clear, at least to me. Lehrer allows (1999, 246), for example, that logical inconsistency does not preclude further acceptance. There are repeated allusions to what is "reasonable," though "reasonable" is left as a primitive term.[6] Curiously, the notion of explanation has all but disappeared from the account.

Thagard's (2000) account of coherence as constraint satisfaction is heavily influenced by computational perspectives. It is a significant work that deserves more attention than I have space here to give it. However, it shares the weakness of other accounts discussed here. It relies on further relations whose nature is unclear. The constraints that figure centrally in the account include those expressed in terms of explanatory and analogical relations. Their import is translated into summable weights that provide a holistic measure of the system's overall coherence (see, e.g., 7, 38, 43). Explanation and analogy

5 It is presumably Lehrer (1990), although the second edition (Lehrer 2000) has only minor changes.

6 An earlier treatment by Lehrer based coherence on a notion of "comparative reasonableness." He suggested that comparative reasonableness could be explicated in terms of comparative expected epistemic utility, "but no such account, including ones I have articulated, strikes me as quite adequate to my purposes" (1989, 253).

are commonly invoked in such discussions, and their principled nature and relation to inductive support remain obscure, as I argued at some length in *The Material Theory of Induction* (Norton 2021, Chapters 4, 8, and 9).

In sum, these accounts of coherentism are compromised by a failure to articulate clearly the core global notion of coherence. Their efforts rely on invoking local relations, notably probabilistic and explanatory relations, while neglecting to give cogent accounts of them or admitting that none is at hand.

The material theory of induction faces no comparable problem. Its core notion is not global but the local relation of inductive support. I elaborated that notion extensively in *The Material Theory of Induction* (Norton 2021). The material theory succeeds just where coherentism fails, for the material analysis of probabilistic, explanatory, and analogical relations, as they figure in evidential support, supersedes the vaguer notions appealed to by BonJour and Lehrer above.

4.3. Coherentism versus Foundationalism

A principal motivation of coherentism is its opposition to foundationalism. The latter, as we saw above, asserts that there are certain basic beliefs foundational in the sense that they do not require further justification for belief. This concept has proven to be the Achilles' heel of coherentism. It leads directly to what BonJour calls the "input objection" (1985, 108) or the "isolation objection" (1999, 127). It is the obvious problem that one can have entirely fictional narratives that exhibit considerable coherence while having nothing to do with the real world. Creating such artifices is the trade of writers of fiction. In BonJour's version, it asserts that

> Nothing about any requirement of coherence dictates that a coherent system of beliefs need receive any sort of input from the world or be in any way causally influenced by the world. . . . Such a self-enclosed system of beliefs, entirely immune from any external influence, cannot constitute empirical knowledge of an independent world. (1985, 108)

The difficulty facing coherentists is that they need to allow for input from the world without conceding to foundationalists. Chapters 6 and 7 of BonJour (1985) contain a labored, extended struggle to allow worldly inputs to beliefs without being forced to make this concession. The result, as summarized by

Murphy (2020, Section 5a), is that these inputs arise through "cognitively spontaneous beliefs" that are nonvoluntary and an "observation[al] requirement" that stipulates that such inputs are required.

This challenge of allowing worldly inputs without conceding to foundationalists has been a defining issue for coherentism. It is based, as far as I can see, on a false dilemma demanding that we choose either to be coherentists or to be foundationalists. Haack (1993, 19) has argued cogently that a serviceable epistemology arises from a combination of foundationalist and coherentist positions. Her "foundherentism" is initially formulated as follows:

> Foundherentism may be approximately characterized thus:
>
> (FHl) A subject's experience is relevant to the justification of his empirical beliefs, but there need be no privileged class of empirical beliefs justified exclusively by the support of experience, independently of the support of other beliefs;

and:

> (FH2) Justification is not exclusively one-directional, but involves pervasive relations of mutual support.

The material theory of induction is almost entirely indifferent to this issue, which has so controlled coherentist thinking. On the local level, the theory proceeds without any need for inputs from the world. The theory can authorize inductive inferences in mathematics.[7] When treating the large-scale structure of inductive support for a science, the analyses of this volume do presume that science is empirical. That is, the content of the science is to be supported by empirically accessible facts of the world. That condition is essential to the argument of Chapter 4 for the uniqueness of the inductive structures of a mature science. According to it, the decision among competing

---

7 Goldbach's unproven but widely believed conjecture is that any even number can be expressed as a sum of two prime numbers. A familiar heuristic argument for it notes that, the larger the number, the more ways in which it can arise as the sum of two numbers, increasing the chances that two of them are prime. This is an inductive argument warranted by the supposition that the two numbers summed are distributed independently enough of the distribution of primes that it is highly likely always to include a pair of primes. For this and more examples, see Franklin (2013, 18).

theories eventually will be made empirically, as long as those theories are genuinely distinct empirically.

This application of the material theory requires only that the overall system of propositions in some way gains inputs from the world in order that the resulting science is empirical. The epistemologies under discussion here are foundationalism, coherentism modified to allow worldly inputs, and foundherentism. All allow for inputs from the world. Insofar as each of their schemes can be reimplemented within relations of inductive support, the material theory of induction can work with all of them.

Indeed, coherentism and material theory differ in the roles that circularities play in their structures. For coherentism, the circularities among relations of justification arise explicitly as a way to block the foundationalist regress argument for basic beliefs.[8] For the material theory, as we saw in Chapter 2, circularities block a different regress argument. Without them, warranting facts would require further warranting facts of ever greater generality.

4.4. Beliefs versus Propositions

A major difference, implicit in the discussion above, can now be made explicit. It concerns the relata of the relations of justification or inductive support. For coherentists, the relata are the *beliefs* consciously held by some agent. For the material theory of induction, the relata are the *propositions* of an inductive logic, independent of any agent's thoughts. There is no presumption that these propositions are the objects of belief in any consciousness. Both BonJour and Lehrer are internalists in their coherentism. That is, the justification of some belief must be accessible cognitively to the agent. In this regard, they are closest to the relations of inductive support of the material theory of induction, for it is also supposed that these relations can be made explicit. In an externalist version of coherentism, if there is such a thing, the justifications of beliefs would not always be accessible.[9] They might arise through some causal process that connects with the world, though that process is not cognitively accessible to the agent.

This difference adds a burdensome extra layer of complications to coherentism. Here is how BonJour expresses it:

8 See, for example, BonJour's (1985, 87) introduction of circularities.

9 BonJour (1985, 101–02) dismisses an externalist coherentism as unacceptable since it would be weaker than an externalist foundationalism.

> But if the fact of coherence is to be accessible to the believer, it follows that he must somehow have an adequate grasp of his total system of beliefs, since it is coherence with this system which is at issue. One problem which we will eventually have to confront is that it seems abundantly clear that no actual believer possesses an *explicit* grasp of his overall belief system; if such a grasp exists at all, it must be construed as tacit or implicit, which creates obvious problems for the claim that he is actually, as opposed to potentially, justified. (1985, 102; BonJour's emphasis)

That the relations of justification must be thought consciously or at least accessible cognitively[10] to some agent does indeed create problems for coherentism (and internalist epistemologies in general). The difficulties are so well known that I need only briefly mention them here. Such epistemologies face a dilemma. Are the justificatory relations those of an ideally rational agent? Or are they those of the actual cognitive processes of real people?

If the relations are those of ideal rationality, then the normative injunctions of the theory are unrealizable by ordinary human cognition. Establishing the logical consistency of even a fairly small set of beliefs is so computationally burdensome that ordinary minds cannot do it.[11] These complications already arise in the easiest case of deductive relations. They are no easier when it comes to securing probabilistic and explanatory consistency as varieties of coherentism require.

If, instead, coherentism pertains to the actual reasoning processes of human agents, then the method of analysis is misplaced. How we actually reason is properly the subject of empirical psychology.[12] A long-standing and well-established tradition in empirical psychology has shown just how poor we are in ordinary deductive and probabilistic reasoning.[13] That we human

10 For example, BonJour writes that "a person for whom a belief is inferentially justified need not have explicitly rehearsed the justifying argument in question — to others or even to himself. . . . What is required is rather that the inference be *available* to the person in question, so that he would be able in principle to rehearse it" (1985, 19; BonJour's emphasis).

11 For details, see Cherniak (1984).

12 Goldman (1985) has investigated the relationship of epistemology and psychology.

13 For a small sample of this enormous literature, see Kahneman, Gilovich, and Griffin (2002).

reasoners conform to the conditions of coherence requires that we have achieved deductive and probabilistic consistency in our belief systems or aspire to it. According to empirical studies in psychology, this goal seems to be beyond the reach of most human agents.

BonJour does not, as far as I can see, directly address this dilemma. Instead, he identifies a reflexive concern of the type that seems to trouble inward-looking philosophers but few others. Believers must have a correct grasp of their own system, and the correctness of this grasp in turn is a further belief that requires justification. The impending regress is blocked, according to BonJour, by a *presumption*, called the "Doxastic Presumption." Its content is developed over several pages and, in one formulation (1985, 105), asserts that "I assume that the beliefs constituting my overall grasp of my system of beliefs are, by and large, correct."

The material theory of induction is merely a codification of inductive logic. It escapes all of these problems. There is no requirement that its relations of inductive support are to figure in their totality in some agent's consciousness. This abstract conception, however, creates the danger that the theory is one of ideal rationality as inaccessible to human agents as the ideal rationality of coherentism. Although that danger is present, the burden taken by material theory is considerably less than that of coherentism.

In the simplest cases, the material theory allows scientists to answer specific questions. Does the evidence of the cosmic background radiation provide more support inductively for big bang cosmology than steady state cosmology? The material theory of induction can answer that question without taking on the burden of establishing the coherence of the entirety of the scientist's belief system.

The more complicated case does concern the entirety of the inductive support for a particular science. Is it, informally speaking, coherent? Coherentism requires a single agent to be able to affirm coherence for the totality of that agent's belief system. The corresponding coherence of the evidential support for a science does not reside in the satisfaction of some overarching concept of coherence. Rather, it is simply the summation of many local relations of support, such that, in mature sciences, each proposition is well supported. Since the overall burden consists merely of many local parts, there is no requirement that any individual scientist has a grasp of their totality. Rather, the task is distributed over the entire community of scientists. For a modern science of any depth, this distribution is inevitable, for full comprehension of all of the details of its evidential support lies outside the cognitive powers of a single scientist. Experts in one wing of the science rely on the affirmations of experts in the other wings and conversely. The process continues over time. Great professional rewards await a scientist who can find evidence of internal inconsistencies that threaten or overturn an existing science. The result is that new generations of scientists scrutinize the consistency and evidential foundations of existing sciences anew. Mature sciences generally survive this scrutiny, indicating their solidity. When they do not, a new science emerges.

4.5. Examples

Associated with these last differences is one of lesser importance that I find to be striking nonetheless. Coherentist epistemology is designed to apply to beliefs of the most mundane variety. Here are a few examples from BonJour (1985) typical of the literature in the epistemology of belief:

> I believe that the piece of paper upon which I am now typing is the very same piece of paper upon which I was typing late yesterday afternoon. (20)
>
> As I sit at my desk (or so I believe), I come to have the belief, among very many others, that there is a red book on the desk. (117)
>
> . . . the car going by is a Lotus. . . . (119)
>
> . . . a figure . . . coming towards me . . . is my friend Frank. . . . (119)
>
> There is a man lurking in the bushes. (120)

The material theory of induction is designed for relations of inductive support in science. There typical propositions that count as empirical evidence are things such as the following.

> Space is filled with electromagnetic radiation of a thermal character with a temperature of 2.7K.
>
> The perihelion of Mercury advances by forty-three seconds of arc per century more than predicted by Newtonian gravitation

theory, after perturbations from other planets are accommodated.

The difference is that the exemplar beliefs of coherentism concern ordinary experience. The corresponding empirical propositions in the material theory of induction are remote from ordinary experience. No one just notices a 2.7K radiation heat bath in the depths of space or that Mercury is moving just a bit too fast over the span of a century. These propositions are secured only after considerable investigation and analysis and are major pieces of science in their own right.

It would be rash to infer from these differences that justification in the epistemology of belief and justification in inductive science are qualitatively different. Indeed, I incline toward the idea that justification in both is the same in their basic natures. Einstein (1936, 349) remarked that "all of science is nothing more than the refinement of everyday thinking." However, there might still be great differences in the refinement: that is, the details and thoroughness of execution. When someone accepts that there is a red book on the table, the justifications might proceed with similar principles as those of the cosmologist who accepts the 2.7K background microwave radiation. Where the first is a snap judgment that happens in moments, the second is underpinned by decades of careful, explicit analysis. These differences matter greatly. In judging that there is a red book on the table, we pay scant attention to the possibility that our experience is the result of some other cause. For the cosmologists, it took decades of measurements at many frequencies before the cosmic radiation could be affirmed to be thermal at 2.7K and not of some other nature.

Furthermore, the differences in the exemplars indicate that the two approaches will prioritize different aspects of the relations. Hence, BonJour frets extensively about the reflexive problem of whether we are justified in our own beliefs about our justifications (addressed in the Doxastic Presumption). In contrast, quantitative methods, such as those found in elaborate statistical testing, are important in inductive inferences in science but do not figure in the simple examples routinely used in coherentist epistemology.

5. Problems of Coherentism

My initial hope, upon recognizing the similarities between coherentism and the large-scale structure of inductive inference, was that coherentist analyses might be a useful resource in resolving problems in the material theory. This hope has not been realized. The two ventures do share similar problems. However, it seems to me that coherentism has fared worse in addressing them, either because of its weaker suppositions or because of its failure to address the problems better.

The most serious problem facing both systems, in my view, is that they harbor circularities of justification and support. As noted in Chapter 3, it is all too common to find that the mere presence of such circles is sufficient for rejection without any further analysis. I have already quoted BonJour's (1985, 91–92) response in Section 3, to which the reader is now referred. The specific response to the threat of circularities in that passage is this:

> And hence, a coherence theory will claim, the apparent circle of justification is not in fact vicious *because it is not genuinely a circle*: the justification of a particular empirical belief finally depends, not on other particular beliefs as the linear conception of justification would have it, but instead on the overall system and its coherence.

BonJour's later analysis gives the same response:

> Justification, when properly understood, is ultimately *nonlinear* or *holistic* in character, with all of the beliefs in the system standing in relations of mutual support, but none being epistemically prior to the others. In this way, it is alleged, any true circularity is avoided. Such a view amounts to making the system itself the primary unit of justification, with its component beliefs being justified only derivatively, by virtue of their membership in an appropriate sort of system. (1999, 123; BonJour's emphasis)

I have quoted this response at length to make its inadequacy clear. The rejection of a linear dependence of relations of justification does not eliminate circularities in the interdependencies within the overall system. We are urged, incorrectly, to think that the potential harm of these circularities evaporates because the justification of particular empirical beliefs depends on the whole system. Murphy's (2020, Section 2b) encyclopedia review recalls BonJour's holistic attempt at escape, finds it lacking, and suggests that circularities are benign if they are within sufficiently strong relations of mutual support. Of course, nothing about strength precludes a vicious circularity or the possibility of arbitrariness.

A stronger response could have been given by coherentists along the lines given in Chapter 3. First, benign circularities are prevalent enough in science that there can be no default supposition that a circularity is harmful. Rather, we have a positive obligation to establish that some specific circularity is harmful and how it is so. The two dangers explored in Chapter 3 were the contradictions of vicious circularities and their opposite, indeterminateness through the possibility of multiple structures that satisfy the circular relationships. In Chapter 3, I argued that both dangers are precluded in the support relations of a material theory of induction by the dynamic character of scientific investigation. Vicious circularities are removed when found, and indeterminateness triggers further investigations that eliminate it (unless we have a true convention). Surely a similar argument can be made concerning circularities among the justifications of beliefs.

This concern about circularity does not figure in BonJour's (1985, 106) list of "three standard and extremely forceful objections":

(I) the alternative coherent systems objection;

(II) the input objection; and

(III) the problem of truth.

BonJour's narrative struggles with all three, whereas the material theory does not.

I have already discussed the second "input" objection in Section 4.3. It is an unnecessary weakness that derives from the damaging conception of coherentism as opposed to foundationalism. There is no corresponding problem for the material theory.

The first objection is that there might be multiple and equally admissible coherent systems that would undermine the justification of the one chosen. The difficulty that BonJour finds in answering derives directly from his conception of coherentism as the alternative to foundationalism. As a result, his coherentist analyses seek to favor coherence over some sort of foundational input from experience. As we saw in Chapter 4, the material theory can argue for the uniqueness of relations of support in a mature science precisely by relying heavily on empirical evidence to decide among competing systems. BonJour (1985, 143) does start to make an argument along these lines. He argues that, once observational input is considered, "it is no longer clear" that multiple and equally admissible systems can be sustained in the long run. That it is no longer clear is not enough. The possibility is there. In Chapter 4, where empirical evidence is given a greater role, this long-term possibility is eliminated through the empirical character of science: if the long-term accumulation of empirical evidence cannot separate two theories, then we have grounds for concluding that they are not distinct in their physical content in the first place.

The material character of inductive inference also provides an added resource: it induces an instability in competition among theories such that, when one theory has gained an advantage evidentially, that advantage is amplified, driving the competition toward resolution in its favor.

The third objection is that mere coherence among beliefs is not enough to establish that they are truths of the world. The obvious counterexamples are the coherent narratives of works of fiction. Mere coherence does not establish truth, unless one is willing to adopt a coherence theory of truth, which BonJour (1985, 109), prudently, is unwilling to do. This problem can be ameliorated if coherentism is not conceived as opposed to foundationalism, for then truths of the world can enter more freely as foundational experiences.

The material theory has no corresponding problem of truth. It does not seek truth conduciveness in some single, global property of the relations of inductive support, such as coherence. Rather, the task is distributed over all of the inductive relations. It is the burden of the individual warranting facts of an inductive inference to be truth conducive for that inference. When a fact warrants an inference to a proposition or warrants its inductive support, what is inferred or supported is the truth of the proposition. The truth conduciveness of the full structure of relations simply results from the accumulated truth conduciveness of the individual relations of support.

6. Probabilistic Accounts of Coherence

In recent decades, there has been vigorous activity in writing on coherentism among Bayesian philosophers of science. There have been two strands of analysis. In the first strand, probabilistic vindication or its negation is sought for the coherentist's notion that coherence among beliefs either constitutes their justification or, more weakly, enhances their justification. Olsson (2005), for example, argues that this notion is not vindicated probabilistically but disproven. Huemer (2011) finds that the probabilistic analysis is inadequate for a disproof while providing an apparently cogent probabilistic implementation of coherentism. Further possibilities of probabilistic implementation are considered in Wheeler (2012). In the second strand, a single probabilistic measure of coherence is sought, such that belief systems that score higher are better justified. Once again the leading results are negative. Bovens and Hartmann (2003) offer a proof that no single probabilistic measure of coherence can serve this function, but they suggest that a quasi-ordering[14] by probabilistically defined coherence is possible. These negative results have been disputed. Schupbach (2011) defends a coherence measure devised by Shogenji. To complicate matters, Shogenji (2013, 2544) then uses probabilistic analysis to argue for an anticoherentism in which coherence reduces the transmission of probabilistic support. These last contributions are only part of a vigorous debate. For a survey, see Olsson (2017, Sections 6–8).

Since there is no consensus among Bayesians on these results, it would be of little value to pursue the details any further. Rather, I will assess at the most general level the relevance of this work to my project. The principal goal of the Bayesian analyses has been to capture the essential intuitions of coherentism within a probabilistic framework and thereby to provide some deeper foundation for it or, pessimistically, a definite refutation of it. That is, they seek to vindicate probabilistically the holistic approach of coherentism or to refute it. As I have indicated above, the material theory does not adopt coherentism's holism. This means that these probabilistic proofs or refutations are of tangential relevance only to the project of this book.

However, since the probabilistic analysis aspires to conclusions concerning justification considered at the large scale, we might wonder whether it can connect somehow with the large-scale conceptions of the material theory

14 That is, the relation is reflexive, transitive, but not complete.

concerning relations of inductive support. My overall assessment is that these Bayesian analyses provide very little of value to someone who does not share the radical Bayesian goal of reducing as much as possible of the epistemology and philosophy of science to repeated applications of Bayes' theorem. I defend this assessment in the next section.

7. Why the Bayesian Analysis of Coherence Fails

Overall, the Bayesian analysis of coherentism has proven to be, at best, an infertile but benign distraction for non-Bayesians and at worst a positive misdirection. There are three reasons for this.

7.1. The Coherence of Coherentism Is not a Probabilistic Notion

Non-Bayesian coherentists have included probabilistic notions in their efforts to explain coherence. Sometimes coherence is manifested within a system of probabilistically related beliefs. However, the notion of coherence itself is not fundamentally a probabilistic notion, such that all of its cases can be reduced to results expressible probabilistically.

This nonprobabilistic character is evident in important examples. One of the best known arises in the competition between Ptolemaic and Copernican astronomy. The two systems could be adjusted so that they provide the same predictions for planetary motions. However, as detailed in Chapter 12, "The Use of Hypotheses in Determining Distances in Our Planetary System," the Copernican system was more coherent than the Ptolemaic system. The latter system needed an independent epicycle-deferent construction for each planet. The former system resulted from the recognition that many of the Ptolemaic circles were not independent motions but actually the superposition of the Earth's orbital circle on those of the other planets.

This greater coherence of the Copernican system was a key argument in its favor. It was widely recognized in the century after the death of Copernicus. Most importantly, it was not probabilistic in nature. It was then expressed and debated without any need for probabilistic conceptions.

7.2. Formalization Is Premature

Coherentism proceeds on the assumption that the coherence of this last example and others like it is the manifestation of a general notion of coherence that can serve to justify the system of beliefs in which it arises. The problem for coherentists is that a general characterization of coherence remains elusive and is one of the recalcitrant problems of coherentism. Just what is coherence?

It is easy to become impatient with the recalcitrance of such a problem. Then one can be tempted by the idea of a formal framework in which the solution to the problem is reducible to a precise mathematical question whose answer is provided by mathematical demonstration. In the seventeenth century, Leibniz offered the prospect of a universal language with this perspective in mind: "When there are disputes among persons, we can simply say: Let us calculate, without further ado, and see who is right."[15] A similar optimism motivates the Bayesian analysis. Olsson (2017, Section 5) writes that

> the arguably most significant development of the coherence theory in recent years has been the revival of C. I. Lewis's work and the research program he inspired by translating parts of the coherence theory into the language of probability.

He proceeds to promise the benefits of the translation:

> The probabilistic translation(s) of coherence theory has made it possible to define concepts and prove results with mathematical precision.

My assessment of this development is that it is retrograde. It is, of course, both satisfying and decisive when mathematical demonstrations in a formal system can resolve vexing, informal confusions. I will celebrate all such successes. However, such a resolution requires that the original problem admits precise mathematical formulation in the first place. This is not the case with coherentism in epistemology. Just what its notion of coherence is remains poorly understood. Instead of a successful clarification of the notion of coherence, we have an intemperate rush to formalization. To superimpose a veneer of probabilities onto an imprecisely understood notion is not to illuminate it but to obscure it and its problems.

The focus of the present probabilistic analyses of coherence has drifted away from clarifying coherence. It has been replaced by extended and apparently fruitless debates on just how to represent probabilistically some simple

15 As quoted by Kulstad and Carlin (2020, Section 3) with the citation *The Art of Discovery* (1685); C 176/W 51.

notions to do with coherence. Because probabilistic independence is a major component of strong results in the probability calculus, there is a premium on giving it a place in the probabilistic analysis. The opportunity to introduce it comes from a notion in coherence theory of the independence of witnesses. How can that independence be expressed probabilistically? Olsson (2005, 25) considers that witness testimonials E_1 and E_2 agree in that both assert H. The independence of the testimony is then represented probabilistically by independencies of the conditional probabilities $P(E_2|E_1, H) = P(E_2|H)$ and $P(E_2|E_1, {\sim}H) = P(E_2|{\sim}H)$. This seems to be odd since we expect it to be more probable that witness 1 and witness 2 will agree on H, if H is the case, than they would if H is not the case. We then learn (Olsson, 2005, 45) that, under correction from other authors, the representation is incorrect since it assumes perfect reliability or unreliability of the witnesses. A more elaborate model is needed. Huemer (2011, 40) argued that the independence of testimonies is too strong a condition in the first place for the probabilistic representation of coherence. All that is needed is that the truth of H makes agreement between the witnesses more likely: $P(E_2|E_1, H) > P(E_2|E_1, {\sim}H)$. That is, the attempt to base a probabilistic treatment on the probabilistic independence of testimonies was too hasty in the first place.

These are just the beginning of familiar problems peculiar to Bayesian probabilistic representations. A non-Bayesian might be willing to admit a probability $P(E_2| H)$ that represents the chance that witness 2 testifies to H when H is the case. But what are we to make of $P(E_2|{\sim}H)$, the probability that witness 2 testifies to H, when H is not the case? We might imagine all sorts of scenarios in which H might be false, so that ${\sim}H$ is true. How likely is the aberrant testimony in each? How likely is each scenario? Have we exhausted all of the scenarios? All of the quantities arising here must be multiplied and summed. And when all of this is done and we sum up all of our quantities, do we have a result with sufficient probabilistic meaning that it can figure in the precise computations that follow? It requires quite an indulgence to imagine so.

The issue here is not just the problem of assigning a precise value to $P(E_2|{\sim}H)$. It is the very idea that we have a quantity here, well represented by any additive measure at all, and that it is one with sufficient commonality of meaning with the additive measure $P(E_2|H)$ for it to be combined freely with it in subsequent computations. It is a standard presumption that all of these maneuvers are admissible, for otherwise Bayes' theorem could not be applied.

Presumption and familiarity bred of necessity, however, are not the same as a well-founded resolution of an enduring problem.

Thagard (2000, Chapter 8, 2004, 2005) has mounted a related critique of the Bayesian treatment of coherence as part of a defense of his account of coherence as constraint satisfaction. Among his many concerns is that Bayesian analysis requires "a host of conditional probabilities that people would be hard pressed to specify" (2005, 311).

Matters become worse when we consider the second strand of the probabilistic analysis, the attempts to define a single numerical measure of coherence in some system that is a function solely of the probabilities in that system. They go beyond problems associated with the mere use of probabilities. It is a risky speculation that any single measure of coherence is possible in the first place. And it is an even riskier speculation that probabilities alone suffice to define it when the notion itself is not probabilistic. It is hardly surprising that no consensus has emerged from the dense fog of elementary theorems in probability and counterexamples that constitute this literature.

For Bayesians committed to the idea that fundamental notions such as coherence are reducible to probabilistic notions, all of these complications are simply work proceeding as usual. The path is not easy, but they are convinced that a happy outcome awaits them eventually. They must persist. They must calculate more.

For those who are not Bayesians, the entire enterprise is one of premature haste in the pursuit of an illusion of precision. The original problem of the nature of coherence has faded away. In its place are exercises in elementary probability theory, endless revisions of them, all with increasingly dubious connections to the original problem. That no consensus has emerged on these probabilistic conjurings is no surprise if one doubts the appropriateness of the formalization in the first place.

Setting Bayesian or non-Bayesian commitments aside, let us recall that we have good reason to think that the notion of coherence is not a probabilistic notion. Is it really wise to persist in efforts to find a probabilistic basis for it?

7.3. A Probabilistic Framework Is not Sufficiently General

The Bayesian analysis of coherence aims at general results. If a probabilistic foundation can be found for coherentism, then it is vindicated universally. If coherentism is refuted in the Bayesian analysis, then it is refuted everywhere.

The problem is that a probabilistic framework is not sufficiently general to support universal conclusions of this type. This is one of the main consequences of *The Material Theory of Induction* (Norton 2021). It asserts that a case-by-case examination of the warranting facts of each domain determines whether relations of inductive support of the domain are probabilistic. There are many domains in which the relations are probabilistic. And there are many in which they are not. Chapters 10–16 of *The Material Theory of Induction* give extended examples. They do not need to be rehearsed here. Insofar as credences are set by strengths of inductive support, the same conclusion applies to them. Thus, one cannot proceed with probabilities as the automatic default when it comes to representing uncertainties in inductive support or credences. There is a positive obligation in each domain to display the facts warranting the probabilistic representation.

When such warranting facts are present, the justification of the results of the probabilistic analysis can ultimately be traced back in part to the facts warranting the probabilities. Figuratively, there should be a disclaimer attached to the Bayesian analysis such as "applies in domains only where probabilities are warranted." To omit this disclaimer is to advance conclusions without proper basis. To suggest that the results have universal applicability is a misrepresentation.

These concerns can be reformulated as a general argument against the possibility of a universally applicable Bayesian epistemology or Bayesian philosophy of science. Any analysis of this type will begin with general framing assumptions. Commonly, the probabilistic analyses of coherentism presume a collection of witnesses testifying to various facts. Further assumptions concern the reliability of the witnesses and the extent to which their testimonies are related, other than through the truth or falsity of facts to which they testify. Any conclusions drawn from these assumptions alone are secure in the sense that they are merely restating what is asserted by the framing assumptions of the analysis.

The delicate part comes when the probabilistic analysis gives us results that go beyond the framing assumptions. If these results are to have universal applicability, then their derivation is flawed since it depends essentially on the assumption that probabilistic relations can apply everywhere. Thus, for any result claimed to have universal validity, we cannot preclude the possibility that it is merely an artifact of an illegitimate assumption.[16] We end up believing that we have proven results when we do not.

Once we are alerted to the danger, examples of this sort of fallacy are easy to find. Here are several. The principle of indifference is a truism of evidence and credence. If you have no basis for distinguishing the support of two outcomes and find them equally favored, then you should accord equal support or equal credence to them. This framing assumption is as anodyne a principle as one can imagine. Yet — famously — it cannot be implemented in a probabilistic system. We end up with the widely known paradoxes of indifference. What is their origin? Perversely, the routine conclusion is that there is some fault in this truism of evidence. The fault lies elsewhere. It is the assumption that probabilities can capture support or credence within this framing. Rather, this framing requires different relations of support or credence (see Norton 2008, 2010).

I provided another example in Chapter 1 through Laplace's rule of succession. The rule tells us that, as a general matter, if we have had 1,826,213 successes, then we should expect a success on the next trial with very high odds of 1,826,214 to 1. As I point out in the chapter, the framing is bare and simply assumes 1,826,213 unrelated successes. Nothing in it warrants the application of probabilities. As a result, nothing more can be inferred using probabilities about future successes. The inference to the very high odds of future successes is fallacious and simply an artifact of applying probabilities without a warrant.

A related example concerns a hypothesis H and its deductive consequences E_1, E_2, E_3, and so on. It is a well-known result of Bayesian analysis, reviewed in Norton (2011, 430–31), that it entails either an excessive pessimism or an excessive optimism concerning the projectability of the hypothesis and its consequences but nothing in between. If we set the prior probability of H to zero as $P(H) = 0$, then conditionalizing on any evidence fails to alter the zero probability. We are dogmatically committed to the failure of any evidence to support H. If, however, we set $P(H) > 0$, even if $P(H)$ is as small

16 Another potential source of error lies in the translation of nonprobabilistic framing assumptions into probabilistic relations. The translation might be erroneous but correctable. More seriously, there might be no good probabilistic translation. This is the case for a state of complete ignorance or completely neutral support. It leads to the failure of probabilistic analysis to accommodate the principle of indifference, as discussed in the main text and Norton (2008, 2010).

as we care to make it, then we commit to an excessive optimism in the projectability of its consequences. That is, if the first 10^n of these consequences have obtained, then we become arbitrarily certain that the next $10^{n+1} - 10^n$ will obtain, as n increases, for it is easy to show that

$$\mathrm{Lim}_{n \to \infty} P(E_{(10^n)+1} + E_{(10^n)+2} + \ldots + E_{(10^{n+1})} \mid E_1 + E_2 + E_3 + \ldots + E_{(10^n)}) = 1$$

That is, we become arbitrarily confident of roughly a tenfold increase in the number of consequences that we expect to obtain. That this is excessively optimistic, even credulous, is clear once we recall that the framing assumptions are sparse. They allow the set $\{E_1, E_2, E_3, \ldots\}$ to be *any* noncontradictory set of propositions whatever (and then H could merely be their conjunction). This excess of optimism or pessimism is an artifact of the application of probabilistic analysis in which the framing assumptions are too sparse to authorize them. They far outstrip what the framing assumptions authorize.

There is an easy escape from this difficulty. Bayesians should renounce the illusion that they are able to deliver results of universal applicability in the epistemology and philosophy of science. Rather, their results apply only in domains in which probabilities are warranted by some positive factual basis. Then strong and interesting domain-specific results will be recoverable, and their basis will not be mysterious. They will rest ultimately on the factual warrant for probabilities in the domain.

8. Conclusion

It appears initially that much is shared by the material account of the large-scale structure of inductive support and coherentism: that is, coherence-based accounts of the justification of beliefs in epistemology. Both require that their respective relations of support are nonhierarchical or nonlinear. That agreement gave some hope of further fruitful connections. That hope was dashed by the rather negative results of this chapter. The similarities in the two projects, we find, scarcely extend beyond the initial agreement on nonhierarchical structures.

The main difference is that coherentism bases justification on a holistic property, the overall coherence of belief systems. The material analysis bases inductive support on local relations, and the overall nonhierarchical structure is derived from them. There are many further differences: coherentism assumes various poorly articulated accounts of justification at the local level, where the material analysis is devoted to a full articulation of a relation of inductive support. Coherentism is formulated as an alternative to foundationalism. That conception creates the enduring problem for coherentism of discerning how the world impinges on beliefs without conceding to foundationalism. The material analysis has no corresponding problem. Coherentism concerns beliefs and thus struggles to accommodate the limits of cognition of any one believer. The material theory concerns abstract relations of inductive support as matters of inductive logic and escapes these problems. Partly through these differences, coherentism has proven to be less able than the material theory to respond to the standard problems that both systems face.

In this chapter, I have reviewed Bayesian analyses of coherentism. Since these analyses accept the holistic conception of coherentism, they are of little relevance to the material analysis of the large-scale structure of inductive support. The review indicates the Bayesian analyses to be overreaching. There is a powerful negative result: insofar as a Bayesian analysis offers universally applicable results on coherentism that go beyond the nonprobabilistic framing assumptions, they are without proper foundation.

Finally, this chapter indicates how the material theory of induction relates to the more general literature in the epistemology of belief. Although I focused on coherentism specifically, much of what I concluded applies more generally. Two more general conclusions can be recovered.

Epistemologists of all varieties treat local relations of justification and inductive support as antecedently understood. They use notions such as probabilistic and explanatory relations in their accounts while leaving the elucidation of such relations to others. The goal of the material theory of induction is to elucidate these very relations and others like them. Hence, the two projects proceed at different levels.

Epistemology concerns beliefs held by cognizers and how these beliefs are justified. The material theory avoids belief as much as possible. It gives an account of what inductively supports what, independent of beliefs and knowing agents, as matters of independent inductive logic. How some agent can use those relations to inform belief is a further problem left largely to the epistemologists.

REFERENCES

BonJour, Laurence. 1985. *The Structure of Empirical Knowledge*. Cambridge, MA: Harvard University Press.

———. 1999. "The Dialectic of Foundationalism and Coherentism." In *The Blackwell Guide to Epistemology*, edited by John Greco and Ernest Sosa, 117–42. Malden, MA: Blackwell.

Bovens, Luc, and Stephan Hartmann. 2003. *Bayesian Epistemology*. Oxford: Clarendon.

Cherniak, Christopher. 1984. "Computational Complexity and the Universal Acceptance of Logic." *Journal of Philosophy* 81: 739–58.

Einstein, Albert. 1936. "Physics and Reality." *Journal of the Franklin Institute* 221: 349–82.

Franklin, James. 2013. "Non-Deductive Logic in Mathematics: The Probability of Conjectures." In *The Argument of Mathematics: Logic, Epistemology and the Unity of Science*, vol. 30, edited by Andrew Aberdein and Ian J. Dove, 11–29. Dordrecht: Springer.

Goldman, Alvin I. 1985. "The Relation between Epistemology and Psychology." *Synthese* 64: 29–68.

Haack, Susan. 1993. *Evidence and Inquiry: Towards Reconstruction in Epistemology*. Oxford: Blackwell.

Huemer, Michael. 2011. "Does Probability Theory Refute Coherentism?" *Journal of Philosophy* 108: 35–54.

Kahneman, Daniel, Thomas Gilovich, and Dale Griffin, eds. 2002. *Heuristics and Biases: The Psychology of Intuitive Judgment*. Cambridge, UK: Cambridge University Press.

Kulstad, Mark, and Laurence Carlin. 2020. "Leibniz's Philosophy of Mind." In *The Stanford Encyclopedia of Philosophy*, fall 2020 ed., edited by Edward N. Zalta. https://plato.stanford.edu/archives/fall2020/entries/leibniz-mind/.

Lehrer, Keith. 1974. *Knowledge*. Oxford: Clarendon.

———. 1989. "Coherence and the Truth Connection: A Reply to My Critics." In *The Current State of Coherence Theory*, edited by J.W. Bender, 253–75. Kluwer Academic Publishers.

———. 1990. *Theory of Knowledge*. Boulder, CO: Westview Press.

———. 1999. "Coherence and Knowledge." *Erkenntnis* 50: 243–58.

———. 2000. *Theory of Knowledge*. 2nd ed. Boulder, CO: Westview Press.

Murphy, Peter. 2020. "Coherentism in Epistemology." In *Internet Encyclopedia of Philosophy*. https://www.iep.utm.edu/coherent/.

Norton, John D. 2008. "Ignorance and Indifference." *Philosophy of Science* 75: 45–68.

———. 2010. "Cosmic Confusions: Not Supporting versus Supporting Not-." *Philosophy of Science* 77: 501–23.

———. 2011. “Challenges to Bayesian Confirmation Theory.” In *Philosophy of Statistics, Vol. 7: Handbook of the Philosophy of Science*, edited by Prasanta S. Bandyopadhyay and Malcolm R. Forster, 391–439. Elsevier.

———. 2021. *The Material Theory of Induction*. BSPS Open Series. Calgary: University of Calgary Press.

Olsson, Erik J. 2005. *Against Coherence: Truth, Probability, and Justification*. Oxford: Clarendon.

———. 2017. “Coherentist Theories of Epistemic Justification.” In *The Stanford Encyclopedia of Philosophy*, spring 2017 ed., edited by Edward N. Zalta. https://plato.stanford.edu/archives/spr2017/entries/justep-coherence/.

Rescher, Nicholas. 1973. *The Coherence Theory of Truth*. Oxford: Oxford University Press.

Schupbach, Jonah. 2011. “New Hope for Shogenji’s Coherence Measure.” *British Journal for the Philosophy of Science* 62: 125–42.

Shogenji, Tomoji. 2013. “Coherence of the Contents and the Transmission of Probabilistic Support.” *Synthese* 190: 2525–45.

Thagard, Paul. 2000. *Coherence in Thought and Action*. Cambridge, MA: MIT Press.

———. 2004. “Causal Inference in Legal Decision Making: Explanatory Coherence vs. Bayesian Networks.” *Applied Artificial Intelligence* 18: 231–49.

———. 2005. “Testimony, Credibility, and Explanatory Coherence.” *Erkenntnis* 63: 295–316.

Wheeler, Gregory. 2012. “Explaining the Limits of Olsson’s Impossibility Result.” *Southern Journal of Philosophy* 50: 136–50.

6

The Problem of Induction

1. Synopsis

Since the problem of induction is so widely known, I expect that many readers will want a simple summary of the main claims instead of the more usual orienting introduction. This synopsis is for those readers.[1]

1.1 The Traditional Problem

I take the problem of induction here to be a specific difficulty in any logic of inductive inference, where "inference" is understood to be a mind- and belief-independent relation of logical support for propositions. Logics prone to the problem are based on universal rules of induction. Traditionally, the rule is enumerative induction: we are authorized to infer from the proposition that some cases bear a property to the proposition that all cases do. Other rules might be abductive: we are authorized to infer to the best explanation or the supposition that relations of inductive support are numerical and conform to the probability calculus.

The problem resides in a short and sharp demonstration that no inductive rule can be justified. The demonstration uses either a circularity or a regress. The rule of enumerative induction itself is justified by some version of that rule: enumerative induction has worked, so we should expect it to continue to work. Hence, its justification is circular. If we consider other rules of inductive inference, then we encounter a similar circularity, if the rule is used to justify itself. Alternatively, the rule might be justified by applying a second rule, and that second rule is justified by a third, and so on in an

1 My thanks to James Norton and Anil Gupta for helpful remarks and reactions.

infinite regress. The regress is fanciful since taking just one or two steps is strained and unrealistic.

Probabilistic accounts of inductive support and analyses of the problem do not escape the fundamental difficulty. In turn, there must be some justification for a logic whose basic rule is that inductive relations of support are probabilistic. Chapter 10 (especially Section 10) of *The Material Theory of Induction* (Norton 2021b) argues that all of the standard justifications for this basic rule are circular.

1.2 The Material Dissolution

The material theory of induction dissolves the problem by denying one of its premises. The problem of induction depends essentially on the presupposition that inductive inference is governed by universal rules. The material theory of induction asserts that there are no universal rules of inductive inference. Inductive inferences are warranted by local facts, not rules. With this understanding, the problem of induction can no longer be set up. It is dissolved.

1.3 Attempts to Recreate It in the Material Theory

A common rejoinder to this dissolution is the proposal that there is an analogous problem for the material theory of induction. It derives from the circumstance that background facts can only warrant an inductive inference if they are true. Thus, they should also be warranted, and the inferences that warranted them must also be warranted. Somehow, lurking in this circumstance, there is supposed to be a regress or circularity as devastating as the original problem of induction. Following are three versions of the problem supposed.

Regress end: Each inductive inference requires a warranting fact of greater generality than the conclusion. The resulting succession of warranting inductive inferences requires a sequence of warranting facts of increasing generality that admits no benign termination.

Regress start: Inductive inference cannot get started: any inductive inference that attempts to go beyond some small, given set of particular propositions requires an unavailable warranting fact of greater generality outside the given set.

Circularity: These successive warranting inferences will eventually form circles of large or small extent. They are supposed to be as harmful as those of the original circularity in the problem of induction.

1.4 Why the Attempts Fail

In earlier chapters, I described the large-scale structure of relations of inductive support in science afforded by the material theory of induction. Briefly, in that theory, inductive inferences are warranted by facts that in turn are supported inductively, those inductive inferences are warranted by further facts, and so on. What results is the massively entangled structure of inductive support relations of a mature science. This structure does not respect any hierarchy of generality. Relations of support routinely cross over one another. It follows that tracing back successively the facts that support some nominated inference leads to a journey through the propositions of the science. There are many forks in the journey's path since its extent grows rapidly and soon might embrace much of science. There is no inexorable and unsustainable ascent to warranting propositions of ever greater generality. Hence, the supposition of *regress end* fails.

This massively entangled structure can be created by hypothesizing provisionally propositions needed to warrant some initial inductive inference. The provisional character of the hypotheses must be discharged by further investigations that provide inductive support for them. This mechanism makes warranting hypotheses of greater generality available when the inductive project of some science is initiated. Hence, the supposition of *regress start* fails.

There are circularities both large and small in this massively entangled structure of relations of inductive support. However, as I argued in Chapter 3, we cannot automatically assume that the mere presence of circularities is harmful. There are benign circularities throughout science. One must establish by positive argumentation that the circularities here are harmful. These harms arise in two ways: as a contradiction of a vicious circularity or as an underdetermination. If either one arises, then it is eliminated by routine adjustments in the science. In place of self-defeating circularities, all that we find in the entangled structure is how one result in a mature science is supported by others, those by others still, and so on. The exercise merely recapitulates, over and over, ordinary relations of inductive support in mature sciences. Hence, the supposition of *circularity* fails.

1.5 Local versus Distributed Justifications of Inductive Inference

The problem of justifying inductive inference has a different character according to whether inductive inference is conceived as warranted by universal rules or by material facts.

If inductive inferences are warranted by universal rules, then the project of justifying them is reduced to that of justifying those few universal rules. All attention is devoted to a small sector of the sciences in which the inductive power is localized. We learn from the endurance of the traditional problem of induction that this localized version of the problem is intractable.

If inductive inferences are warranted by facts, then the justification for inductive inferences is not localized. It is distributed over all of the sciences. In a mature science, the justification for some chosen inductive inference lies in the applicable warranting facts. It is an unproblematic application of the material theory to a particular case. This is true for *every* inductive inference in the mature science, and *that is all* there is to the justification for induction, understood materially. The totality of the justification for inductive inference lies in the accumulation of many such unproblematic justifications and thus itself is unproblematic. When the justification is so distributed, the difficulty is reversed. Efforts to set up the problem of induction fail repeatedly.

2. Introduction

The synopsis above is merely a sketch of the analysis to be developed in greater detail in this chapter. My hope is that readers who might be unsatisfied by its brevity will be satisfied by the lengthier analysis below. Its first step is a more precise statement of the original problem. Although the problem of induction is widely recognized, I have no confidence that we all address the same problem. Before a claim of a dissolution of the problem of induction can be sustained, the problem itself must be clearly delineated. That delineation is the task that I will undertake in Sections 3–10 of this chapter. The task is largely historical, and readers who are confident that they know the history might want to skip ahead to Section 10.

Since inductive inference traditionally has been regarded as generally troublesome, in Section 3 I will seek to sweep away some preliminary distractions that might be taken mistakenly to be the problem of induction. In recalling a collection of what I call "inductive anxieties," the section identifies what the problem of induction is not. It is not, for example, the problem that enumerative induction is capricious. An inference from some *A*s are *B* to all *A*s are *B* sometimes can be sustained by only a few cases of *A*s that are *B*, or it can fail to be sustained even by many cases.

In Section 4, I review David Hume's own presentation of the celebrated argument. It was a masterpiece of philosophical writing, still justly admired today. His argument was narrower than the version that modern authors have taken from his analysis: Hume limited all inductive inferences to causal inferences. And it was broader since he posed the problem largely in psychological terms. He characterized inferences as mental processes, as the "operation of thought." In his celebrated fork, Hume divided all such operations as concerning relations of ideas or matters of fact. Neither could justify inductive inferences about the future, he urged. The first cannot since we can imagine it failing. The second cannot since it requires that we presume in advance the very thing to be justified, that the future will resemble the past.

In Sections 5 and 6, I review the early reception of Hume's analysis. After an initial response, notably from Immanuel Kant and even possibly Thomas Bayes, the analysis faded and merited only passing mention in nineteenth-century discussions of induction. The term "the problem of induction" did not univocally have its modern meaning. Rather, it was a marker for more general inductive anxieties. For Mill, it denoted the capriciousness of inductive inference. In Section 7, I review the twentieth-century revival of Hume's problem, first in the writing of Bertrand Russell and then, with greater focus, in that of Hans Reichenbach and his student Wesley Salmon. They advocated a "circularity" version, reminiscent of Hume's own. Briefly, inductive inference, now understood in Salmon's formulation as any form of ampliative inference, cannot be justified by deduction, since then it would not be inductive, and it cannot be justified inductively, for that would be circular. In Section 8, I recall the "regress" version, delineated most thoroughly by Karl Popper. Instead of the circularity of a rule of inductive inference justifying itself, Popper imagined a rule of inductive inference being justified by another rule, and that by another rule, and so on in an unsustainable infinite regress.

In Section 9, I report that both Russell and Salmon insisted that their modern version of the problem of induction drops the psychological clothing that Hume gave it. The problem is purely one of inductive logic, which pertains to relations with propositions, independent of our thoughts and beliefs. Although modern epistemologists run together logical inference and mental operations, I was pleased to find that this rarely caused confusion. The exception is noted in Section 10, where I review failed attempts to argue that an externalist epistemology of beliefs can solve the problem of induction.

The material dissolution of the problem of induction is presented again in Section 11. In Sections 12 and 13, I respond to the concern that the harmful regresses and circularities of the problem of induction reappear in the tangle of relations of inductive support for the material theory of induction. I argue in Section 12 that the regress of the problem of induction is already fanciful and dubious in its first steps, whereas that of the material theory is merely the recapitulation of ordinary relations of inductive support in familiar science. I argue in Section 13 that the circularities of the problem of induction are harmful since they leave its rules of induction indeterminate. Drawing from the analyses of Chapters 3 and 4, I argue that the circularities of the material theory do not create analogous problems of indeterminacy.

Elliott Sober and Samir Okasha have given responses to the problem of induction that are close to this material dissolution. I review their work briefly in Section 14. Since I claim that there is no problem for the material theory in justifying inductive inference, in Section 15 I give a short summary of the character of the positive justification.

Finally, the present material dissolution of the problem of induction appeared in its earliest form in my first paper on the material theory of induction (Norton 2003). It has attracted some small though continuing critical attention. This attention has been stimulating and led to refinements of the material dissolution. In Section 16, I review the critical reception of the material dissolution in the literature and show how the refinements respond to and answer the negative criticism. Section 17 is a short conclusion.

3. What the Modern Problem of Induction Is Not: Inductive Anxiety

The very idea of inductive inference has been a long-standing target of hesitation and vilification. The dissolution of the problem of induction advocated here is not designed to address all hesitations about induction. To preclude confusion, in this section I report two of these other hesitations. One is simply the observation that inductive inference is not deductive inference and thus must admit the possibility of failure. The second is that a particular form of induction, enumerative induction, is capricious. Sometimes it works well.

Sometimes it does not, and then it encourages ill-advised hastiness. Beyond these two identifiable hesitations, for many, induction is surrounded by an unfocused but nonetheless menacing miasma. In it, induction simply is a problem. I will call the totality of these hesitations "inductive anxiety."

The first hesitation already has clear expression in the ancient tradition of skepticism. As part of his broadly spread critique of all forms of justification, the skeptic Sextus Empiricus himself gave a terse statement that still serves us well today:

> It is easy, I think, to reject the method of induction. For since by way of it they want to make universals convincing on the basis of particulars, they will do this by surveying either all the particulars or some of them. But if some, the induction will be infirm, it being possible that some of the particulars omitted in the induction should be contrary to the universal; and if all, they will labour at an impossible task, since the particulars are infinite and indeterminate. Thus in either case it results, I think, that induction totters. (Annas and Barnes 2000, 123)

Earlier in his text, Sextus Empiricus gives a colorful illustration of how induction totters: "Since most animals move their lower jaw but the crocodile alone moves its upper jaw, the proposition 'Every animal moves its lower jaw' is not true" (120).

We need not linger over this first hesitation. It is constitutive of (ampliative) inductive inference that it can sometimes fail. That fact does not impugn its utility as long as the inferences are secure enough that their failures are tolerably rare. To abandon inductive inference entirely would destroy science, all of whose major results are supported inductively.[2]

For the second hesitation, Mill, in his monumental *System of Logic*, recounts several inductive inferences, some of which proceed securely from a few particulars, whereas others are never judged secure. They lead to a synoptic lament of the capriciousness of induction:

2 Popper's ([1959] 2002) attempt to account for scientific practice solely with deductive inference fails. Salmon (1981) has shown that close adherence to Popper's strictures precludes science from making predictions.

> Why is a single instance, in some cases, sufficient for a complete induction, while in others, myriads of concurring instances, without a single exception known or presumed, go such a very little way toward establishing a universal proposition? Whoever can answer this question knows more of the philosophy of logic than the wisest of the ancients, and has solved the problem of induction. (Mill 1882, 228)

In arguing for the cautious inductive ascent of his preferred method, Francis Bacon provided a celebrated riposte, which seems to be a combination of both of the hesitations listed above:

> The induction which proceeds by simple enumeration is puerile, leads to uncertain conclusions, and is exposed to danger from one contradictory instance, deciding generally from too small a number of facts, and those only the most obvious. (Bacon [1620] 1902, 83)

This second hesitation also need not detain us. Many accounts of inductive inference have taken up the task of accounting for why enumerative induction works when it does and why it fails when it does. This was explicitly the task of Harman's (1965) paper in which the term "inference to the best explanation" was introduced. My material account of inductive inference in Chapter 1 of *The Material Theory of Induction* (Norton 2021b) identifies the warrant for this form of inductive inference in background facts. Generalizations are warranted or not according to whether these background facts are favorable or not. No doubt a Bayesian will find some combination of prior probabilities and likelihoods to fit the expected behavior of even the most capricious of inductive generalizations.

For further details of the troubled history of enumerative induction and a compilation of striking counterexamples mentioned in the traditional literature, see Norton (2010).

4. Hume's Critique

Hume's celebrated critique of inductive inference elevated these traditional anxieties about induction from answerable concerns to what became the model of a recalcitrant philosophical problem in the twentieth century. His critique needs some refinement before we recover the modern version of the problem of induction. Two refinements are notable.

First, Hume restricted all ampliative, nondemonstrative inferences to those mediated by relations of cause and effect:

> All reasonings concerning matter of fact seem to be founded on the relation of *Cause and Effect*. By means of that relation alone we can go beyond the evidence of our memory and senses. (1777a, 26; Hume's emphasis)

This restriction needs to be loosened.

Second, Hume did not separate cleanly two things that should be kept separate. First are thoughts, beliefs, and mental processes, such as is properly the subject of a theory of mental action. They are distinct from logical relations among propositions, such as is the subject of an abstract logic, formulated independently of thoughts and beliefs. For example, Hume's fork, the celebrated distinction of "Relations of Ideas" and "Matters of Fact," is introduced in terms of mental processes. The first "Relations of Ideas" are discoverable, Hume insists (1777a, 25), "by the mere operation of thought, without dependence on what is anywhere existent in the universe." This possibility is contrasted with a "Matter of Fact" whose contrary (negation) is possible. That is, "it can never imply a contradiction, and is conceived by the mind with the same facility and distinctness, as if ever so conformable to reality" (25). Elsewhere, however, his language could easily be mistaken by the unwary as conforming to an analysis of purely logical relations among propositions. On the supposition that present regularities might fail in the future, he asks "what logic, what process of argument secures you against this supposition?" (38). I will urge below that the distinctive Humean problem of induction resides in the inductive logic and can be formulated only indirectly in terms of mental processes.

With these complications noted, we can follow Hume's development of the problem.[3] First, Hume affirms that demonstrative reasoning cannot give us knowledge of these relations of cause and effect:

3 Comparable arguments can also be found more tersely in Hume's earlier *Treatise* (1739, 89–90).

> I shall venture to affirm, as a general proposition, which admits of no exception, that the knowledge of this relation is not, in any instance, attained by reasonings *a priori*, but arises entirely from experience, when we find that any particular objects are constantly conjoined with each other. (1777a, 27)

His argument is based on the immediately following claim:

> Let an object be presented to a man of ever so strong natural reason and abilities; if that object be entirely new to him, he will not be able, by the most accurate examination of its sensible qualities, to discover any of its causes or effects. (27)

The claim is illustrated by examples (27–28) that, Hume asserts, outstrip demonstrative reasoning. He imagines Adam, presumably new to the world and innocent of experiences of it. Adam cannot infer that water suffocates from its fluidity and transparency or that fire consumes from its heat and warmth. Someone innocent of natural philosophy could not infer that polished marble blocks will adhere tightly, that gunpowder is explosive, that lodestones attract, and more. An example earlier in the text, we shall see, reappears later in the text:

> *That the sun will not rise to-morrow* is no less intelligible a proposition, and implies no more contradiction than the affirmation, *that it will rise*. (25–26; Hume's emphasis)

Hume then looks for other possibilities for arriving at knowledge of cause and effect. There is only one candidate, "moral reasoning," for he recalls his fork:

> All reasonings may be divided into two kinds, namely, demonstrative reasoning, or that concerning relations of ideas, and moral reasoning, or that concerning matter of fact and existence. (1777a, 35)

Yet, he continues, moral reasoning cannot provide a firm basis for such knowledge. He justifies this failure by identifying a circularity within efforts to use moral reasoning for this purpose:

> We have said that all arguments concerning existence are founded on the relation of cause and effect; that our knowledge of that relation is derived entirely from experience; and that all our experimental conclusions proceed upon the supposition that the future will be conformable to the past. To endeavour, therefore, the proof of this last supposition by probable arguments, or arguments regarding existence, must be evidently going in a circle, and taking that for granted, which is the very point in question. (35–36)

Since this is the celebrated circularity upon which the modern problem is based, we can pause for another trenchant statement of it:

> It is impossible, therefore, that any arguments from experience can prove this resemblance of the past to the future; since all these arguments are founded on the supposition of that resemblance. (38)

5. The Reception

While Hume fretted that his earlier *Treatise* (1739) fell "dead-born from the press" (1777b, 8), there was still some fairly immediate and noteworthy reaction. It had a profound impact on Kant ([1783] 1909, 7), who famously credited Hume for "interrupt[ing] my dogmatic slumber." Hume's contemporary, Thomas Reid, mounted efforts to refute Hume's skepticism.[4] It is even plausible that his skepticism was one of the motivations for Thomas Bayes' analysis of inverse probabilities. Zabell (1989, 292) notes that the timing of the initiation of Bayes' research on inverse probabilities coincided with Hume's publication in 1748 of his *Enquiry*. After Bayes' death, his result was published and annotated by Richard Price (Bayes 1763). Zabell (1989, 294) and Earman (2002, Section 1) note that much in Price's annotations indicates a response to Hume, even though Hume is not mentioned by name. For example, Price writes (in Bayes 1763, 371–72) that

4 See Landesman and Meeks (2003, Chapter 29).

> Common sense is indeed sufficient to shew us that, from the observation of what has in former instances been the consequence of a certain cause or action, one may make a judgment what is likely to be the consequence of it another time. . . .

Price also considers "the case of a person just brought forth into this . . . world" (409) (reminiscent of Hume's mention of Adam) who makes successive observations of the sunrise and forms odds of its return. The example is one that Hume had used but to skeptical ends.

6. The Nineteenth-Century Hiatus

In the nineteenth century, any recognition that Hume might have received for identifying the problem of induction faded. He was instead generally tolerated as a troublesome skeptic concerning topics such as causation and miracles. His analysis was not lauded then, unlike today, as the revered *locus classicus* for the modern problem of induction. In that century, the phrase "the problem of induction" appeared frequently. However, its focus was diffuse, and it appeared mostly to designate some version of the "inductive anxieties" sketched in Section 3.

Whatever role Hume's critique might have had in the initiation of Bayes' work on inverse probabilities, there is little trace of it in subsequent work. Laplace's development of the rule of succession in his *Essay* (1814), sketched here in Chapter 1, used Hume's example of successive sunrises but made no mention of Hume. The *Essay* includes an entire chapter (1902, Chapter 17) on induction and similar ampliative inferences. It recounts some history of such inferences, including mentions of the English writers Newton and Bacon but not the Scottish writer Hume.

Perhaps it is unsurprising that logic texts of the nineteenth century make scant mention of Hume's critique. Their charter is to delineate the structure of the logics, not to rehearse skeptical assaults against them. Kirwan's (1807, 231) early logic treatise does cite Hume but to dispute his assertion that chance is the absence of a cause. Munro's (1850, 233–340) *Manual of Logic* decrees that induction is material and thus "extralogical" insofar as the induction is not complete. That means that its premises fail to include all instances of the generalization, so the inference is not deductive. Whately's *Elements of Logic* (1856) includes a lengthy chapter on induction (Book 4, Chapter 1) and struggles with many hesitations but never clearly articulates Hume's argument or mentions Hume in the context of induction. Creighton's *An Introductory Logic* (1898) has a section entitled "The Problem of Induction" (Chapter 13, Section 47). However, the term "problem" is less the identification of a difficulty than the setting of a task: how are we to pass from chaotic experience to scientific knowledge?

As late as Schiller's (1912) discussion of formal logic, the phrase "the problem of induction" did not have its modern meaning. The work has a chapter entitled "The Problem of Induction" (Chapter 17). The problem identified is the difficulty of determining the truth of premises used in deductive syllogisms. Hume's concern appears only briefly some eight pages into the meandering chapter as the unanswered question "how do we know that the future will resemble the past?" (239).

One might have expected more from W. Stanley Jevons, notable for his nineteenth-century writing on scientific methodology. His two logic texts (1888, 1902) make no mention of Hume or any problem of induction, although both discuss induction extensively. His major work of methodology, *Principles of Science* (1874), similarly covers induction extensively and advocates for a Bayesian inverse approach. It too has no mention of Hume or any trace of the possibility that Bayes himself might have been motivated by Hume's challenge.

John Stuart Mill might have been the preeminent writer of his age on scientific methodology. We saw in Section 3 that he labeled the capriciousness of inductive inference as "the problem of induction" and declared hyperbolically that to solve it is to "know more of the philosophy of logic than the wisest of the ancients."

The third book of the six forming his *System of Logic* is devoted to induction. In it, Mill presents his methods, whose content remained a core of presentations of scientific methodology into the mid-twentieth century. Buried in this third book among its twenty-five chapters is Chapter XXI. It addresses what, in effect, is Hume's circularity argument. Its subsidiary treatment indicates that Mill regarded the problem as a minor nuisance, a philosopher's sophistry, that can be dispatched forthwith by his sharp wit. Mill notes (1882, 398) that his inductive methods depend on the law of causality, that every event has an invariable, antecedent cause. We are assured of this law by processes of induction that join those cases in which causation is not yet apparent with those in which it is. The inevitable circularity appears:

> If, then, the processes which bring these cases within the same category with the rest, require that we should assume the universality of the very law which they do not at first sight appear to exemplify, is not this a *petitio principii*? Can we prove a proposition, by an argument which takes it for granted? And if not so proved, on what evidence does it rest? (398)

In stating this Humean circularity, Mill makes no mention of Hume. It is not for lack of knowledge of his work, for Hume's controversial analysis of miracles is discussed at length elsewhere in Mill's *System of Logic*. That Hume's analysis had indirect or even direct influence on Mill, however, is suggested by his distinctively Humean choice of examples:

> It would be absurd to say, that the generalizations arrived at by mankind in the outset of their experience, such as these — food nourishes, fire burns, water drowns — were unworthy of reliance.[5] (401)

Mill's dismissal of the circularity fares as poorly as any that underestimates its gravity. His dismissal allows that we first arrive at the law of causality by a fragile, simple, enumerative induction but that our inductive methods are subsequently reinforced by applying the law to itself so that a certainty results:

> The law of cause and effect, being thus certain, is capable of imparting its certainty to all other inductive propositions which can be deduced from it. . . . And hence we are justified in the seeming inconsistency, of holding induction by simple enumeration to be good for proving this general truth, the foundation of scientific induction, and yet refusing to rely on it for any of the narrower inductions. (403)

Mill has staked here the entirety of his inductive enterprise on the certainty of the law of cause and effect, which in his writing amounts to a principle of determinism. The irony, of course, is that this certainty was about to be falsified by the discovery of quantum theory in the 1920s.

5 That bread nourishes is an example that Hume uses repeatedly in his *Enquiry* (1777a, 28 ff.).

In any case, authors contemporary to Mill were not so easily bluffed. Lachelier devoted Section 2 of his 1871 doctoral dissertation, *Du fondement de l'induction*, to Mill's argument. No matter how artful Mill's analysis, Lachelier concluded that a purely empiricist view like Mill's cannot derive conclusions for the future from the knowledge of the past (1907, 25; translated from Ballard 1960, 13):

> If we see nature as nothing more than a series of impressions without reason and without connection, we can indeed record, or rather undergo, these impressions at the moment they are produced, but we cannot predict them nor even conceive of their production in the future.

Lachelier's own ideas inclined toward a Kantian, rationalistic idealism, so Lachelier regarded this empiricist failure merely as motivation for his preferred approach. Although Hume's circularity would have provided powerful further direction, Lachelier mentions it but immediately abandons analysis of it (Lachelier 1907, 17; Ballard 1960, 9):

> The principle of induction itself, then, must be the product of an induction . . . (we leave aside the circle suspected to be in this reasoning).

Similarly, the British idealist F.H. Bradley had little interest in induction and any problems that Hume might have found in it. In his *Principles of Logic* (1883, 342), the treatment of inductive inference is deeply buried in the text and passed over dismissively: "[Mill's methods of inductive logic] will not work unless they are supplied with universals. They presuppose in short as their own condition the result they profess alone to produce." Bradley concludes that "we may set down Inductive Logic as a *fiasco*." Although this conclusion is reminiscent of Hume's circularity, Hume is not credited with any insight and is not mentioned by name anywhere in the 534 pages of the text.

Perhaps prominent recognition of Hume's argument has slipped past this sampling of nineteenth-century writing. If his critique had prominence in the nineteenth century, then we would expect it to register in survey writing. In light of this expectation, it is revealing that Thomson's (1887) philosophical dictionary has an entry for "The Problem of Inductive Logic," but it simply defines the problem as the capriciousness of inductive inference by giving the quotation from Mill above in Section 3. This while elsewhere in the dictionary Hume appears copiously as something of a disreputable gadfly. Hume's skeptical nihilism, Thomson reports, "gave . . . offence so serious to the British public" (xxx).

Still more remarkable is that the introduction of twenty-five pages to the 1894 edition of *Enquiry*, written by Lewis Amherst Selby-Bigge in 1893, makes no mention of Hume's charge of circularity concerning inductive inference. Rather, what attracts the editor's attention concerns causation (xv). It is Hume's affirmation "that there is nothing at the bottom of causation except a mental habit of transition or expectation, or, in other words, a 'natural relation.'" Selby-Bigge then turns to other concerns and reports similar skeptical remarks by Hume on the relation of resemblance (xvi).

7. Twentieth-Century Revival: The Circularity Formulation

With the start of the twentieth century, "the problem of induction" was a phrase used variously to represent a variety of inductive anxieties or even just as a caption to introduce a wide-ranging discussion of induction.[6] The phrase did not indicate the short, sharp problem posed by Hume that any justification for a rule of induction must be inductive and thus circular.

Matters soon changed. Russell's *Problems of Philosophy* (1912) gave terse and readily accessible accounts of a series of philosophical problems. The chapter "On Induction" developed a clear and compelling version of Hume's original problem. Although Hume is not mentioned by name, the chapter's Humean inspirations are clear by its use of familiar Humean examples. The running example asks what justifies our belief that the Sun will rise tomorrow. Russell asks, for example,

> Do *any* number of cases of a law being fulfilled in the past afford evidence that it will be fulfilled in the future? If not, it becomes plain that we have no ground whatever for expecting

6 Ernst Cassirer (1910) has a long chapter entitled "On the Problem of Induction" ("Zum Problem der Induktion"). The phrase "the problem of induction" seems to designate no sharply defined difficulty for induction, such as that posed by Hume. Rather, it serves as a general heading under which Cassirer can develop complaints about empiricism and defend Kantian perspectives on induction.

> the sun to rise to-morrow, or for expecting the bread we shall eat at our next meal not to poison us, or for any of the other scarcely conscious expectations that control our daily lives. (96; Russell's emphasis)

The inevitable circularity emerges. Russell develops and refines the circularity until it becomes one of justification for what he calls the "principle of induction" (103). It is expressed in several cautious clauses. Its overall import, however, is that past association of things of sorts A and B make probable that this association will continue. Justification for this principle itself inevitably falls victim to Hume's circularity. The chapter concludes darkly:

> The inductive principle, however, is equally incapable of being *proved* by an appeal to experience. Experience might conceivably confirm the inductive principle as regards the cases that have been already examined; but as regards unexamined cases, it is the inductive principle alone that can justify any inference from what has been examined to what has not been examined. All arguments which, on the basis of experience, argue as to the future or the unexperienced parts of the past or present, assume the inductive principle; hence we can never use experience to prove the inductive principle without begging the question. Thus we must either accept the inductive principle on the ground of its intrinsic evidence, or forgo all justification of our expectations about the future. (106; Russell's emphasis)

Hans Reichenbach proved to be a more tenacious and exacting proponent of the cogency of Hume's critique. In his contribution to the first issue of the new journal *Erkenntnis*, Reichenbach argued on Humean grounds that there can be no justification for probabilistic forms of inductive inference. It is just that we have no choice but to use them:

> There is no other justification for our belief in logic than to point to the fact that we cannot think at all otherwise. We can however give the analogous [justification] for the laws of probability: we cannot do anything else at all other than to believe in the laws of probability. (1930, 187)

The point is soon given an even stronger form:

> It is exactly the same with probabilistic logic [as with deductive logic]; we cannot justify it, but we can affirm that we just cannot think of any alternative. (188)

Reichenbach concluded that

> Our reply, then, to the problem of validity does not consist in an answer to Hume's question. Rather, the attempt to find a logical foundation for probabilistic assertions seeks an impossible goal, comparable to the squaring of the circle. (188)

The idea that we have no choice but to think probabilistically in inductive terms now seems to be unreflective and unimaginative.[7] Perhaps Reichenbach recognized the weakness of this idea, for he shortly replaced the "no choice but" defense of the use of probabilistic induction with a stronger and now celebrated pragmatic argument. In Section 38, "The Problem of Induction," of his *Experience and Prediction* (1938), Reichenbach formulated a "principle of induction" (340). Loosely speaking, it tells us to expect that the observed frequency of some property in a sequence of events will persist at this value, approximately, within error bounds, as the sequence proceeds. Hume, Reichenbach continued, had mounted a most significant challenge to the principle. He summarized it as follows:

> 1. We have no logical demonstration for the validity of inductive inference.
>
> 2. There is no demonstration a posteriori for the inductive inference; any such demonstration would presuppose the very principle which it is to demonstrate.
>
> These two pillars of Hume's criticism of the principle of induction have stood unshaken for two centuries, and I think they will stand as long as there is a scientific philosophy. (342)

7 That seems so especially to me after having written several chapters in *The Material Theory of Induction* (Norton 2021b) that explore calculi of inductive inference that are alternatives to the probability calculus.

Reichenbach then roundly chastised the philosophers and logicians of the nineteenth century for their failure to recognize the gravity of Hume's challenge:

> It is astonishing to see how clear-minded logicians, like John Stuart Mill, or Whewell, or Boole, or Venn, in writing about the problem of induction, disregarded the bearing of Hume's objections; they did not realize that any logic of science remains a failure so long as we have no theory of induction which is not exposed to Hume's criticism. (342)

Reichenbach's *The Theory of Probability* (1949) gave a similar formulation derived from Hume's original. In Section 91, "The Justification of Induction," citing Hume's *Enquiry*, Reichenbach asks Hume's question. What grounds the inference that the same causes will still be followed by the same effects in the future? Following Hume, Reichenbach divides the negative answer into two parts: there can be no deductive justification and no inductive justification:

> 1. The conclusion of the inductive inference cannot be inferred *a priori*, that is, it does not follow with logical necessity from the premises; or, in modern terminology, it is not tautologically implied by the premises. Hume based this result on the fact that we can at least imagine that the same causes will have another effect tomorrow than they had yesterday, though we do not believe it. What is logically impossible cannot be imagined — this psychological criterion was employed by Hume for the establishment of his first thesis.
>
> 2. The conclusion of the inductive inference cannot be inferred *a posteriori*, that is, by an argument from experience. Though it is true that the inductive inference has been successful in past experience, we cannot infer that it will be successful in future experience. The very inference would be an inductive inference, and the argument thus would be circular. Its validity presupposes the principle that it claims to prove. (470)

Reichenbach proceeded in both works to his well-known answer to Hume's problem: we are justified in using induction pragmatically. Although we have no guarantee that it will work, if anything can work, then it will work.

Wesley Salmon, one of Reichenbach's most successful students, continued the Reichenbachian analysis. His *Foundations of Scientific Inference* (1967) gave, in my view, the most incisive development of Hume's objection.[8] Salmon's version of Hume's objection is slightly more general than Reichenbach's version; it proceeds to a systematic and gently ruthless refutation of each escape proposed in the then present literature; it then concludes with Reichenbach's pragmatic answer.

The inductive inferences of earlier formulations of Hume's problem are replaced by Salmon with the considerably more general notion of "ampliative" inference. Such an inference is defined negatively by Salmon (1967, 8) merely as an inference that is not demonstrative —

> . . . an ampliative inference, then, has a conclusion with content not present either explicitly or implicitly in the premises.

Loose as this definition is, Salmon has no difficulty recreating Hume's charge of circularity against it:

> Consider, then, any ampliative inference whatever. . . . We cannot show deductively that this inference will have a true conclusion given true premises. If we could, we would have proved that the conclusion must be true if the premises are. That would make it necessarily truth-preserving, hence, demonstrative. This, in turn, would mean that it was nonampliative, contrary to our hypothesis. Thus, if an ampliative inference could be justified deductively it would not be ampliative. It follows that ampliative inference cannot be justified deductively.
>
> At the same time, we cannot justify any sort of ampliative inference inductively. To do so would require the use of some sort of nondemonstrative inference. But the question at

8 Wes Salmon was highly respected and the kindest senior colleague in my junior years on the faculty of the University of Pittsburgh. I regret that time robbed me of the opportunity to show him my analysis, for his approval would have meant the world to me. Then again, his disapproval would have been devastating.

issue is the justification of nondemonstrative inference, so the procedure would be question begging. Before we can properly employ a nondemonstrative inference in a justifying argument, we must already have justified that nondemonstrative inference.

Hume's position can be summarized succinctly: We cannot justify any kind of ampliative inference. If it could be justified deductively it would not be ampliative. It cannot be justified nondemonstratively because that would be viciously circular. (11)

8. Twentieth-Century Expansion: The Regress Formulation

Explicit notions of induction, when Hume wrote, were limited to some version of generalization. The simplest was the long-standing form, enumerative induction: from some *A*s are *B*, we infer that all are. Bacon's method of tables provided a more sophisticated, if still limited, version of inductive practice. Nonetheless, writing after Bacon, Hume was comfortable reducing all inductive inferences to one simple form: the same causes will continue to have the same effects. With similarly limited conceptions of inductive inference, Russell and Reichenbach[9] worked with comparably simple conceptions of inductive inference, as codified in their respective "principles of induction" sketched above. The simplicity of these conceptions makes it possible for Hume's critique to be expressed in terms of a circularity. There is one simple notion of inductive inference, and the only way to justify it inductively is to apply that notion to itself.

As the twentieth century unfolded, this simple conception of inductive inference ceased to be viable, if ever it was. It became all too clear that there are many forms of ampliative inference in addition to the few considered by Hume, Russell, and Reichenbach. By the start of the twenty-first century, the variety was so great that I found it a challenge to write a survey of accounts

9 I have excluded Salmon's analysis from the list since his analysis is not limited to the narrow conceptions of inductive inference of Russell and Reichenbach. His ampliative inferences include all nondemonstrative inferences. However, his formulation of the problem as one of circularity omits the possibility of an infinite regress.

of inductive inference that would capture and usefully systematize them. My best effort is Norton (2005).

With many such accounts available, the circularity of Russell's and Reichenbach's analyses ceased to be sufficiently expansive. What if their principles of induction are just not justified by applying the principles to themselves? What if they are justified by some *other* form of ampliative inference? Harman's (1965) revival of abductive inference as "inference to the best explanation" was offered explicitly as providing a warrant for enumerative induction. Justifying one form of inductive inference inductively by another does not settle the matter. Now we must ask what inductively justifies this second form, Harman's schema of inference to the best explanation, and when another form of inductive inference is invoked we must ask what justifies that further form.

The resulting succession of justifications for inductive inference schemas either leads back to a schema already used, in which case we have a circularity, or triggers an infinite regress. This last possibility is the "regress" form of the problem of induction.

The earliest clear articulation that I have found of this regress form of the problem of induction comes in Karl Popper's *Logik der Forschung* (1935), translated as *Logic of Scientific Discovery* ([1959] 2002). Popper formulates the problem of induction as the problem of justifying a principle of induction, the fact that authorizes inductive inferences. He dismisses the possibility that such a principle could be analytic or a tautology: that is, a purely logical truth. Rather, it is a proposition whose truth is known from experience by induction. This immediately leads to the infinite regress:

> To justify it [the first principle of induction], we should have to employ inductive inferences; and to justify these we should have to assume an inductive principle of a higher order; and so on. Thus the attempt to base the principle of induction on experience breaks down, since it must lead to an infinite regress. (5)[10]

10 Popper's *Logik der Forschung* is noted for its decisive rejection of inductive inference. His deeply skeptical view of induction was not so novel in 1935. We have seen that Reichenbach's *Erkenntnis* paper in 1930 abandoned the project of justifying induction on Humean grounds. Popper cites Reichenbach's paper, mentions Reichenbach's endorsement of probabilistic

This is a terse but serviceable formulation of the regress version of the problem. A more developed version can be found in what Popper (2009, preface) describes as drafts and preparatory writings of 1930–33 for *Logik der Forschung*. They were first published in German in 1979 and then in English translation in 2009. In the translation (Book 1, Chapter 3), we find that Popper prefers the regress form of the problem of induction because the circularity form would be open to the objection that the mere assertion of the circularity involves self-reference, which Russell had shown to raise the possibility of vicious circularity. Popper continues:

> The concept of "infinite regression" is not open to these objections, but otherwise it accomplishes the same task, namely that of demonstrating the existence of an impermissible operation.

Popper continues the chapter, slowly developing the infinite regress and eventually providing this summary:

> In this way, a hierarchy of types emerges:
>
> *Natural laws* (these may be understood as statements about singular empirical statements, and as of a higher type than the latter). The induction of a natural law requires a
>
> *First-order principle of induction*, which as a statement about natural laws is of a higher type than the latter; the induction of a first-order principle of induction, in turn, requires a
>
> *Second-order principle of induction*, which as a statement about first-order principles of induction is, in turn, of a higher type than the latter; and so on.
>
> Every universal empirical statement requires a principle of induction of a higher type than the *inductum*, if it is to possess any *a posteriori* validity value at all (either true or false) as an inductum.
>
> Therein consists the infinite regression. (Popper's emphasis)

inferences, but does not note Reichenbach's deep skepticism about justifying probabilistic inferences (5–6).

9. Logic of Induction, Not Epistemology of Belief

We saw above that Hume's formulation of his critique of induction mixed logical and psychological notions. Hume identified deductive necessities as those discoverable by "the mere operation of thought," and contingencies are characterized as freely conceivable by the mind. As a result, his account leaves open whether the problem that he identified arises in inductive logic or in the psychological processes of belief formation. The first context, inductive and deductive logic, is independent of human thoughts and beliefs. It consists of propositions and inferences that arise as relations among propositions. The second context resides within the operation of the mind. Its relata are not propositions but beliefs, and reasoning[11] is a mental process that carries us from some beliefs to the formation of other beliefs.

The modern version of the problem of induction, the version that I wish to address, resides within the first context, the logic of induction, and not within the second context, the epistemology of belief. The problem is formulated in terms of rules governing inductive inferences and what happens when these rules are applied to themselves or to other rules. They are defined within the context of logic. These rules and the resulting problem of induction appear only indirectly in the epistemology of beliefs, after the problem has been formulated in the logical context. It arises in this second context in the specific case in which a reasoner uses these rules to direct reasoning from a belief in some propositions to a belief in others.

It is not possible, as far as I can see, to define the problem of induction within the epistemology of belief without first formulating it in the logical context. There is no problem of induction if a reasoner merely passes from one belief to another. The problem arises only when that passage is authorized by some rule of inductive inference, and we then ask what justifies that rule.

That the problem of induction is best formulated within the logical context is explicitly part of the twentieth-century revival of the problem. Russell makes the point:

11 It is common to describe this mental process as "inference" in the epistemological literature. Here I restrict the term "inference" to the first context, in which it denotes mind- and thought-independent relations over propositions. (This strictly logical operation is often called "implication" in the epistemological literature.)

> Now in dealing with this question we must, to begin with, make an important distinction, without which we should soon become involved in hopeless confusions. Experience has shown us that, hitherto, the frequent repetition of some uniform succession or coexistence has been a *cause* of our expecting the same succession or coexistence on the next occasion. (1912, 96–98; Russell's emphasis)

He continues with some examples. They include the well-known but dark chicken remark.[12] Russell concludes that

> We have therefore to distinguish the fact that past uniformities *cause* expectations as to the future, from the question whether there is any reasonable ground for giving weight to such expectations after the question of their validity has been raised. (98; Russell's emphasis)

Salmon (1967, 6) is similarly explicit. The problem, he stresses, is "a *logical* problem" (Salmon's emphasis). "It is the problem of understanding the logical relationship between evidence and conclusion in logically correct inferences." He then concludes thus:

> The fact that people do or do not use a certain type of inference is irrelevant to its justifiability. Whether people have confidence in the correctness of a certain type of inference has nothing to do with whether such confidence is justified. If we should adopt a logically incorrect method for inferring one fact from others, these facts would not actually constitute evidence for the conclusion we have drawn. The problem of induction is the problem of explicating the very concept of *inductive evidence*. (Salmon's emphasis)

---

12 "The man who has fed the chicken every day throughout its life at last wrings its neck instead, showing that more refined views as to the uniformity of nature would have been useful to the chicken" (98).

10. Epistemology Does Not Solve the Problem of Induction

In principle, misidentifying the problem of induction as deriving from the epistemology of belief could be troublesome. After a review of the literature on epistemology that was not especially diligent, my impression is that the danger has not been realized. Although the literature has made no special efforts to separate the two contexts, the failure seems not to have been troublesome. In internalist epistemologies, what justifies a belief is cognitively accessible to the reasoner. When a belief is justified by inductive inference, the reasoner knows it and knows that a rule of inductive inference was used. Thus, the problem that Hume identified can be spelled out in appropriate logical terms. In externalist epistemology, cognizers have no access to what justifies some beliefs. If they include the justifications for reasoning that corresponds to inductive inferences, then Hume's problem cannot be set up. We are unaware by stipulation of what justifies our reasoning and how it effects the justification. It follows that we cannot know whether these external justifications can be applied to themselves or even what it is for these external justifications to be applied to themselves.

There is only one case that I found of a clear confusion of logical and epistemological issues. In a widely known paper,[13] van Cleve (1984) sought to give an externalist solution to the problem of induction. It is evident from the start that the project cannot succeed. The challenge is to provide an explicit justification for inductive inference. Such a thing cannot be supplied by an epistemology in which the means of justification, by definition, are inaccessible to us.[14]

Van Cleve is undeterred. In the briefest sketch, he identifies two related inductive inference schemas:

> *x*% of the *A*'s I have examined were *B*'s.
> Hence, *x*% of *all A*'s are *B*'s.

13 I learned of this paper through correspondence with Job de Grefte.

14 For a critique of the capacity of externalist epistemologies to answer a broad range of skeptical challenges, see Fumerton (1995, Chapter 6). He notes (163, 171) that philosophers do not have the neurophysiological expertise to assess the efficacy of externalist justifications: "If I had wanted to go mucking around in the brain trying to figure out the causal mechanisms that hook up various stimuli with belief, I would have gone into neurophysiology."

and

> Most of the *A*'s I have examined were *B*'s.
> Hence, The majority of *all A*'s are *B*'s. (1984, 555–56; van Cleve's emphasis)

Somehow, through an external process inaccessible to us, we know that these are good inference schemas, and we know how to restrict application of these rules so that grue-like problems are avoided. This schema is then applied to our history of inductive reasoning to form what van Cleve calls "Argument A":

> Most of the inductive inferences I have drawn in the past from true premises have had true conclusions.
> Hence, The majority of *all* inductive inferences with true premises have true conclusions. (557; van Cleve's emphasis)

With the conclusion of Argument A, we have arrived at some form of justification for inductive inference.

This analysis cannot withstand scrutiny. There are two problems. The first problem is that the analysis is entirely too optimistic about the accuracy of our spontaneous human attempts at inductive reasoning. We human reasoners are naturally rather poor at it. Our natural inclinations are toward inductive fallacies.[15] If we could find some way to quantify the "majority of all inductive inferences" in the premise of Argument A, then we would likely find that the premise is false. That we are disposed to infer in some specific way, without any explicit justification for that disposition, is a poor justification for the correctness of the argument form implemented.

Indeed, a strong motivation for modern scientific methodology lies in the need to correct our natural inclination toward inductive fallacies. We see patterns where there is none. We too easily scan some collection of numerical data and come to the wrong conclusion. Too many of us judge a chance remission of some ailment as caused by whatever dubious therapy happened to be tried at that moment. Too many find an occasional cold day a basis for denying our warming climate. These misapprehensions are corrected by explicit statistical analysis. Similarly, we are too easily misled by anecdotal reports to

15 How is it that we survive? Our natural inductive inclinations are toward safety and the exaggeration of threats, not toward accuracy. There is ample redundancy in our interactions with the world, so that our many errors are individually correctible and mostly not fatal.

believe in the efficacy of some faulty treatment. The impulse to believe must be reined in by requiring controlled studies.

One can well imagine that an externalist justification is viable for narrowly specific beliefs, such as "Jones believes he has just seen a mountain-goat," to use Goldman's (1979, 10) example. However, it is much harder to see how such external mechanisms could reliably implant within us the sorts of universal logical schema sought by inductive logicians. Rather, we should expect most of the schemas spontaneously occurring to us to be incorrect. We will need explicit methods, such as those of science, to separate the few good ones from the many bad ones. Since these internal methods decide which schemas we should accept, any real advantage that externalist epistemologies could provide is lost.

The second problem is more serious. If externalism can solve the problem of induction, then we should expect the analysis to display a justified inductive inference schema. A principal consequence of the material theory of induction is that this end is unachievable if the goal is a universal schema of the type offered by van Cleve (1984). Inevitably, the particular schemas displayed by van Cleve, to put it charitably, are incomplete. That most of the *A*s that I have examined are *B*s is quite insufficient to authorize the conclusion that the vastly greater majority of *all A*s are *B*s.

Van Cleve simply avers that "I shall assume that we know how to restrict the predicates involved in these inferences so as to avoid Goodman's paradox about the grue emeralds" (1984, 556). That brash display of wishful thinking only begins to address the troubles that van Cleve has to suppose away. Even without the trickery of grue-ified predicates, inductive inference schemas, such as van Cleve displays, most commonly fail unless the *A*s and the *B*s are chosen very selectively under the guidance of background facts specific to the domain. This is the extended lesson drawn in Chapter 1 of *The Material Theory of Induction* (Norton 2021b). Even then additional facts might be needed. For example, depending on the case, we might need some assurance that the *A*s at issue have been sampled appropriately. That requires further background assumptions, such as the specification of a random sampling protocol.

These two concerns leave little of van Cleve's (1984) analysis intact. With the inductive inference schemas so incompletely specified, we have no assurance that they can be applied to our history of inductive reasoning to recover the core Argument A. And there is little motivation to do so since the premise of Argument A is likely false.

Papineau (1992, Section 2) gives a briefer analysis, similar to that of van Cleve (1984). We carry out an induction, premised on the successes of our past history inductions, to conclude that inductions lead to true conclusions. The correctness of this larger induction is based on the supposed reliability of induction as an inference scheme. It is sufficient here to note that the criticism above of van Cleve's analysis applies equally to it.

De Grefte (2020) has more modest ambitions. He disavows van Cleve's (1984) attempt to justify inductive inference. Rather, he argues that there is no problem of induction for a reliabilist externalist epistemologist:

> My present aim is only to establish that a reliabilist would not be troubled by the problem of induction. And that follows from the fact that reliabilists maintain that reliability is sufficient for justification, and that inductive inference may be reliable even if it is impossible to provide an argument for its inductive validity. We thus do not need to make the controversial assumption that inductive inference is, in fact, reliable. (102)

Here I agree with de Grefte: the modern problem of induction does not arise in a context in which there are no rules of inductive inference. However, he is wrong to conclude from this "... that externalist epistemologies are generally able to dissolve the problem of induction" (100). The problem of induction is a problem of inductive logic. It is not solved or dissolved by pointing out that the problem does not arise in another context.

There is a related problem that leaves reliabilist externalist epistemologists in a worse position than inductive logicians. That some epistemic process has been reliable in the past is no guarantee that it will continue to be reliable. Since these processes are invisible to externalists, they cannot even identify the processes justifying beliefs and thus have no means of controlling and assessing them.

11. The Material Dissolution of the Problem of Induction

The material theory of induction dissolves the problem of induction. The reason is simple and has already been given in the synopsis at the start of this chapter: the problem of induction is formulated in terms of universally applicable rules or schemas for inductive inference. There are no such rules or schemas in the material theory. It follows that the problem of induction cannot be set up. That is, there is no problem of induction within the material theory.

The analysis could stop with that. However, a common but mistaken reaction to this dissolution is that it is too easy. Surely a recalcitrant problem like the problem of induction cannot be dispatched so simply. In other failed solutions to the problem, the core difficulty remains but is somehow sufficiently disguised that it is no longer immediately apparent. Any claim of a solution to the problem of induction is then taken as an invitation to dig deeper to expose the trick and defeat the solution. It is the default reaction of philosophers to any claim of a solution to the problem of induction. This understandable intuition, mistaken in this case, directs us to seek a comparably troublesome regress or circularity in the justifications for inductive inferences within the material theory. There are both — regresses and circularities — within the material theory of induction. However, they are benign, unlike their counterparts in theories of induction with universally applicable schemas. Demonstrating this is the goal of the next two sections.

12. Regresses

Consider first the regresses within the material theory of induction. Each inductive inference is warranted by background facts in the applicable domain. If they are to provide a warrant, then they must be facts — that is, truths — so we expect that in turn they are supported by further inductive inferences. And these inductive inferences in turn require further facts to warrant them. And so on. What results is a regress of facts of some sort. However, it is a benign regress that merely recapitulates the mundane relations of inductive support that arise routinely within the sciences. It is unlike those troubling universally applicable inductive inference schemas of the problem of induction.

12.1. In the Traditional Problem

To see this, we begin with the troublesome case. For universally applicable inductive inference schemas, the traditional starting point of the regress is some version of enumerative induction. The regress is already troublesome at the outset, for, as we just saw, schemas of enumerative induction are incomplete. If applied mechanically, then they lead mostly to false conclusions. All too often, when we have some *A*s that are *B*, it is *not* the case that all *A*s are *B*. Hence, the first step of the regress, using another rule of inductive inference to justify the schema, has been set an impossible task. Still, we might follow Harman's (1965) lead and seek to use inference to the best explanation to vindicate enumerative induction. The effect is merely to add another layer of trouble. As argued at some length in Chapters 8 and 9 of *The Material Theory of Induction* (Norton 2021b), the schema of inference to the best explanation itself is incomplete. We have no agreement in the literature on what counts as an explanation, let alone just how to judge which is the best explanation. Indeed, I have argued, a distinctive notion of explanation seems to play no role in the standard examples of inference to the best explanation in science.

The regress cannot stop, however. We press on. How, as a general matter, are we to justify inference to the best explanation? Often explanations require simplifications that intentionally introduce idealizing falsehoods. Explanation and truth need not coincide. Nonetheless, perhaps we can find a third rule to justify this second rule. Might we suppose that the general use of this argument form has passed some sort of severe test so that it is justified by the rule of severe testing? Has the rule of inference to the best explanation been tested severely enough to justify its universal use?

Finally, might we tap instead into the unbridled optimism of Bayesians that their system can account for everything? Might there be a Bayesian vindication of inference to the best explanation, even if we remain unsure of just what an explanation is? Or might a Bayesian vindication succeed for any of the other rules that we might seek to justify in the regress? Whatever the prospects of success here for Bayesian vindications, we have still only postponed the difficulty. We must now ask what justifies the Bayesian system? In Chapters 10 and 11 of *The Material Theory of Induction* (Norton 2021b), I argued that all of the many attempts to justify probabilism are circular. This does halt the regress but at the cost of circularity.

We have explored only a few steps of the regress, and our store of distinct, universally applicable schemas of inductive inference is depleted. The prospect of sustaining an *infinite* sequence of such steps is not just distant but also obviously impossible. Our inferences have become as brittle as glass. We must feign some grasp of the application of inductive inference schemas in all generality, and then pretend to grasp clearly just what it is to apply still further inductive inference schema to them, and then more to them. We rightly judge this infinite regress of rules applied to rules and to rules as fanciful and unsustainable.

12.2. In the Material Theory

The regress of factual warrants in the material theory of induction is different. Where the regress of rules applied to rules is incomplete, speculative, and dubious, the regress of factual warrants is distinctive precisely because of its lack of distinction. It is simply the recapitulation of the grounds given in a mature science for its results. In Chapter 2, I described the nonhierarchical, massively entangled relations of inductive support within a mature science and argued that the totality of these relations is self-supporting. Another example can remind us of just how routine the regress of factual warrants is in a mature science. In Chapter 2, I used the illustration of the impossibility of a perpetual motion machine in the case of the EmDrive.

Consider now the general proposition that a perpetual motion machine of any kind is impossible. Our certainty of its impossibility is warranted by the fact of the conservation of energy. We can now begin the regress of warrants. What supports our confidence in the conservation of energy? I indicated in Chapter 2 that the totality of support for this fact is so immense that it extends well beyond what can be specified here. However, it is sufficient to say a little more to make the key point.

The conservation of energy — then commonly known as the "conservation of forces" — was one of the proud triumphs of mid-nineteenth-century physics. The result derived from the joint achievements of many, including James Joule, Julius Mayer, and Hermann Helmholtz. It was established through the accumulation of many smaller results, for the conservation of energy applies to all physical transformations. What needed to be shown was that, in each physical transformation, where a capacity was lost in some component, it was restored in another, and the restoration was such that a quantitative measure of the capacity was preserved.

When the result was still a scientific novelty, Helmholtz gave a popular lecture in Karlsruhe, sometime in the winter of 1862–63, summarizing its basis. Helmholtz (1885) proceeded methodically through the various transformation processes that contribute to the general result.

> *Simple mechanical processes*, such as bodies moving under gravity. They include the motion of pendula powering clocks, the falling weights that powered such clocks, mills powered by falling water, and the operation of diverse lever and pulley systems.
>
> *Processes involving and powered by elastic bodies.* They include springs and crossbows, along with bodies moved by the expansive powers of heated gases, such as those produced in a gun barrel by exploding gunpowder or the steam within a steam engine.
>
> *The many transformations of heat.* They include its transmission among solids, liquids, and gases and by radiative processes; its latency in phase transitions such as the melting of ice; and its production and absorption in chemical processes. Of great historical importance was the novel recognition that heat and motive power are intertransformable. Motive power can be converted to heat by friction, and heat can be converted to motive power in a heat engine.
>
> *Chemical transformations.* They include all manner of heat-generating, combustion reactions and fermentation reactions that produce pressurized gases.
>
> *Electrical processes.* They include the creation of combustible gases by electrolysis, the use of chemicals to generate an electrical current in the cells of a battery, and the interconvertibility of motive power into electrical currents in electric motors and dynamos. These electric currents can then produce chemical changes or, in resistances, create heat.

As Helmholtz worked his way, step by step, through all of these processes, the same result was recovered over and over: "Thus, whenever the capacity for work of one natural force is destroyed, it is transformed into another kind of activity" (359).

We see here the first steps of the regress of inductive support. Each result claimed by Helmholtz required further support. To recover them, he could indicate a long history of experimental work preceding his work in each of the sciences touched on by his inventory of processes. The best known of them in this context was the experimental work of Joule. He painstakingly measured the exact conversion between heat and motive power, the mechanical equivalent of heat. His was just one of many experiments touching all of the sciences. They include Regnault's painstaking measurements of the physical properties of steam and Faraday's many researches into electrochemistry. Following this path, the regress takes us on a tour of earlier nineteenth-century experimental work in the physical sciences. These next steps of Helmholtz's regress are not limited to experimental work. They also engage with established physical sciences. The conservation results pertaining to the motions of bodies under gravity could be drawn directly from well-established Newtonian mechanics and the conservation of heat itself from results in calorimetry and from what could be preserved of the caloric theory of heat.

Helmholtz's lecture gives an early portrait of the regress of inductive support shortly after the initial recognition of the conservation of energy. The regress continued for decades with ever growing strength. Each item listed in Helmholtz's inventory identified a distinct science: conservative mechanics, the mechanics of fluids, thermodynamics, chemistry, and electrical theory. As each developed, it affirmed the conservation of energy within the processes peculiar to its domain. Might we fear that the mysteries of electricity, magnetism, and radiation harbor a violation of the law of energy conservation? With the perfection of Maxwell's electrodynamics as the century progressed, the conservation of energy was issued as a simple theorem, a deductive consequence of his equations. There were also interactions among the sciences. The joint sciences of electrochemistry and thermochemistry emerged, for example. In each, the conservation of energy was maintained. Overall, the conservation of energy proves to be affirmed multiply in each of the sciences and in many experiments. Its affirmation in one area then provides support for its affirmation in another and conversely.

The law found new strength with the arrival of novel physics in the twentieth century. With Einstein's $E = mc^2$, energy and mass are identified. The law of conservation of mass had figured prominently in Lavoisier's establishment of the oxygen theory of combustion and his tabulation of elements. The law of conservation of mass is now merged with the conservation of energy. Evidential support for one is also evidential support for the other. As relativistic mechanics developed, a similar merger of conservation laws appeared. In the four-dimensional account of special relativity developed by Hermann Minkowski, the laws of conservation of energy and of momentum proved to be a manifestation of a single law of conservation of energy-momentum. The conservation of momentum supports the conservation of energy and conversely. The standard Hilbert space formulation of quantum theory emerged in the late 1920s and early 1930s. It gave energy conservation a special role. Physical systems were routinely represented by conservative Hamiltonian operators whose action on quantum states generates their time evolution. The resulting temporal dynamics then automatically conserves the energy of a system with determinate energy. The success of quantum dynamics depended on the conservation of energy and conversely.

This recounting of the evidential support for energy conservation and the necessary failure of all perpetual motion machines is likely not a moment of great excitement for the reader. It reads like a dull recitation of an introductory chapter in a dreary science text. That, of course, is precisely the point. When we ask what justifies a fact warranting some inference in a mature science, we begin a regress that recounts relations of inductive support upon relations of inductive support. We rapidly find that tracing these relations takes us on a tour of much science, and we find the relations entangled in many mutually reinforcing interactions that give rigidity and strength to the structure. At each moment in our tour, we encounter a piece of ordinary, unremarkable science. What we do not find is what we found in the regress of universal schemas of induction: an accumulation of incompletenesses that terminates in dubious speculation after only a few steps of regress.

The justificatory regress of universal schemas of inductive inference is almost immediately ruinous and presents a severe challenge to any account of such schemas. The regress of inductive support in the material theory of induction is merely the recapitulation of mundane science. It just recalls how science is done.

13. Circularities

To recall a theme stressed repeatedly in this volume, the massive tangle of relations of inductive support in a mature science includes circularities, both large and small in compass. We have just seen several of them. Einstein's $E = mc^2$ merged the conservation of energy and the conservation of mass. Through the mediation of this fact of merger, it now follows that the earlier establishment of the conservation of mass in chemistry provides support for the conservation of energy in physics, and the earlier establishment of the conservation of energy in physics provides support for the conservation of mass in chemistry. Similar relations of mutual support arise for the laws of conservation of energy and conservation of momentum, through the fact that these laws are expressed as a single law of conservation of energy-momentum in the four-dimensional formulation of special relativity.

I argued at some length in Chapter 3 that the mere presence of a circularity in some system is not an automatic condemnation of the system. Many circularities, like the ones just noted, are common in unobjectionable science. Rather, if we are to assert that a circularity is troublesome, then we have a positive obligation to demonstrate that the specific circularity is so. The chapter provided two means for doing this. The most serious case is a vicious circularity. In it, the circular relations lead to a contradiction. The less serious case is a circularity that leaves the structure indeterminate. If that indeterminacy is not transient but ineliminable, then the common resolution is to judge the structures involved as not factual. That is, they can be set conventionally, much as we are free to set our units of measurement.

In the circularity forms of the problem of induction, we seek to use a universal schema of inductive inference to justify itself. This circularity is immediately troublesome, for it forms a tight circle that leaves the schema indeterminate. It is easy to show that there are too many dubious universal schemas of inductive inference that are self-justifying. The trouble is that self-justification is too permissive. Salmon's (1967, 12) preliminary example is of a psychic who makes predictions by gazing into a crystal ball:

> When we question his claim he says, "Wait a moment; I will find out whether the method of crystal gazing is the best method for making predictions." He looks into his crystal ball and announces that future cases of crystal gazing will yield

> predictive success. . . . "By the way, I note by gazing into my crystal ball that the scientific method is now in for a very bad run of luck."

Another of Salmon's examples is a counterinductive rule that is self-justifying. It mimics the familiar attempt to allow inductive inference to be self-justifying. Salmon defines an inductive rule R_3:

> To argue from
>
> *Most instances of A's examined in a wide variety of conditions have been non-B*
>
> to (probably)
>
> *The next A to be encountered will be B.* (15; Salmon's emphasis)

Salmon then takes as a premise that most applications of rule R_3 (*A*) have been unsuccessful (not-*B*). Rule R_3 then assures us that it will be successful on its next application. More formally, Salmon writes that

> R_3 has usually been unsuccessful in the past.
>
> *Hence (probably):*
>
> R_3 will be successful in the next instance. (15; Salmon's emphasis)

Douven (2017, Section 3.2) provides an amusing variant of Salmon's counterinductive rule:

> For suppose that some scientific community relied not on abduction but on a rule that we may dub "Inference to the Worst Explanation" (IWE), a rule that sanctions inferring to the worst explanation of the available data. We may safely assume that the use of this rule mostly would lead to the adoption of very unsuccessful theories. Nevertheless, the said community might justify its use of IWE by dint of the following reasoning: "Scientific theories tend to be hugely unsuccessful. These theories were arrived at by application of IWE. That IWE is a

> reliable rule of inference — that is, a rule of inference mostly leading from true premises to true conclusions — is surely the worst explanation of the fact that our theories are so unsuccessful. Hence, by application of IWE, we may conclude that IWE is a reliable rule of inference."

I stressed above that we have a positive obligation to demonstrate that circularity is troublesome. Salmon's and Douven's analyses show just this trouble for the self-justifying schema.

That there is no such demonstration of troublesome circularity in the material theory of induction I argued in Chapters 3 and 4. Contradictions can arise provisionally in the tangle of mutual relations of inductive support of a developing theory. They are merely an indication that we have an error somewhere in our structure. They are routine and provide a helpful guide to finding the error and its subsequent elimination. Indeterminacies can also arise. If they are ineliminable, then we have good reason to conclude that what is left indeterminate is not factual but something that can be set by convention, for we have found something beyond the reach of evidence. Finally, if the indeterminacies admit multiple theories but remain within the reach of evidence, then we find that the resulting competition among those theories is unstable. An advantage for one strengthens it at the expense of the others. Under this instability and the accumulation of further evidence, inductive support is driven to favor just one of the competing theories.

The circularities that arise when universal schemas of inductive inference seek to justify themselves are self-defeating. The circularities of inductive support that arise in the material theory of induction are merely symptoms of a massively interconnected network of relations of inductive support. They are part of what gives strength and rigidity to the evidential support of mature sciences.

14. Sober and Okasha

It would be a surprise if a response to Hume's problem this simple had been entirely overlooked in the literature. As far as I know, there are two older versions of this escape. Neither is complete since each omits at least one key piece, but they have enough for me to characterize them as close to the material dissolution.

Sober (1988) notes that Humean skepticism about our knowledge of the future is equally a problem for historical sciences, such as evolutionary biology, for they try to discern the past from evidence in the present. In these inferences, invocations of simplicity can play a prominent role. Sober, however, understands them materially:

> Whenever observations are said to support a hypothesis, or are said to support one hypothesis better than another, there must be an empirical background theory that mediates this connection. It is important to see that this principle does not evaporate when a scientist cites simplicity as the ground for preferring one hypothesis over another in the light of the data. *Appeal to simplicity is a surrogate for stating an empirical background theory.* (64; Sober's emphasis)

Sober then applies this material understanding of induction to Hume's problem. According to the problem, as Sober recalls it, inductive inference depends on an inductive principle that cannot be justified by reason alone. In place of this failure, he finds a regress:

> What we do find in any articulated inductive argument is a set of empirical assumptions that allow observations to have an evidential bearing on competing hypotheses. These background assumptions may themselves be scrutinized, and further observations and background theory may be offered in their support. When asked to say why we take past observations to support the belief that the sun will rise tomorrow, we answer by citing our well-confirmed theory of planetary motion, not Hume's Principle of the Uniformity of Nature. If challenged to say why we take this scientific theory seriously, we would reply by citing other observations and other background theories as well. (65–66)

All that is needed for this analysis to coincide with the material dissolution is for Sober to affirm a benign termination of the regress. Here he falters. Through an obliquely answered rhetorical question, he concludes that there is no "stage where an empirical belief that is not strictly about the here and now is sufficiently supported by current observations, taken all by themselves" (66).

Such a stage is incompatible with his earlier conclusion that observations can support a hypothesis only relative to a background theory. Sober concludes that

> The thesis that confirmation is a three place relation sustains Hume's skeptical thesis, but not the argument he constructed on its behalf. (66)

Sober's objection to a benign termination to the regress, we can now see, depends on a tacit adherence to the hierarchical structure of relations of inductive support denied in Chapter 2 of this volume. Without it, we are freed from the requirement that a warranting fact must be drawn from somewhere in a later stage of the regress. A benign termination is possible merely using warranting propositions supported elsewhere.[16]

Okasha (2001) recounts the key idea of the material dissolution of the problem of induction in a section headed "IV. No Rules of Induction, No Humean Argument." The section ends thus:

> To conclude, a Humean sceptical argument will only work if our inductive behaviour can be characterized as a process of rule-governed ampliation. There is no necessity that our inductive behaviour can be so characterized. I have offered reasons for thinking that it cannot be. If this is correct, then Hume's argument cannot be converted from a valid one into a sound one, and the threat of inductive scepticism is successfully parried. (324)

Okasha also recognizes that an inductive rule is applicable only if the background factual conditions are hospitable. In the material theory, it is inferred from this circumstance that rules of inductive inference can be applied only locally in suitably hospitable domains. Here, unfortunately, Okasha takes a different course that precludes a full dissolution of the problem of induction. He treats inductive rules as universally applicable and finds this to require us to abandon all rules of inductive inference. That is, he writes,

16 See Okasha (2005) for an account of Sober's analysis and the material dissolution as presented in Norton (2003).

> To use an inductive rule is to assume that the world is arranged in a particular way, as I have stressed. . . . So following any particular inductive rule does seem less than fully rational. It embodies a fixed commitment to the world's being in a certain state; but *qua* empiricists we should undertake such commitments only provisionally, not hold on to them at all costs. (321)

The result is that Okasha must seek some other account of the inductive practices of science. He explores Bayesianism, understood as the dynamics of opinion change, and Popper's deductivist elimination of induction. Hume's problem is escaped but at the cost of denying that science infers inductively.

15. What Justifies Induction in the Material Theory

Showing that there is no problem of induction in the material theory might seem to leave the fundamental question unanswered. What, one might still wonder, justifies the practice of inductive inference, according to the material theory? Although the answer was implicit in the discussion in the previous section, it might be helpful to make it more explicit.

The question can appear to be unanswered if it is accompanied by a false presumption. In asking "What justifies . . . ?" the presumption might be that we can identify a particular thing that does the justifying. That was the sort of answer that Mill tried to give with his principle of uniformity of nature. In the material theory of induction, there is no single identifiable thing that justifies inductive inference. Rather, the justification for inductive inference is distributed over the entirety of the complicated network of relations of inductive support that comprises a mature science. In the early stages of a new science, when these networks are not fully in place, justification might be only partial. At least some of the justificatory work is done by propositions, introduced hypothetically, without themselves having proper support. The goal, as the science develops, is to provide support for each of these hypotheses so that no proposition of the resulting mature science is without inductive support.[17]

Perhaps an analogy will help to illustrate the sufficiency of this answer. The vitalists of the eighteenth and nineteenth centuries sought in vain for

17 This notion of distributed support has already appeared in variant forms in Chapters 2 and 5 of this volume.

the animating spirit that distinguishes living matter from dead matter. As biology advanced into the twentieth century and our knowledge of the details of life processes became increasingly detailed, the futility of the search for this *élan vital* became clear. However, there was no simple answer to the question of what makes something alive. A biologist could examine in great detail any portion of a living organism and find only inanimate chemical and electrochemical processes, even if of great complexity. We can point to no single thing that animates matter. The best — in fact the only — answer to the question of what makes some organism alive is just this: it is no one piece of the organism. Its life derives from the synthesis of all of the many processes of its many parts.

16. Critical Responses to the Material Dissolution

Section 6 of my first paper on the material theory of induction (Norton 2003) described how that theory eluded the problem of induction. I have described in the preface to this volume how this dissolution of the problem of induction generated a critical response out of proportion to its place in the original paper. However, the criticism revealed that I had not developed the details of the dissolution well enough. It needed to be sharpened. Here I will recall that criticism and show how subsequent refinements have responded to it. There were two broad areas of concern, indicated below.

16.1. From Particulars to Generalities

First, I had correctly identified the regress of justifications in the material theory as benign and merely recapitulating ordinary relations of support in standard science. However, I had not identified the nonhierarchical structure of these relations and the role of hypotheses in its erection. Rather, in Norton (2003, Section 6), I merely asserted that the regress is benign and gave some inconsequential speculation on the possibilities for its termination. They included a termination in "brute facts of experience."

Both John Worrall (2010) and Tom Kelly (2010) found this inadequate. As Worrall correctly noted,

> However, if we follow this backward direction, we clearly meet what seems to be an insuperable problem: the accreditational buck has to stop somewhere: it cannot be an infinite chain (or

> rather tree . . .). . . . [W]e know that nodes in the tree must contain, at some stage, universal claims — and so we would still have to account for some initial act (or acts) of generalization. And given that we want each node to be justified, we would seem to be back at the same old problem. (746)

And then

> I am unsure what a "brute fact" of experience is. But presumably brute facts for Norton here had better be singular: if so, then the problem has not been solved since the tree needs to go universal at some point; . . . (747)

Kelly set up his objection by defining *E*:

> . . . [C]onsider that time immediately before we acquired our first piece of inductive knowledge. Let E represent the totality of our knowledge at that moment. (760)

Trouble, Kelly continued, then ensues:

> Suppose that we try to take a first, minimal step beyond E. Again, intuitively, this proposition will be our first piece of inductive knowledge. In that case, we must have recourse to at least one known material postulate. Of course, that material postulate has to be a part of E, since it has to be known, and E represents the totality of our knowledge at the time. . . . My worry is that, given that the only empirical knowledge that one has at that point is observational knowledge and its deductive consequences, there would not be anything suitable around to play the role of material postulate. (761)

In brief, the concern is that we start knowing only particular facts. To extend our knowledge inductively to generalities of vastly greater scope, we need a material postulate of vastly greater scope. By supposition, we have no such fact in our starting point.

This is an objection that needs a response, and I am grateful to Worrall and Kelly for pressing me on it. The response to these worries came in Norton (2014) (and is elaborated in Chapter 2 here). Their objection fails. It neglects the use of hypotheses as a way of extending the inductive reach of evidence well beyond its initially limited scope. We can and routinely do take a first faltering step in inductive inference by hypothesizing the warranting fact needed. This warranting fact can be of generality greater than the facts from which we initially proceed. The key is that its use is provisional. We have a positive obligation to return to the hypothesis and show in subsequent investigations how it is supported inductively. When we succeed, we commonly end up with cogent but massively entangled relations of inductive support. If we do not succeed, then we must concede that the inference has no warrant and should be abandoned.

It is a lesson hard won by authors of philosophy papers that their solutions to problems can be overlooked. Such has happened with works by Schurz (2019) and Schurz and Thorn (2020). They mischaracterize the material approach to induction as a "uniformity account" (Schurz 2019, 17; Schurz and Thorn 2020, 89): that is, an account based on uniformity assumptions. Then they assume that the regress of inductive support depends on a sequence of uniformity assumptions of increasing generality that cannot terminate satisfactorily.[18] Both texts provide instances of such sequences. Readers should be forewarned that these sequences are proposals by Schurz and Thorn and not part of my account. Their supposition is based on a mistaken assumption about how the warranting facts of an inductive inference themselves are to be warranted. The supposition is that the successive warranting facts in this process inexorably must become ever more general. In this way, relations of inductive support are supposed to be adapted to a hierarchy of increasing generality.

That inductive support, materially understood, avoids just such sequences was an important consequence of my identification (Norton 2014, Section 10) of the nonhierarchical structure of relations of inductive support, further elaborated here in Chapter 2. Rather, relations of inductive support cross over one another in a massively entangled structure that respects no such hierarchy of generality. Schurz and Thorn (2020) draw their treatment of the

18 "A closer look at Norton's example shows that the uniformity assumptions that justify inductive inferences become more and more general" (Schurz 2019, 17); ". . . the uniformity assumptions that justify material inductive inferences become unavoidably more and more general" (Schurz and Thorn 2020, 90). Independent of any consideration of the material theory, Bird (1998, 111) characterizes the regress form of the problem of induction in terms of an unsustainable regress of ever more general, justifying facts.

material escape from Norton (2003), supplemented by references to Worrall (2010) and Kelly (2010). They do not cite Norton (2014) and make no accommodation for its assertions. My fuller response to them is in Norton (2021a).

16.2. Logic versus Epistemology

Second, I had not separated questions of inductive logic from those of the epistemology of beliefs, as I have now done in Section 9 of this chapter. That this should have happened in my response (Norton 2014) to Kelly (2010) was almost inevitable since his critique mingled the two throughout. Kelly presented a core claim of the material theory in epistemological terms:

> In what sense are inductive inferences "grounded in" material facts? . . . [W]hat is required is that the person drawing the inference knows (or at least, reasonably believes) that they obtain. (759)

This Kelly soon reinforces as the key supposition that will lead to his objection to the material dissolution:

> . . . Norton's view is that knowledge of the underlying material postulate is what is required: "In order to learn a fact by induction, the material theory says that we must already know a fact, the material postulate that licenses the induction" (2003, 666).[19]
>
> Let us call this commitment of the material theory:
>
> **Prior Knowledge**: in order to learn a fact by induction, one must have prior knowledge of the material fact that licenses the induction. (2010, 760; Kelly's emphasis)

Kelly's narrative here takes a central claim of the material theory of induction from the context of the logic of induction and reconstitutes it as a claim in the epistemology of belief. With this revision, as quoted above, Kelly sets up E: the totality of our knowledge at the moment immediately before we acquire

19 The remark quoted from me ("In order to learn a fact . . . know a fact . . .") reports a consequence of the material logic of induction for the epistemology of belief. The "knowing" is not constitutive of inductive inference relations in the material theory. Kelly mistakenly makes it so.

our first piece of inductive knowledge. He can now pose what appears to be an insurmountably difficult problem. How can we proceed from E to make the first induction to a generalization of vastly greater scope?

Understood as a problem of inductive logic, it is not so formidable. We have some body of particular facts. Which inductive inferences can it support? As I recounted in Chapter 2, once we abandon the unnecessary hierarchical restrictions on applicable material postulates, we find that there is no barrier to them grounding an extensive science with propositions of general scope, as long as the propositions of E themselves are varied enough. We can even recover the inductive structure from a sequence of inductive inferences that employ hypotheses provisionally.[20]

If, however, we conceive E epistemologically, as some sort of exhaustive specification of the beliefs of a fictional primitive human, then now we have posed a new and more difficult problem. We have somehow to imagine the unimaginable. What is it to be such a human, fully grasping many particulars but no generalities? What would such a human do next? Would such a human have any confidence that generalities are somehow within inferential reach? What might motivate such a human even to want to try?

This epistemological formulation of the problem led me (Norton 2014, Sections 6–7) to give some epistemological analysis of what I called "the historical-anthropological objection." I agreed with John Worrall about the spuriousness of the epistemological problem posed. We have no reason to believe that our forebears were ever in the cognitive state represented by E. Even while objecting that the problem as posed engaged in wild speculation, I sought to make the point by responding with more speculation of my own on the prospects of primitive cognition in what I called a "counter-fable."

Looking back, I stand by the content of my analysis. However, I now regret not choosing a more cautious response. The material theory of induction has no trouble dealing with the inductive logic of the problem. Once the problem is enmeshed with fabrications of fictitious primitive humans in the epistemology of belief, it can no longer be addressed responsibly by armchair philosophers. Even though this was the basic point that I sought to make, it

20 One might worry that this use of hypotheses strays into the epistemology of beliefs. The use of hypotheses, as described in Chapter 2, is akin to the positing of a hypothesis in ordinary deductive logic as part of a *reductio ad absurdum*. In both cases, the hypotheses figure in explicit logical relations over propositions. Beliefs need not enter the analysis.

was a mistake to engage in any more detail,[21] for it invites the misapprehension that the material theory of induction has some responsibility to make sense of primitive humanoid cognition. It does not. Its compass is restricted to inductive logic defined over propositions and especially those that enter into routine science. It has no responsibility to the inchoate speculations of a primitive Adam when he first stumbles out of his cave.

De Grefte (2020) entangles logic and epistemology in a sequence of dubious arguments. First, he argues that "proponents of the material theory of induction are in fact committed to an externalist epistemology." Here I resist all attempts to enmesh the material theory in issues of epistemology and have no interest in connecting the material theory of induction with any particular epistemology. Lest the point pass, however, I should report that de Grefte's efforts to establish a commitment to an externalist epistemology are weak. As far as I can see, internalists can employ the material theory of induction simply by being aware of the material facts authorizing the inductive inferences behind their reasoning.

Second, de Grefte (2020, 100) argues "that externalist epistemologies are generally able to dissolve the problem of induction." In Section 10 of this chapter, I argued that this is a mistake. That there is no problem of induction in an externalist epistemology does not solve a problem in inductive logic. Moreover, reliabilist externalist epistemologies are felled by a problem analogous to the problem of induction.

Hence, finally, with these two failures, there is no foundation for de Grefte's claim:

> Like extant forms of externalism, Norton's material theory of induction dissolves the problem of induction. But since the material theory *entails* an externalist epistemology, one may suspect it is this externalism that does the epistemological work here. (2020, 104; de Grefte's emphasis)[22]

Weintraub's (2016, Section 4) appraisal of the material dissolution illustrates again the dangers of mixing logic and epistemology incautiously.

---

21 This regret also applies to remarks in Norton (2003, as in 668n9), in which I assert (correctly) that brute facts like "the ball is red" already presupposed universal knowledge.

22 See also my response in Norton (2021a, Section 6).

After recounting the much-cited "bismuth" example of Norton (2003, 649),[23] she writes

> But it is extremely implausible to suppose that if bismuth is *in fact* an element, but we justifiably believe that it *isn't* or have no opinion about the matter, our belief that it melts at 271 C is justified, our sample of positive instances notwithstanding. (72; Weintraub's emphasis)

That is, Weintraub supposes that we have mistakenly come to believe falsities of the background domain or perhaps have no suitable background beliefs. Then she correctly notes that we would be unable to justify the appropriate conclusion concerning the melting point of bismuth. There is no fault here in the inductive logic. The fault lies in the translation of logic into belief states. The cognizer proceeds by supposition from false or inadequate beliefs. It is a failure outside the compass of the material theory of induction.

Her dismissal of the material dissolution of the problem of induction seems to rest on a misreading of the material theory. Weintraub (2016, 72) characterizes the material theory as "an attempt to eliminate induction," grouped with Popper's inductive eliminativism. I understand her to hold that the material theory treats inductive inferences as enthymemes. That is, they will be rendered deductive with the addition of the material postulate as another premise.[24] Weintraub reports correctly some truisms of deductive logic, such as "that all observed instances of bismuth were elements doesn't entail that *all* instances of bismuth are elements" (72; italics in the original). However, these truisms are insufficient to support her conclusion: "Norton's attempt to dissolve the problem of induction, I conclude, fails (again) because its characterization of our practice is erroneous" (72). Weintraub's critique is based on an erroneous characterization of the material theory.

---

23 From "some samples of the element bismuth melt at 271C," we infer that "all samples of the element bismuth melt at 271C" using the warranting fact that "all samples of bismuth are uniform just in the property that determines their melting point, their elemental nature. . . ."

24 Here Weintraub overlooked the disclaimer in Norton (2003, 651): "Chemical elements are generally uniform in their physical properties, so the conclusion of the above induction is most likely true." A footnote explains the inductive risk taken: "Why 'generally'? Some elements, such as sulfur, have different allotropic forms with different melting points."

Finally, Skeels (2020) somehow manages to convince himself that there are two "Nortons" who advocate two different material theories. They correspond to the real logical version and Skeel's invented epistemological version of the material theory. In the first, justifications derive from facts and, in the second, from knowledge. Skeels then seeks to use his misidentification to impugn the material dissolution of the problem of induction. See Norton (2021a, section 14) for my response.

16.3. More Treatments

For completeness, I recall some other treatments of the material dissolution of induction in the recent literature.

Livengood and Korman (2020) accept the material dissolution of the problem of induction as a matter of inductive logic. However, they urge that rational entitlement to future beliefs goes beyond consideration of evidence and inductive logic. The entitlement fails in the absence of a suitable explanatory relationship between the belief and the fact to be believed. As I indicate in my response (Norton 2021a, Section 9), this problem goes beyond the concerns of the material theory of induction. It is an issue of the epistemology of belief formation, and I hope that epistemologists can resolve it.

Jackson (2019, 164) disputes the material dissolution of Hume's problem by disputing a key condition of the material theory itself, that warranting facts must be facts: that is, truths. He argues, erroneously, that this precludes proper warrant for eighteenth-century predictions that employed Newton's laws of motion. There is no problem here. Our best theory of gravity, general relativity, returns Newton's entire theory in the weak gravitational fields pertinent to eighteenth-century physics. Jackson also worries that "scientifically ignorant people" might no longer have a warrant for inferring that night will follow day. Having learned my lesson, I will not be lured again into speculating about the inductive practices of fictitious or vaguely specified "scientifically ignorant" people. If inventions and fictions are to be avoided, then Jackson is well advised to do the same.

Peden (2019) offers a friendly amendment to the material dissolution of the problem of induction. He argues that it would benefit from supplementation by the combinatorial justification for induction of Williams and Stove, in conjunction with what is sometimes called "direct inference," "statistical syllogisms," or "proportional syllogisms." Whether this supplement is helpful is a topic that needs to be dealt with elsewhere. However, I am wary of such gifts since my fear is that they create more problems than they solve.

17. Conclusion

Hume's problem of induction has the reputation of being one of the most fearsome and intractable problems of philosophy. In her synoptic article, Henderson (2020) reports Russell's dark warning: "If Hume's problem cannot be solved, [Russell laments, then] 'there is no intellectual difference between sanity and insanity.'" Henderson finds a huge range of solutions in the present literature and an enduring belief by many that none succeeds. When such diversity persists, we can only conclude that, so far, we are doing a poor job of protecting ourselves from the lamentable conclusion that Russell feared.

Of the many solutions currently on offer, in my view, the best is Reichenbach's pragmatic solution. It is a dominance argument. We should infer inductively, even if we cannot justify induction as leading to the truth, since, pragmatically, if any method can work, then induction will work. The pragmatic solution has its best exposition and elaboration in Salmon (1967). Over half a century after its publication, I still find it to be one of the best treatments of Hume's problem. Ingenious as it is, Reichenbach's pragmatic solution is unsatisfying. It puts us in the same position as a drowning man, clutching at straws. Both we inductive inferers and the drowning man would like some further assurance of the efficacy of our desperate measures. We should like something a little stronger than "What have you got to lose?!" That this pragmatic answer and clever formal elaborations of it should retain a firm position in the literature is a sure index of the literature's failure to treat the problem well.

This despondent view was my view until I began work on the material theory of induction. It became clear then that even the most intractable problems are defined within a framework. What can make them intractable is precisely that we seek solutions within the framework. If we can break out of that framework, then perhaps the problem can be solved. In the best case, the problem can no longer even be set up. That proves to be the case when we adopt a material theory of induction. The problem of induction, in its most intractable modern form, is a problem for universal rules of induction. Once we adopt a material theory of induction, we abandon universal rules of induction. We break out of the confining framework. The problem of induction can no longer be set up. It is dissolved.

REFERENCES

Annas, Julia, and Jonathan Barnes, trans. and eds. 2000. *Sextus Empiricus: Outlines of Scepticism*. Cambridge, UK: Cambridge University Press.

Bacon, Francis. (1620) 1902. *Novum Organum*. Edited by Joseph Devey. New York: P.F. Collier and Son.

Ballard, Edward G., trans. and ed. 1960. *The Philosophy of Jules Lachelier*. Dordrecht: Springer.

Bayes, Thomas. 1763. "An Essay towards Solving a Problem in the Doctrine of Chances." *Philosophical Transactions of the Royal Society of London* 53: 370–418.

Bird, Alexander. 1998. *Philosophy of Science*. N.p.: Routledge.

Bradley, Francis H. 1883. *The Principles of Logic*. London: Kegan Paul, Trench, and Company.

Cassirer, Ernst. 1910. *Substanzbegriff und Funktionsbegriff*. Berlin: Bruno Cassirer.

Creighton, James E. 1898. *An Introductory Logic*. New York: Macmillan.

de Grefte, Job. 2020. "Epistemic Benefits of the Material Theory of Induction." *Studies in History and Philosophy of Science*, Part A, 84: 99–105.

Douven, Igor. 2017. "Abduction." In *The Stanford Encyclopedia of Philosophy*, summer 2017 ed., edited by Edward N. Zalta. https://plato.stanford.edu/archives/sum2017/entries/abduction/.

Earman, John. 2002. "Bayes, Hume, Price and Miracles." *Proceedings of the British Academy* 113: 91–109.

Fumerton, Richard A. 1995. *Metaepistemology and Skepticism*. Lanham, MD: Rowman and Littlefield.

Goldman, Alvin I. 1979. "What Is Justified Belief?" In *Justification and Knowledge: New Studies in Epistemology*, edited by G.S. Pappas, 1–23. Dordrecht: Reidel.

Harman, Gilbert H. 1965. "The Inference to the Best Explanation." *Philosophical Review* 74: 88–95.

Helmholtz, Hermann. 1885. "On the Conservation of Force." In *Popular Lectures on Scientific Subjects*, translated by E. Atkinson, 317–97. New York: D. Appleton and Company.

Henderson, Leah. 2020. "The Problem of Induction." In *The Stanford Encyclopedia of Philosophy*, spring 2020 eds., edited by Edward N. Zalta. https://plato.stanford.edu/archives/spr2020/entries/induction-problem/.

Hume, David. (1739) 1896. *A Treatise of Human Nature*. Edited by Lewis Amherst Selby-Bigge. Oxford: Clarendon.

———. (1777a) 1894. *An Enquiry Concerning the Human Understanding, and an Enquiry Concerning the Principles of Morals*. Edited by Lewis Amherst Selby-Bigge. Oxford: Clarendon Press.

———. 1777b. *The Life of David Hume, Esq, Written by Himself.* London: W. Strahan and T. Cadell.

Jackson, Alexander. 2019. "How to Solve Hume's Problem of Induction." *Episteme* 16: 157–74.

Jevons, W. Stanley. 1874. *The Principles of Science: A Treatise on Logic and Scientific Method.* New York: Macmillan.

———. 1888. *Elementary Lessons in Logic: Deductive and Inductive.* London: Macmillan.

———. 1902. *Science Primers: Logic.* London: Macmillan.

Kant, Immanuel. (1783) 1909. *Prolegomena to Any Future Metaphysics.* Edited by Paul Carus. Chicago: Open Court.

Kelly, Thomas. 2010. "Hume, Norton and Induction without Rules." *Philosophy of Science* 77: 754–64.

Kirwan, Richard. 1807. *Logick: Or an Essay on the Elements, Principles and Different Modes of Reasoning.* 2 vols. London: Payne and Mackinlay.

Lachelier, Jean. 1907. *Du fondement de l'induction.* Paris: Felix Alcán.

Landesman, Charles, and Roblin Meeks, eds. 2003. *Philosophical Skepticism.* Oxford: Blackwell.

Laplace, Pierre Simon. 1902. *A Philosophical Essay on Probabilities.* 6th ed. Translated by Frederick Wilson Truscott and Frederick Lincoln Emory. New York: Wiley.

Livengood, Jonathan, and Daniel Z. Korman. 2020. "Debunking Material Induction." *Studies in History and Philosophy of Science*, Part A, 84: 20–27.

Mill, John Stuart. 1882. *A System of Logic, Ratiocinative and Inductive.* 8th ed. New York: Harper and Brothers.

Munro, H.H. 1850. *A Manual of Logic, Deductive and Inductive.* Glasgow: Maurice Ogle and Son.

Norton, John D. 2003. "A Material Theory of Induction." *Philosophy of Science* 70: 647–70.

———. 2005. "A Little Survey of Inductive Inference." In *Scientific Evidence: Philosophical Theories and Applications*, edited by Peter Achinstein, 9–34. Baltimore: Johns Hopkins University Press.

———. 2010. "A Survey of Inductive Generalization." http://www.pitt.edu/~jdnorton/homepage/cv.html#L2010.

———. 2014. "A Material Dissolution of the Problem of Induction." *Synthese* 191: 671–90.

———. 2021a. "Author's Responses." *Studies in History and Philosophy of Science* 85: 114–26.

———. 2021b. *The Material Theory of Induction.* BSPS Open Series. Calgary: University of Calgary Press.

Okasha, Samir. 2001. "What Did Hume Really Show about Induction?" *Philosophical Quarterly* 51: 307–27.

———. 2005. "Does Hume's Argument against Induction Rest on a Quantifier-Shift Fallacy?" *Proceedings of the Aristotelian Society* 105: 237–55.

Papineau, David. 1992. "Reliabilism, Induction and Scepticism." *Philosophical Quarterly* 42: 1–20.

Peden, William. 2019. "Direct Inference in the Material Theory of Induction." *Philosophy of Science* 86: 672–95.

Popper, Karl. (1959) 2002. *The Logic of Scientific Discovery.* London: Routledge Classics.

———. 2009. *Two Fundamental Problems of the Theory of Knowledge.* London: Routledge.

Reichenbach, Hans. 1930. "Kausalität und Wahrscheinlichkeit." *Erkenntnis* 1: 158–88.

———. 1938. *Experience and Prediction.* Chicago: University of Chicago Press.

———. 1949. *The Theory of Probability.* 2nd ed. Translated by Ernest H. Hutten and Maria Reichenbach. Berkeley: University of California Press.

Russell, Bertrand. 1912. *The Problems of Philosophy.* New York: Henry Holt and Company.

Salmon, Wesley C. 1967. *The Foundations of Scientific Inference.* Pittsburgh: University of Pittsburgh Press.

———. 1981. "Rational Prediction." *British Journal for the Philosophy of Science* 32: 115–25.

Schiller, Ferdinand C.S. 1912. *Formal Logic: A Scientific and Social Problem.* London: Macmillan and Company.

Schurz, Gerhard. 2019. *Hume's Problem Solved: The Optimality of Meta-Induction.* Cambridge, MA: MIT Press.

Schurz, Gerhard, and Paul Thorn. 2020. "The Material Theory of Object-Induction and the Universal Optimality of Meta-Induction: Two Complementary Accounts." *Studies in History and Philosophy of Science*, Part A, 82: 88–93.

Skeels, Patrick. 2020. "A Tale of Two Nortons." *Studies in History and Philosophy of Science*, Part A, 83: 28–35.

Sober, Elliott. 1988. *Reconstructing the Past: Parsimony, Evolution, and Inference.* Cambridge, MA: Bradford/MIT Press.

Thomson, J. Radford. 1887. *A Dictionary of Philosophy in the Words of Philosophers.* London: Reeves and Turner.

van Cleve, James. 1984. "Reliability, Justification, and the Problem of Induction." *Midwest Studies in Philosophy* 9: 555–67.

Weintraub, Ruth. 2016. "The Problem of Induction Dissolved: But Are We Better-Off?" *American Philosophical Quarterly* 53: 69–83.

Whately, Richard. 1856. *Elements of Logic.* New York: Harper and Brothers.

Worrall, John. 2010. "For Universal Rules, against Induction." *Philosophy of Science* 77: 740–53.

Zabell, Sandy. 1989. "The Rule of Succession." *Erkenntnis* 31: 283–321.

PART II

Historical Case Studies

7

The Recession of the Nebulae

1. Introduction

In 1929, the astronomer Edwin Hubble announced what would become the single most important observation of modern cosmology.[1] Hubble reported that the extragalactic nebulae[2] are receding from us with a velocity proportional to their distance, a result that soon came to be known as "Hubble's law."[3] The establishment of this linear relation seems to be one of the simplest of generalizations. Hubble needed only to compare the velocities of recession and distances to a selection of nebulae, note their linear relation, and declare the result. This is how his affirmation of the linear relationship is often reported in summary. McKenzie's *Major Achievements of Science* describes it thus:

> In 1929 Hubble compared Slipher's determinations of the recession of the nebulae with his own determinations of

1 I thank Siska De Baerdemaeker for helpful comments on an earlier draft.

2 Hubble's "extragalactic nebula" or just "nebula" are the older terms for "galaxy." In 1929, the term "galaxy" referred unambiguously only to our star system, the Milky Way. The Latin *nebula* (plural *nebulae*) means "cloud" and was used by astronomers of Hubble's time to denote the luminous clouds visible in astronomical telescopes. As Hubble explained (1936, 16–17), some of these clouds proved to be gas and dust within our Milky Way. These he called "galactic nebulae." Others were more distant star systems in their own right — "extragalactic nebulae" — that he would just call "nebulae." Hubble defended his reluctance to label these other nebulae "galaxies": "The term *nebula* offers the values of tradition; the term *galaxies*, the glamour . . . of romance" (18; Hubble's emphasis).

3 In 2018, the members of the International Astronomical Union voted to rename the law the "Hubble-Lemaître law."

> distances and he discovered a simple relation now called Hubble's law, that the velocity is proportional to the distance. (1960, 333)

This simple determination seems to be a good illustration of a natural hierarchical structure for inductive support. In it, inductive inferences may proceed only from a lower, more particular level to a higher, more general level.

> *Inductive Hierarchy*
> *Lower level*: velocity and distance assignments to particular nebulae.
> *Higher level*: general relation connecting the velocities and distances of all nebulae.

Hubble's inference, it seems, merely proceeds up the hierarchy. The particulars of a few individual nebulae at the lower level provide inductive support for the general law at the higher level.

Simple as this inference might seem, Hubble's celebrated paper of 1929 showed no respect for this inductive hierarchy. Rather, a multiplicity of inductive inferences moved up and down the hierarchy in an intricate arrangement of interlocking parts, much like those of a complicated geometric puzzle.

To begin, in 1929, Hubble had access to measurements of the velocities of recession of forty-six extragalactic nebulae, but he had independent distance estimates for only twenty-four of them. For these twenty-four, in what initially appears as a simple generalization, he found a linear velocity-distance relation within statistical uncertainties. However, the inference was not a simple generalization since the determination of most of the distances among these twenty-four nebulae depended on assuming hypotheses still needing further support. They are the hypotheses of *Brightest Star Magnitude* and *Clustering of Nebular Luminosity* detailed in Section 3. These hypotheses cannot be located uniquely in the inductive hierarchy above. In the inferences, they are presumed by the determinations of distance, so they are prior to the lower level: that is, still lower. However, the hypotheses accrue support once the inferences of the paper of 1929 are complete. That means that they come at the end of the inferential chain, so they should be placed higher in the inductive hierarchy.

The remaining twenty-two nebulae were more problematic. For them, Hubble had measurements of velocities and apparent luminosities but not distances. He was determined somehow to make use of these data. In doing so, he introduced relations of support that further cut across the inductive hierarchy. This happened in two related ways.

First, Hubble averaged the apparent luminosities of the twenty-two nebulae and computed the average distance associated with them, assuming the *Clustering of Nebular Luminosity* hypothesis and a mean absolute luminosity found in his second, inverted inference (described below). The mean velocity and the mean distance fell within the expectations of the linear relation that he had found for the first twenty-four nebulae, providing further support for that relationship.

Second, Hubble inverted the direction of evidential support. He used the velocity-distance relation itself, in conjunction with the velocities of recession of these twenty-two nebulae, to infer their distances. This inference proceeds down the inductive hierarchy from the higher level to the lower level. He then used the distances computed to determine the absolute luminosities of the twenty-two nebulae. The results provided direct support for the *Clustering of Nebular Luminosity* hypothesis, already used in the earlier analyses.

The overall outcome was a tangle of inductive inferences that failed to respect any simple linear, inductive hierarchy, such as the one indicated above. We shall see that Hubble remarked repeatedly on the agreement among and later the consistency of the results of the inferences as providing the strongest support for his general conclusions. His notion of consistency was much stronger than mere logical compatibility. Rather, it reflects the mutual agreement among the many entangled relations of support. What might be evidence that supports a result in one relation becomes the result supported by evidence in another relation. This agreement among relations of mutual support gives the structure its inductive solidity.

Hubble's analysis also illustrates the use of hypotheses in initiating inductive investigations. The two hypotheses above were used provisionally as warrants since they themselves were not yet fully supported evidentially. Part of Hubble's overall project became the successful discharging of this inductive debt by providing support for these hypotheses.

In Section 2, I will describe how Hubble came to be concerned with the velocities of the nebulae. In Section 3, I will outline the hypotheses that he used in his determinations of the distances to the nebulae. In Sections 4 and 5, I will review the inference to the linear velocity-distance relation for the first twenty-four nebulae. In Section 6, I will review the inverted inferences for the remaining twenty-two nebulae. In Section 7, I will reflect briefly on the strength of support that Hubble could display in 1929 for the linear relationship. In the concluding Section 8, I will summarize the interwoven relations of support in Hubble's paper of 1929. An appendix to this chapter includes technical details of the computations relating absolute and apparent nebular luminosities, known tersely as "magnitudes."

2. Background to Hubble's Investigations

It is now a commonplace of astronomy that space is filled with many immense star systems akin to our own Milky Way. They are the galaxies, as they are now called, or the extragalactic nebulae, as Hubble called them. Yet whether the stars were so distributed in space remained unsettled in the early 1920s. A landmark in the decision was a debate held between the astronomers Harlow Shapley and Heber Curtis on April 26, 1920, at the Smithsonian Museum of Natural History. Shapley defended the view that our Milky Way is the unique great star system of the universe. Curtis, however, argued that our Milky Way is just one of many such "island universes,"[4] as they were then called. The matter was settled fairly quickly. According to Trimble (1995, 1142), it was Hubble himself who provided a cleaner resolution. Starting with observations in 1923,[5] he was able to discern Cepheid variable stars in two nearby nebulae, most notably Andromeda. As we shall see below, this enabled a determination of the distances to these nebulae. They were located outside our Milky Way, he found.

Our solar system has a motion within the Milky Way. With the recognition that our Milky Way is just one of many nebulae, a prosaic question arises: what is the motion of our solar system with respect to these other nebulae? In his later work, *The Realm of the Nebulae*, Hubble (1936, 106–18) recalled how the answer to this question developed. The velocities of nebulae relative to the Earth were known from red shift measurements in the 1910s. The motion of the solar system was then estimated as around 420 mi/sec. The expectation was that, once this motion was subtracted from the motions of the nebulae,

---

4 The cases each made are published in Shapley and Curtis (1921). See Trimble (1995) for further details.

5 As reported in Hubble (1929b). The results also appeared in a *New York Times* article on December 23, 1924 (Anonymous 1924), and were communicated orally by H.N. Russell at the December–January 1924–25 meeting of the American Astronomical Society (Anonymous 1925).

those motions would be small and random. In particular, there would be as many velocities of approach as of recession. Using a statistical analysis to average away these random motions, we should recover the motion of our solar system with respect to the mean rest state of the nebulae in our vicinity.

As early as 1918, it was already clear that the statistical project was not proceeding smoothly. Wirtz (1918) found the need to add a "*k* term" that corresponded to an overall recession of the nebulae. It meant that the motions of the nebulae visible to us were not distributed randomly about some nebular state of rest. In place of the state of rest was some sort of expansion. The *k* term represented a constant motion of recession from our solar system of 656 km/sec. The motions of the individual nebulae were distributed randomly around that constant motion of recession. Wirtz wrote that,

> If we give this value a verbal interpretation, it is that the system of spiral nebulae disperses [*auseinandertreibt*] with a speed of 656 km [per second] in relation to the momentary position of the solar system as a center. (115)

Over the next decade, Wirtz and others refined the correction term, allowing it to be a function of distance from our solar system. Hubble's celebrated paper of 1929 was a direct contribution to this literature. Its first paragraph identifies the issue to be addressed:

> Determinations of the motion of the sun with respect to the extra-galactic nebulae have involved a K term of several hundred kilometers [per second] which appears to be variable. Explanations of this paradox have been sought in a correlation between apparent radial velocities and distances, but so far the results have not been convincing. The present paper is a re-examination of the question, based on only those nebular distances which are believed to be fairly reliable. (1929a, 168)

The result announced (170–71) was that a statistical fit gave the overall motion of the nebulae as distributed, with some considerable deviations, around a velocity of recession that increases linearly with distance from us. In more detail, the best estimate of the motion of our solar system is 280 km/sec, and, when this is subtracted from the motions of the nebulae, their motions are scattered around an average recessional velocity of 500 km/sec for each million parsec (Mpc) of distance.[6]

A prosaic question about the motion of our solar system had led Hubble to the single most important observational result of modern cosmology.

3. The Determination of Distances

To carry out the analysis of his paper of 1929, Hubble needed determinations of both velocities of and distances to the nebulae. For the forty-six nebulae of his analysis, the velocity determinations proved to be relatively unproblematic. They were determinable from frequency shifts in the spectra of light from the nebulae. The shifts were immediately interpreted as the results of radial velocities: that is, motions along the lines of sight to each nebula.[7] As Hubble (1936, 102–05) recounts, Vesto Slipher, working at the Lowell Observatory, had begun the arduous work of measuring these shifts in 1912. By 1925, he had provided the velocities of twenty-five nebulae.

The locus of difficulty in the analysis was the determination of distances. Two means were available for determining these distances. One was the angular size of the nebula. Nearby nebulae are large: Andromeda extends over 3° in the sky, six times the extent of the full Moon. If we know the absolute size of the nebula in, say, light years, then the distance to the nebula is immediately determined by elementary geometry.

This means of determining distance to the nebulae is not mentioned in Hubble's paper (1929a).[8] Rather, Hubble explicitly reports only luminosity-based determinations. They depend on the fact that the intensity of light emitted by a celestial object diminishes with the inverse square of distance.

---

6 This value of 500 km/sec.Mpc of what we now call the Hubble constant proved to be about an order of magnitude too large as a result of systematic errors in Hubble's determinations of distances. By 1958, the value had been reduced by Sandage to a more modern value of 75 km/sec. Mpc, which corresponded to a Hubble age of the universe of 1.3×10^{9} years.

7 Slipher (1912, 56) wrote that, "the velocity-like displacement might not be due to some other cause, but I believe we have at the present no other interpretation for it. Hence we may conclude that the Andromeda Nebula is approaching the solar system with a velocity of about 300 kilometers per second." Hubble (1936, 34) held the same view but more cautiously: "Although no other plausible explanation of redshifts has been found, the interpretation as velocity-shifts may be considered as a theory still to be tested by actual observations."

8 Hubble and Humason (1931, 52) recount that the difficulty with the method is that the brightnesses of the nebulae fade as we move away from their centers, so that different photographic exposures of the same nebula give different sizes.

Thus, if we know the absolute magnitude of the object's luminosity, then we can determine its distance: we compare this absolute magnitude with the apparent magnitude that we perceive, either visually or photographically.

The weakness of this approach is that the absolute magnitudes are hard to determine; direct measurements give us only apparent magnitudes. Without some independent means of determining the absolute magnitude, the approach cannot be applied. In his paper of 1929, Hubble relied on three methods of determining absolute magnitude.

1. *Cepheid Variable Stars.* Henrietta Leavitt (1912) had reported that certain stars in the Magellenic Clouds varied periodically in magnitude and that there was a definite relationship between the period and the magnitude. Subsequent parallax measurements to other Cepheid variable stars enabled determinations of their distances and thus their absolute magnitudes. Combining these results meant that an observation of the period of one of these variable stars enabled a determination of its absolute magnitude and thus its distance. Hubble himself used this method in 1923 in his determination of the distance to the nebula Andromeda. The distinctive shape of the curve[9] plotting the change of visual magnitude with time enabled him to identify the variable stars that he found in Andromeda as Cepheid variable stars. This was, Hubble (1936, 16) reported, the first reliable method of determining distances to nebulae. It was also the most reliable of the three methods of the paper of 1929 but could be applied only if a Cepheid variable star could be resolved in the nebula.

2. *Brightest Star Magnitude.* It seemed reasonable to assume that different nebulae are constituted of the same sorts of stars, with the same range of possible magnitudes. That led to the expectation that the brightest stars in each nebula have the same absolute magnitude.[10] Hubble (1929a, 168) offered

9 Shown in Hubble (1936, 95).

10 Hubble footnoted an earlier paper (1926) in which he had already advanced the hypothesis (357–61), although only hesitantly, as a "reasonable assumption, supported by such evidence as is available" (357).

an absolute magnitude determined photographically of $M = -6.3$ (see the appendix for a review of the system of units used for apparent and absolute magnitudes). This assumption is important in untangling the evidential relations displayed in Hubble's paper. So I will display it as a hypothesis to which we will return:

> *Brightest Star Magnitude.* The brightest stars in each nebula have the same absolute magnitude.

Hubble approached the hypothesis with optimism and caution:

> The apparent luminosities of the brightest stars in such nebulae are thus criteria which, although rough and to be applied with caution, furnish reasonable estimates of the distances of all extra-galactic systems in which even a few stars can be detected. (1929a, 168–69)

Hubble conceded the limitation that the method could be applied only to nebulae close enough for individual stars to be resolved telescopically. The third method was untroubled by this limitation.

3. *Clustering of Nebular Luminosity.* Drawing from his earlier survey of nebulae (Hubble 1926), he suggested that the absolute magnitudes of nebulae were similar insofar as they were distributed randomly but not too distant from their average. The average value offered (1929a, 169) is a visually determined magnitude of $M = -15.2$ (recall from the appendix that smaller magnitudes correspond to greater brightnesses. A magnitude of -15 is very bright). Actual values, Hubble reported, are "exhibiting a range of four or five magnitudes about [this] average" (169). Once again, this assumption will play an important role in the evidential relations and is displayed thus:

> *Clustering of Nebular Luminosity.* The absolute magnitudes of nebulae cluster in a small interval of four or five units of magnitude about a single mean common to all nebulae.

Four to five units of magnitude amount to a considerable error if we are trying to estimate the distance to just one nebula. It is shown in the appendix that this uncertainty in the absolute magnitude of any particular nebula introduces an uncertainty in the determination of distance of roughly one order of magnitude: that is, the extremes of the full range differ by a factor of 10.

These deviations can be averaged away if we aggregate data from many nebulae so that we can recover more reliable distance determinations for averages. This is especially helpful in getting a more accurate distance estimate to a cluster of nebulae whose members are assumed to be grouped around the same location in space. Hubble (1929a, 169) explains that he would use this averaging technique:

> The application of this statistical average [$M = -15.2$] to individual cases can rarely be used to advantage, but where considerable numbers are involved, and especially in the various clusters of nebulae, mean apparent luminosities of the nebulae themselves offer reliable estimates of the mean distances.

Hubble (1929a) says little more on the use of this technique. Hubble and Humason (1931) is a lengthier and more detailed exposition, using considerably more data. There we find how effective the averaging can be. They report clusters consisting almost always of several hundred nebulae, up to a maximum of 800.[11]

To determine the distance to some particular nebula in a cluster, they would survey the full range of apparent magnitudes of the nebulae in the cluster. The aggregation of survey data greatly reduces errors. For example, consider a cluster of 400 nebulae whose magnitudes are spread over an interval of 4 or 5 magnitudes around the true mean of –15.2. The spread of the average of the magnitudes of the cluster around that true mean is reduced by

11 A table in Hubble and Humason (1931, 74) lists the numbers of nebulae in named clusters as Virgo-(500), Pegasus-100, Pisces-20, Cancer-150, Perseus-500, Coma-800, Ursa Major-300, and Leo-400. Whatever hesitation is flagged by the parentheses for the Virgo cluster, Hubble (1936, 54) reports "several hundred" nebulae in it.

a factor of 400 = 20. We find in the appendix that this reduces the interval to 0.25 magnitudes and corresponds to an error in distance estimates in which the farthest distance is merely 12% greater than the nearest distance. This provides a good determination of the absolute magnitude of and distance to a nebula whose brightness matches the average.[12] That distance is then also the estimate of the distance to the particular nebula of interest.

4. From Particulars to Generalities

Although forty-six nebulae were included in Hubble's (1929a) analysis, Hubble was able to estimate individual distances to only twenty-four of them. He inferred the linear relation between their distances and velocities by directly comparing distances and velocities. He reported the results of two ways of arriving at the linear relation.

The first, and most direct, way took the velocities and distances of the individual nebulae and used standard statistical methods to find the best fit of a relation written in more modern vector notation as

$$\mathbf{v}_i = \mathbf{r}_i K + \mathbf{V}_0$$

Here $\mathbf{v}_i$ is the vector velocity of the ith nebula located a vector displacement $\mathbf{r}_i$ from us and $\mathbf{V}_0$ is the vector velocity of our solar system. The constant K is now known as the "Hubble constant" and is the parameter of greatest interest to us now. It converts a scalar distance r to a nebula to its scalar velocity of recession $v = Kr$. The velocity $\mathbf{v}_i$ is not the velocity observed from the Earth through the red shift, for those observations are taken from a vantage point itself moving at $\mathbf{V}_0$. The velocity that we observe for the ith nebula is the difference $\mathbf{v}_i - \mathbf{V}_0$. Hubble reported that the best fit gave

$$K = 465 \pm 50 \text{ km/sec.Mpc} \quad V_0 = 306 \text{ km/sec} \quad A = 286^\circ \quad D = 40^\circ$$

The second way proceeded by first reducing the data for the twenty-four nebulae to nine groupings and first averaging within each grouping. Hubble indicated only that the groupings were selected "according to proximity in direction and in distance" (1929a, 170). Presumably, the effect of the

12 Hubble and Humason (1931, 56) summarize the strategy as "the mean or most frequent apparent magnitude of the many members [of a cluster] is a good indication of the distance of a cluster, and hence clusters offer the greatest distances that can definitely be assigned to individual objects."

averaging, once again, was to reduce the effect of random deviations from linearity, this time prior to finding the statistical best fit of the above relation. The index i would now refer to the ith group. Hubble reported that best fit as

$$K = 513 \pm 60 \text{ km/sec.Mpc} \quad V_0 = 247 \text{ km/sec} \quad A = 269^\circ \quad D = 33^\circ$$

For his final result, Hubble selected values intermediate between these two sets and rounded them:[13]

$$K = 500 \text{ km/sec.Mpc} \quad V_0 = 280 \text{ km/sec} \quad A = 277^\circ \quad D = 36^\circ$$

Since the solar velocity $\mathbf{V}_0$ is comparable in size to the nebular velocities $\mathbf{v}_i$, Hubble's analysis had to pass through the more indirect route of finding the best fit of the above relation. Merely computing the ratio of observed velocity and distance for each nebula would have omitted the essential correction for the Earth's motion. Hubble's figure, redrawn here as Figure 7.1, gives a sense of the large size of the residuals that deviate from his best-fit relations. It displays the velocities of nebulae, after the velocity of our solar system has been subtracted, in relation to their distances.

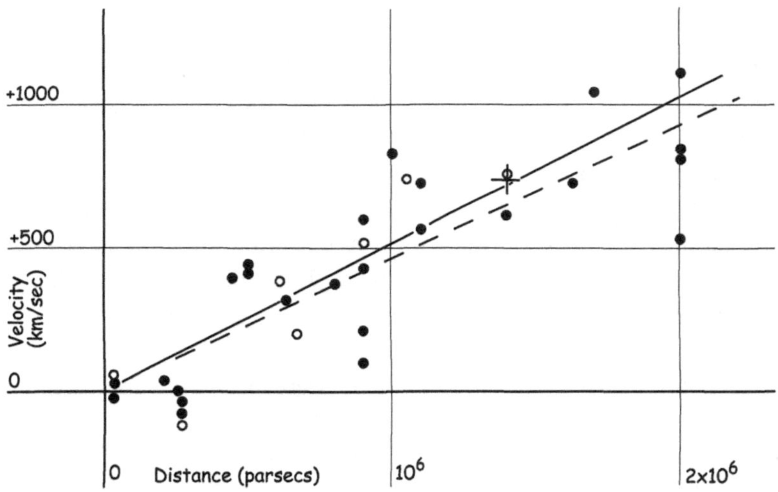

Figure 7.1. Hubble's "Velocity-Distance Relation among Extra-Galactic Nebulae"

13 Hubble converted the celestial coordinates into galactic coordinates: longitude 32°, latitude +18°.

An extended caption explains the data presented. Hubble (1929a, 172) writes that

> The black discs and full line represent the solution for solar motion using the nebulae individually; the circles and broken line represent the solution combining the nebulae into groups. . . .

There are twenty-four black discs, and they correspond loosely[14] to the data in Table 7.1 for twenty-four nebulae whose distances can be determined. Hubble concluded that

> . . . the cross represents the mean velocity corresponding to the mean distance of 22 nebulae whose distances could not be estimated individually.

I will return to the treatment of these twenty-two nebulae in Section 6.

5. Hubble's Hypotheses

The appearance of this last inference is of a traditional generalization that proceeds from the particulars of the lower level to the covering generality at the higher level of the hierarchy indicated in Section 1. The appearance is deceptive, for most of the distance determinations in the particulars depend on the hypotheses indicated in Section 3. Since the subsequent generalizations depended on them, the generalization was not secure until Hubble provided further evidence in support of the hypotheses. This stage of his investigation took on an inductive debt. We shall see that Hubble continued the analysis in a way intended to discharge some of that debt.

The data for these twenty-four nebulae were presented in Table 1 of Hubble's paper (1929a), reproduced here as Table 7.1.

To arrive at the distances in this table, Hubble used all three of the methods discussed above. He did not lay out the specifics of the determinations in each case. All of the details would be lengthy and not fit into the short announcement that he offered. Hubble and Humason (1931) provide a similar

14 We should not expect the velocities in the figure to match those of Table 7.1 up to a constant subtractive factor. The correction for solar motion is a vector subtraction whose scalar effect will vary according to the differences in the directions of the vectors in the subtraction.

Table 7.1. Hubble's "Nebulae Whose Distances Have Been Estimated from Stars Involved or from Mean Luminosities in a Cluster"

	Object	m_s **photographic magnitude of brightest stars**	r **distance* in megaparsecs**	v **velocity km/sec**	m_t **visual magnitude**	M_t **absolute visual magnitude computed† from** r, m_t
1	Small Magellenic	..	0.032	+170	1.5	-16.0
2	Large Magellenic	..	0.034	+290	0.5	-17.2
3	NGC 6822	..	0.214	-130	9.0	-12.7
4	NGC 598	..	0.263	-70	7.0	-15.1
5	NGC 221	..	0.275	-185	8.8	-13.4
6	NGC 224	..	0.275	-220	5.0	-17.2
7	NGC 5457	17.0	0.45	+200	9.9	-13.3
8	NGC 4736	17.3	0.5	+290	8.4	-15.1
9	NGC 5194	17.3	0.5	+270	7.4	-16.1
10	NGC 4449	17.8	0.63	+200	9.5	-14.5
11	NGC 4214	18.3	0.8	+300	11.3	-13.2
12	NGC 3031	18.5	0.9	-30	8.3	-16.4
13	NGC 3627	18.5	0.9	+650	9.1	-15.7
14	NGC 4826	18.5	0.9	+150	9.0	-15.7
15	NGC 5236	18.5	0.9	+500	10.4	-14.4
16	NGC 1068	18.7	1.0	+920	9.1	-15.9
17	NGC 5055	19.0	1.1	+450	9.6	-15.6
18	NGC 7331	19.0	1.1	+500	10.4	-14.8
19	NGC 4258	19.5	1.4	+500	8.7	-17.0
20	NGC 4151	20.0	1.7	+960	12.0	-14.2
21	NGC 4382	..	2.0	+500	10.0	-16.5
22	NGC 4472	..	2.0	+850	8.8	-17.7
23	NGC 4486	..	2.0	+800	9.7	-16.8
24	NGC 4649	..	2.0	+1,090	9.5	-17.0
	NGC = nebula number in the New General Catalog					mean -15.5

* These distances are systematically low. Hubble reports 0.275 Mpc for the distance to nearby Andromeda, whereas the more recent estimate is 0.780 Mpc.

† Using formula (A3) of the appendix. The table has distances in units of megaparsecs, whereas distance in (A3) are entered in parsecs.

analysis, with more data and details, which has to be considerably lengthier and more complicated in its reporting. In his report, Hubble (1929a, 170) limited himself to general statements:

> The first seven distances are the most reliable, depending, except for M32 [= NGC 221] the companion of M31 [= Andromeda, NGC 224], upon extensive investigations of many stars involved.

For Andromeda (M31 = NGC 224), we know from Hubble (1929b) that he used Cepheid variable stars for the distance determination. Presumably, the *Brightest Star Magnitude* hypothesis was not used in the distance estimates for these first seven objects since there are no brightest star magnitude entries for them. Subsequent distance estimates did consider the magnitudes of the brightest stars since they are given for rows seven to twenty. Hubble continued:

> The next thirteen distances,[15] depending upon the criterion of a uniform upper limit of stellar luminosity, are subject to considerable probable errors but are believed to be the most reasonable values at present available. (1929a, 170)

The use of mean nebular magnitudes for distance determination is finally mentioned for rows twenty-one to twenty-four:

> The last four objects appear to be in the Virgo Cluster. The distance assigned to the cluster, 2×10^6 parsecs, is derived from the distribution of nebular luminosities, together with luminosities of stars in some of the later-type spirals, and differs somewhat from the Harvard estimate of ten million light years. (170)

Here the *Clustering of Nebular Luminosity* hypothesis was employed. That it had a larger role is suggested by the title given to the table as a whole: "Distances . . . from Mean Luminosities in a Cluster."

15 Presumably, he means the "next fourteen," rows seven to twenty.

6. From Generalities to Particulars

Hubble then turned to the remaining twenty-two nebulae for which velocities were known but not distances. He was intent on recovering some evidential import from the data. The data with which he worked are presented in Table 7.2, which reproduces his Table 2 (1929a). The column v is the velocity determined by red shifts for the nebula with the indicated NGC number. The next column v_s indicates the correction that must be subtracted from the observed velocity to correct for solar motion.

With these data in hand, Hubble proceeded with two approaches. The first was the crudest. It simply worked out the velocity-distance relation for the average behavior of all of the twenty-two nebulae. Since the velocity-distance relation is presumed to be linear, it should hold for the average of the velocities and distances. Hubble found an average velocity of 745 km/sec and an average distance of 1.4 Mpc. These averaged data then give an estimate for the constant $K = 745/1.4 \approx 530$ km/sec.Mpc. Given the magnitude of errors likely (see below), the agreement was likely well within error limits for the value of 500 km/sec.Mpc estimated in the earlier part of the paper.

For my purposes, it is interesting to see that even here Hubble's analysis relied on the *Clustering of Nebular Luminosity* hypothesis. It was not needed to recover the average velocity. That was simple arithmetic.[16] The hypothesis was needed to determine the average distance. According to the hypothesis, the absolute magnitudes of the individual nebulae varied at an interval of 4 to 5 magnitudes around the common mean value. This range would then be reflected in the apparent magnitudes reported in the column m_t of Table 7.2. However, taking the average of the apparent magnitudes reduces the interval by a factor of $1/\sqrt{22} = 1/4.69$ to an interval of roughly the size of a single magnitude. We find in the appendix that the farthest distance in the associated distance interval is 58% greater than the nearest distance. The average apparent magnitude of 10.5 is far from the absolute magnitude of −15.3 assumed.[17] The diminution is entirely the result of the great distance associated with the average. That distance is computed[18] from (A3) and is 1.445 Mpc.

16 (Average $v = 748.4$) – (average correction $v_s = 2.95$) = 745.4 km/sec.

17 This absolute magnitude of −15.3 is recovered from the next stage of calculations on these twenty-two nebulae.

18 That is $\log_{10} d = 0.2(10.5 + 15.3) + 1 = 6.16$, so that $d = 10^{6.16} = 1.445 \times 10^6$ pc.

The more elaborate of the two approaches involved using the velocity-distance relation in reverse. Starting with the corrected velocity, $v - v_s$, for each of the twenty-two nebulae, Hubble computed the distance r that the linear velocity-distance relation required, where he assumed a value for the K constant of 500 km/sec.Mpc. The results are reported in the r column of Table 7.2 and conform to the formula $r = (v - v_s)/500$. Since these distances were computed using the very relation under scrutiny, by themselves they could provide no evidence for the relation. To extract some useful evidential import, Hubble used these distances r to calculate[19] the absolute magnitude M_t of each nebula from the measured, apparent magnitude, m_t. The results are reported in the last column of Table 7.2. Hubble computed the mean to be –15.3.

What he found notable was that the mean absolute magnitude computed for these twenty-two nebulae matched almost exactly the mean –15.5 computed for the first twenty-four nebulae using their independently known distances. Similarly, their ranges agreed: 4.9 for the twenty-two nebulae of Table 7.2[20] and 5.0 for the twenty-four nebulae of Table 7.1. The most direct reading is that the new results from the twenty-two nebulae provide another instance of the *Clustering of Nebular Luminosity* hypothesis, using the same mean and range as the earlier analysis. This provides direct support for the hypothesis. Hubble was more celebratory and expansive in his assessment:

> The two mean magnitudes, –15.3 and –15.5, the ranges, 4.9 and 5.0 mag., and the frequency distributions are closely similar for these two entirely independent sets of data; and even the slight difference in mean magnitudes can be attributed to the selected, very bright, nebulae in the Virgo Cluster. This entirely unforced agreement supports the validity of the velocity-distance relation in a very evident matter. Finally, it is worth recording that the frequency distribution of absolute magnitudes in the two tables combined is comparable with those found in the various clusters of nebulae. (1929a, 172–73)

---

19 The calculation employed formula (A3) of the appendix. Note that d in that formula is in parsecs, whereas r in Table 7.2 is in megaparsecs.

20 I find the range to be 4.8, extending from –12.8 for NGC 1700 to –17.6 for NGC 4594.

Table 7.2. Hubble's "Nebulae Whose Distances Are Estimated from Radial Velocities"

	NGC nebula number	***v* Velocity km/sec**	**v_s Velocity correction subtracted for solar motion**	***r* Distance Mpc**	**m_t Apparent magnitude**	**M_t Absolute magnitude computed from r, m_t**
	278	650	-110	1.52	12.0	-13.9
	404	-25	-65	..	11.1	..
	584	1,800	75	3.45	10.9	-16.8
	936	1,300	115	2.37	11.1	-15.7
	1023	300	-10	0.62	10.2	-13.8
	1700	800	220	1.16	12.5	-12.8
	2681	700	-10	1.42	10.7	-15.0
	2683	400	65	0.67	9.9	-14.3
	2841	600	-20	1.24	9.4	-16.1
	3034	290	-105	0.79	9	-15.5
	3115	600	105	1	9.5	-15.5
	3368	940	70	1.74	10	-16.2
	3379	810	65	1.49	9.4	-16.4
	3489	600	50	1.1	11.2	-14.0
	3521	730	95	1.27	10.1	-15.4
	3623	800	35	1.53	9.9	-16.0
	4111	800	-95	1.79	10.1	-16.1
	4526	580	-20	1.2	11.1	-14.3
	4565	1,100	-75	2.35	11	-15.9
	4594	1,140	25	2.23	9.1	-17.6
	5005	900	-130	2.06	11.1	-15.5
	5866	650	-215	1.73	11.7	-14.5
Mean		748.4	2.95		10.5	-15.3

7. How Strong Was the Evidence for Linearity?

My concern here is the tangled structure of the relations of inductive support. Although it is independent of this concern, it is worth noting that Hubble's evidence in 1929 for the linear relation was weak. This was so even though his paper of 1929 is routinely celebrated as the origin of the linear relation between the velocities of recession of the nebulae and their distances. A glance at Figure 7.1 shows just how weak was the establishment of the linearity. The data points are so broadly scattered about the straight lines fitted that all that can be inferred securely is that the velocities are increasing with the distances. The difficulty is that nebulae close to our Milky Way have particular motions in random directions of the order of the overall velocity of recession. These motions confound the linear motion of recession. To reveal the linear relation more clearly requires examination of more distant nebulae for which the particular motions become successively smaller in relation to the velocity of recession.

As long as Hubble's interest lay in the original project of determining the motion of our solar system, the weakness of the evidence for linearity is a smaller concern. We might reasonably expect that other velocity-distance relations compatible with the data would have only a minor effect on the estimates of solar motion. The threat is more serious, however, if Hubble's paper is to underwrite the founding empirical observation of modern cosmology: the linearity of the velocity-distance relation.

Hubble had a response to this threat in his paper. He allowed that his data merely "establish a roughly linear relation" (1929a, 173). The solution lay in an extension to more distant nebulae and was already under way. Hubble reported a result for NGC 7619, whose distance he estimated at roughly 7 Mpc. That greatly exceeded the distance of 1 or 2 Mpc of nebulae investigated so far. Its speed of recession still fit well enough with his *K* factor of 500. Shortly after, in joint work, Hubble and Humason (1931) reported on velocities of recession of still more distant nebulae. Their Figure 5 (77) plotted data for nebular clusters, one of which is more than 30 Mpc distant. In this plot, the linearity of the paper of 1929 survives. Hubble and Humason had become so confident of the linear relationship that they proposed its use to determine distances. It was, they boasted,

> . . . a new method of determining distances of individual objects in which the percentage errors actually diminish with distance. (76)

This remark foreshadowed the recent practice of identifying the locations of distant galaxies merely by citing their red shift factors directly. Red shift has become the surrogate for distance.

By the time of his more popular work in 1936, Hubble reasserted his confidence that the linearity of the relation had been vindicated. He wrote of the success of the extension of the investigation to more distant nebulae:

> The relation is plausible but not unique. The true relation might be a curve which was nearly linear within the range covered by the observations, but which departed widely from a straight line in the regions beyond the faintest nebulae in the group. This possibility was investigated by extrapolating the adopted relation extending it far out into the hitherto unobserved regions and testing it by new observations. Such a procedure often leads to minor, or even to major, revisions in the relation first selected: it has been said that research proceeds by successive approximations. However, in the investigation of red-shifts, no revision was definitely indicated. The linear relation has survived repeated tests of this nature and is known to hold, at least approximately, as far out into space as the observations can be carried with existing instruments. (3–4)

8. Conclusion and Summary

In the introduction, I sketched the inductive hierarchy to which one might assume that Hubble's inferences of 1929 conformed. We have now seen that his inductive inferences did not respect this hierarchy. Rather, his inferences are interwoven nonhierarchically through the following sets of propositions.

(a) Sets of velocities of recession assigned to nebulae
(b) Sets of distances assigned to nebulae
(c) Linear relations asserted between their velocities and distances

(d) Hypothesis of *Brightest Star Magnitude*
(e) Hypothesis of *Clustering of Nebular Luminosity*

The inferences were as follows:

(i) In Sections 4 and 5, we saw inferences from the sets of velocity (a) and distance (b) assignments to a linear relationship (c), in which many of the distance assignments already presumed the two hypotheses (d) and (e).

(ii) In Section 6, we saw an inference from the means of the velocities (a) and distances (b) to an instance of the linear relationship (c). The determination of the mean distance once again presumed hypothesis (e) as well as a mean absolute magnitude for nebulae determined by the inferences of (iv).

(iii) In Section 6, we saw an inference from sets of velocity assignments (a) and the linear relationship (c) to sets of distance assignments (b).

(iv) In Section 6, Hubble proceeded from the distances computed in (iii) and inferred to a set of absolute magnitudes that affirmed hypothesis (e).

Use of the velocity-distance relation in (iii) to infer back to distances became a fixture in astronomy. In his more popular work, Hubble (1936, 115) was confident enough of this inference that he wrote

> The velocity-distance relation, once established, could evidently be used as a criterion of distance for *all* nebulae whose velocities were known.

This inference appears initially as the mere recovery of a deductive consequence of the velocity-distance relation. It also has an inductive component. I have emphasized the "all" since it includes the nebulae originally used to establish the velocity-distance relation. We gain inductive support for an independently determined distance to some nebula if we find that it conforms to the velocity-distance relation. Alternatively, if conformity fails, then we have a check and a correction for the original distance determination.

The cogency of Hubble's inferences required that strong evidential support be provided for hypotheses (d) and (e), or the distance determinations of his analysis would be compromised. Discharging this inductive debt was an obligation taken seriously in the later analysis of Hubble and Humason (1931). Of its thirty-eight pages, six were devoted to a section on the "Upper Limit of Stellar Luminosity as a Criterion of Distance" (46–51), and another five pages were devoted to a section on the "Total Luminosity of Nebulae as a Criterion of Distance" (52–56). That is, almost 30% of the paper was spent elaborating and establishing these two hypotheses.

More generally, Hubble repeatedly offered the agreement among the results of all of these inferences as giving general support to his analysis. We have already seen his remark that "this entirely unforced agreement supports the validity of the velocity-distance relation in a very evident matter" (1929a, 172–73).[21] Hubble and Humason (1931, 43) commence their paper by defending their methods of determining nebular distance, whose initiating assumption is "supported in a general way by the consistency of the results to which it leads." Later they announce that, "since the two investigations were based upon different criteria of distance, the close agreement emphasizes the internal consistency of our present ideas concerning luminosities of nebulae" (76).

In his more popular narrative, Hubble (1936, 101) reflected on the various criteria used to determine nebular distances, including the velocity-distance relation itself, and he concluded that

> The exploration of the realm of the nebulae was carried out with the aid of these criteria. The early work was justified largely by the internal consistency of the results. The foundations were firmly established, but the superstructure represented considerable extrapolations. These were tested in every way that could be devised, but the tests for the most part concerned internal consistency. The ultimate acceptance of the

21 Hubble (1929a) does not provide further evidence explicitly and specifically supporting the *Brightest Star Magnitude* hypothesis. Perhaps this unforced agreement provides independent support for the nebular distances determined using this hypothesis and thus, indirectly, support for the hypothesis itself.

> superstructure was due to the steady accumulation of consistent results rather than to critical and definitive experiments.

A few pages later Hubble reflected on the use of distances derived from the mean and range of the absolute luminosities in establishing the velocity-distance relation:

> The consistency of these results was additional evidence of the validity of the velocity-distance relation. (115)

The consistency so important to Hubble is not the consistency of deductive logic, in which it merely designates a lack of contradiction. This deductive sense of consistency by itself provides no inductive support. The Hubble law of expansion of the nebulae in our universe is logically consistent with the existence of another, parallel universe, isolated from ours, in which nebulae approach each other. The fact of logical consistency supplies no inductive support for the existence of such a parallel universe.

The consistency alluded to by Hubble was the agreement among the many entangled relations of inductive support of his analysis. The Hubble law itself in one part is inductively supported by other results and in another part is used to provide inductive support. The hypotheses of *Brightest Star Magnitude* and *Clustering of Nebular Luminosity* are used, in one part, to warrant inductive inferences to other results, and in another part the results are supported by inductive inferences. The overall import is that no proposition within Hubble's analysis is left without inductive support, and that fact gives his analysis its inductive solidity.

Appendix: Luminosity and Magnitude

Hubble's accounts above discuss the brightness of stars and nebulae using the standard system of magnitudes. His paper of 1929 was written for experts, so Hubble had no need there to explain the system. His more popular *The Realm of the Nebulae* (1936, 9–13), however, describes the system. The luminosity L of an object is the rate at which it emits luminous energy. Our perception of brightness associates equal increments in brightness to equal multiples of luminosity. Thus, the brightness of an object is given by a logarithmic function of the luminosity. That is, the apparent magnitudes m_1 and m_2 of two objects at the same distance from us are related to their luminosities L_1 and L_2 by

$$m_1 - m_2 = -2.5 \log_{10} (L_1/L_2) \tag{A1}$$

The minus sign in the relation means that a *brighter* object has a *smaller* magnitude.

This particular logarithmic relation was chosen to preserve continuity with the ancient visual system of reporting star brightnesses, already found in Ptolemy's *Almagest*. There stars were grouped by their brightnesses into six magnitudes. The first magnitude was the brightest and the sixth the dimmest visible. If the associated luminosities are $L_1, L_2, \ldots L_6$, then stepping through them represents equal increases in apparent brightness as long as

$$L_1/L_2 = L_2/L_3 = L_3/L_4 = L_4/L_5 = L_5/L_6 = 2.5$$

The ratio of 2.5 arises from the stipulation that the full range of luminosities spans 100 to 1: that is, $L_1/L_6 = 100$. Thus, each of the five steps corresponds to a multiplicative factor of $100^{1/5} = 2.512$, rounded down to 2.5. The magnitudes are labeled "visual" or "photographic" according to the media with which they are measured. The distinction is important since the two media have different sensitivities to different frequencies of light.

The apparent brightness of an object diminishes with the inverse square of distance from us. If the two objects in formula (A1) were removed to distances d_1 and d_2 respectively, then the ratio (L_1/L_2) must be replaced by the ratio $(L_1/d_1^2) / (L_2/d_2^2)$. The relation among apparent magnitudes becomes

$$m_1 - m_2 = -2.5 \log_{10} (L_1/L_2)(d_2^2/d_1^2) \tag{A2}$$

The absolute magnitude of an object M is stipulated to be the apparent magnitude that the object would have were it placed 10 parsecs from us.[22] Using only the distance dependency in (A2), it follows that the apparent magnitude m of an object of absolute magnitude M at a distance of d parsecs is[23]

$$m = M + 5 \log_{10} d - 5 \quad \text{or} \quad \log_{10} d = 0.2(m - M) + 1 \tag{A3}$$

Hubble (1929a) supposes that the intrinsic brightnesses of all nebulae are within 4 to 5 absolute magnitudes of each other. Assuming a mean absolute magnitude for some nebula will lead to errors in distance estimates. To take

22 A parsec is the distance at which the mean Earth-Sun distance subtends one second of arc. It is a convenient astronomical unit since distances to nearby stars are revealed by their parallax during the Earth's annual motion around the Sun. 1 parsec = 3.258 light years. A megaparsec or Mpc is 1 million parsecs.

23 Set $d_2 = 10$ and $d_1 = d$; note that $\log_{10} (d^2/10^2) = 2 \log_{10} d - 2 \log_{10} 10 = 2 \log_{10} d - 2$.

the most extreme case, an apparent magnitude m can derive from an object with absolute magnitude M_1 at distance d_1 or another object with absolute magnitude M_2 at distance d_2, where $M_1 - M_2 = 5$. Thus, we have from (A3) that

$$M_1 + 5 \log_{10} d_1 = M_2 + 5 \log_{10} d_2$$

and then

$$5 = M_1 - M_2 = 5 \log_{10} (d_2/d_1)$$

It follows that $\log_{10} (d_2/d_1) = 1$, so that $d_2/d_1 = 10$. That is, the uncertainty in the absolute magnitudes of nebulae corresponds to an uncertainty of one order of magnitude in their spatial distances.

If, however, we follow Hubble's technique of averaging, then this uncertainty is greatly reduced in estimating the value of the true mean.[24] For a cluster of 400 nebulae, the spread of the mean is reduced by a factor of $1/\sqrt{400} = 1/20 = 0.05$. So the spread is $5 \times 0.05 = 0.25$. Thus, we have from (A3) as before

$$0.25 = M_1 - M_2 = 5 \log_{10} (d_2/d_1)$$

We now have for the corresponding distances that $\log_{10} (d_2/d_1) = 0.05$, so that $d_2/d_1 = 1.122$. That is, the farthest distance of the associated interval of distances is merely 12% greater than the nearest distance.

For a group of twenty-two nebulae, the spread of the mean reduces by a factor of $1/\sqrt{22} = 1/4.69$. If we approximate the spread of 4 to 5 magnitudes to be reduced to one order of magnitude, then we have from (A3) that

$$1 = M_1 - M_2 = 5 \log_{10} (d_2/d_1)$$

We now have $\log_{10} (d_2/d_1) = 0.2$, so that $d_2/d_1 = 1.585$. That is, the farthest distance of the associated interval of distances is 58% greater than the nearest distance.

24 Assume that we have $n = 400$ independent samples from the same distribution with variance σ^2. The variance of the mean is σ^2/n. Hence, the standard deviation is $\sigma/\sqrt{n}$.

REFERENCES

Anonymous. 1924. "Finds Spiral Nebulae Are Stellar Systems: Dr. Hubbell [*sic*] Confirms View that They Are 'Island Universes' Similar to Our Own." *New York Times*, December 23, 1924, 6.

———. 1925. "Thirty-Third Meeting of the American Astronomical Society." *Popular Astronomy* 33: 158–60.

Hubble, Edwin. 1926. "Extragalactic Nebulae." *Astrophysical Journal* 64: 321–69.

———. 1929a. "A Relation between Distance and Radial Velocity among Extra-Galactic Nebulae." *Proceedings of the National Academy of Sciences* 15: 168–73.

———. 1929b. "A Spiral Nebula as a Stellar System, Messier 31." *Astrophysical Journal* 69: 103–58.

———. 1936. *The Realm of the Nebulae.* Oxford: Oxford University Press; London: Humphrey Milford.

Hubble, Edwin, and Milton L. Humason. 1931. "The Velocity-Distance Relation among Extra-Galactic Nebulae." *Astrophysical Journal* 74: 43–80.

Leavitt, Henrietta. 1912. "Periods of 25 Variable Stars in the Small Magellenic Cloud." *Harvard College Observatory Circular* 173: 1–3. Reported by Edward C. Pickering.

McKenzie, A.E.E. 1960. *The Major Achievements of Science.* Vol. 1. Cambridge, UK: Cambridge University Press.

Sandage, Allan. 1958. "Current Problems in the Extragalactic Distance Scale." *Astrophysical Journal* 127: 513–27.

Shapley, Harlow, and Heber D. Curtis. 1921. "The Scale of the Universe." *Bulletin of the National Research Council*, vol. 2, part 3, no. 11: 171–217.

Slipher, Vesto M. 1912. "The Radial Velocity of the Andromeda Nebula." *Lowell Observatory Bulletin* 58, II, no. 8: 56–57.

Trimble, Virgina. 1995. "The 1920 Shapley-Curtis Discussion: Background Issues and Aftermath." *Publications of the Astronomical Society of the Pacific* 107: 1133–44.

Wirtz, Carl Wilhelm. 1918. "Ueber die Bewegung der Nebelflecke." *Astronomische Nachrichten* 206: 109–15.

8

Newton on Universal Gravitation

1. Introduction

Isaac Newton's reasoning in his seventeenth-century *Mathematical Principles of Natural Philosophy* ([1726] 1962) remains to this day a model of tight, carefully controlled argumentation. Its inductive centerpiece lays out the evidential case for his theory of universal gravitation with exemplary caution and discipline. Within his argumentation, there are two cases of pairs of propositions in which relations of inductive support cross over each other, in analogy to the relations of structural support in an arch. The first pair comprises the two core propositions of Newton's celebrated "Moon test." The second pair comprises the propositions of an inverse square law of gravity and of the elliptical orbits of the planets.

In both cases, the individual relations of support have the following structure: the observed evidence supports a proposition by means of a warranting hypothesis. Schematically, this can be written as

Observed evidence

(warrant) Hypothesis

________________(deduce)

Proposition

The crossing over of relations of support arises in both cases in the following way. We have two propositions, proposition_1 and proposition_2, such that

Observed evidence | Observed evidence
(warrant) Proposition_1 | (warrant) Proposition_2
_____________(deduce) | _____________(deduce)
Proposition_2 | Proposition_2

Finally, each of the individual inferences above is deductive. They combine to give a totality in which the observed evidence inductively supports both propositions. That is, the relations of support are locally deductive but inductive in their combination.

Observed evidence

_____________(induction)

Proposition_1 & Proposition_2

The two examples are treated in turn in the sections that follow.

2. The Moon Test

One of Newton's more remarkable discoveries in his theory of universal gravitation is the identity of two forces. The first is the celestial force that deflects planets into orbit around the Sun and deflects moons into orbits around their planets. The second is the force of gravity that leads to the fall of free bodies at the Earth's surface, such as hurled stones. That these forces are the same is now a commonplace. It was a major discovery in the seventeenth century, for the ancient tradition had been that the physics of terrestrial bodies differs from the physics of celestial matter. Newton needed a strong argument to establish the identity.

The identity of the two forces was established early by Newton in Book III of his *Principia* ([1726] 1962). That book presents a sequence of propositions laying out his argument for universal gravitation. The first three propositions establish that the celestial force of attraction acting on an orbiting body varies with the inverse square of distance from the center of the attracting body in three cases: the orbits of Jupiter's moons about the center of Jupiter, the orbits of the planets about the Sun's center, and the orbit of the Moon about the Earth's center. The fourth proposition asserts the identity of terrestrial gravity and the celestial force acting on the Earth's Moon.

To arrive at this fourth proposition, Newton determined the acceleration of the Moon toward the Earth. It is this acceleration that deflects the Moon from its linear, inertial motion and brings it into orbit around the Earth. We would now represent this acceleration directly as so many feet/second^2 or meters/second^2. Newton proceeded indirectly. A body falling with constant acceleration *a* from rest will cover a distance $at^2/2$ in time *t*. Newton used this distance as the measure of acceleration.

As a result of its orbital motion, Newton noted, the Moon falls 15 Paris feet 1 inch 1 4/9 lines (1/12 of an inch) in one minute. That is, it falls 15.0934 Paris feet in one minute. The Moon is roughly 60 times farther away from the center of the Earth than a point on the Earth's surface. Hence, if the celestial force acting on the Moon is governed by an inverse square law all the way down to the Earth's surface, then it would be 60^2 times greater on the Earth's surface. This means that a body falling under its action at the Earth's surface would fall 15.0934 x 60^2 Paris feet in one minute. One minute is a time unfamiliar in our experience for bodies to fall above the surface of the Earth. So Newton scaled the time of fall to one second. Conveniently, one second is 1/60th of a minute. Since the distance fallen varies with the square of time *t*, a body falling under the celestial force at the Earth's surface for one second would fall $1/60^2$ of 15.0934 x 60^2 Paris feet: that is, 15.0934 Paris feet. This matches well how bodies fall on the surface of the Earth under gravity, as measured by experiments on pendula. Newton ([1726] 1962, 408) concluded:

> And therefore the force by which the Moon is retained in its orbit becomes, at the very surface of the Earth, equal to the force of gravity which we observe in heavy bodies there. And therefore (by Rule 1 & 2) the force by which the Moon is retained in its orbit is that very same force which we commonly call gravity; for were gravity another force different from that, then bodies descending to the Earth with the joint impulse of both forces would fall with a double velocity. . . .

The case that Newton made here is a powerful one. In recollections recorded much later, he asserted that he had found the arguments of these first four propositions in 1666. He noted (1888, xviii) of the Moon test that

> At the same year [1666] I began to think of gravity extending to the orbit of the Moon, . . . and thereby compared the force requisite to keep the Moon in her orb with the force of gravity at the surface of the earth and found them answer pretty nearly.

3. The Inferences Summarized

The inference above can be summarized as follows:

Observed acceleration of fall of terrestrial bodies and the Moon.

(warrant) $H_{\text{inv. square}}$: The celestial force acting on the Moon is strengthened by an inverse square law with distance at the Earth's surface.

____(deduce)

Intermediate conclusion: Equality of accelerations at the Earth's surface due to gravity and the celestial force.

(warrant) Rules 1 and 2 of Newton's Rules of Reasoning in Philosophy

---

H_{identity}: Terrestrial gravitation and the lunar celestial force are the same.

The last step might seem to be superfluous. Newton found that the acceleration because of gravity and the celestial force match at the Earth's surface. Is that not enough to show the identity of the two forces? It is very close, but there is a loophole. It might just be that the force of gravity does not act on celestial matter such as comprises the Moon and that the celestial force does not act on ordinary, terrestrial matter. Newton closed the gap with the rules of reasoning that he had declared earlier in *Principia* ([1726] 1962). The relevant idea is that we are to assign the same cause to the same effect. I will not pursue this use of the rules further. In Chapter 6, "Simplicity," of *The Material Theory of Induction* (Norton 2021), I described my discomfort with the rules and indicated how they can be replaced in this case by a simple material fact: that the matter of the Moon would behave like terrestrial matter were it brought to the Earth's surface. What results is the simpler inference:

Observed acceleration of fall of terrestrial bodies and the Moon.

(warrant) $H_{\text{inv. square}}$: The celestial force acting on the Moon is strengthened by an inverse square law with distance at the Earth's surface.

____(deduce)

Intermediate conclusion: Equality of accelerations at the Earth's surface due to gravity and the celestial force.

(warrant) Terrestrial and lunar matter respond to the same forces.

____(deduce)

H_{identity}: Terrestrial gravitation and the lunar celestial force are the same.

For my purposes here, what matters is that the inverse square law, $H_{\text{inv. square}}$, is used as part of the inference to the identity result, H_{identity}. This usage forms half of the arch shown in Figure 8.1.

There is a second inference here that Newton did not make explicit. He inferred that the celestial force is governed by an inverse square law in other parts of the solar system. But how did he know that this inverse square dependence on distance would continue to hold when he moved out of the celestial realm down to the terrestrial realm? It is striking that the inference sketched above works so well. That the two forces "answer pretty nearly," as Newton remarked, gives one confidence that the inverse square law, introduced as a hypothesis above, is also supported by the successful outcome. Perhaps this was why Newton reported the agreement as a memorable phase in his discovery of universal gravitation. Although not given explicitly by Newton, we can summarize this naturally suggested argument as follows:

Observed acceleration of fall of terrestrial bodies and the Moon.

(warrant) $H_{identity}$: Terrestrial gravitation and the lunar celestial force are the same.

____(deduce)

Intermediate conclusion: Celestial/gravitational accelerations at the Earth's surface and the Moon's orbit are in the ratio of an inverse square of distances to the Earth's center.

(warrant) Terrestrial and lunar matter respond to the same forces.

____(deduce)

$H_{inv.\ square}$: The celestial force acting on the Moon is strengthened by an inverse square law with distance at the Earth's surface.

This second inference forms the second half of the relations of support displayed in Figure 8.1.

For my purposes here, we have two inferences each of whose conclusions is used as a warrant in the argument for the other. The corresponding arch can be drawn as follows.

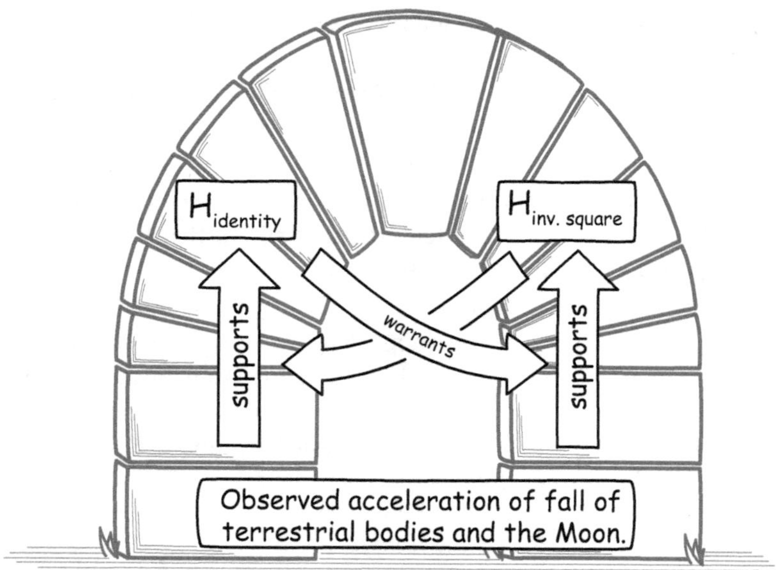

Figure 8.1. The arch for the Moon test

Although the component relations of support are deductive, the combined result is that the observed accelerations provide inductive support for the two hypotheses:

Observed acceleration of fall of terrestrial bodies and the Moon.

____(induction)

$H_{identity}$: Terrestrial gravitation and the lunar celestial force are the same.

$H_{inv.\ square}$: The celestial force acting on the Moon is strengthened by an inverse square law with distance at the Earth's surface.

4. Elliptical Orbits and the Inverse Square Law

The next pair of mutually supporting propositions asserts that the planets move along elliptical orbits and that their motion is governed by an inverse square law of gravity. Planetary astronomy poses a curve-fitting problem. We have many observed positions of the planets. Which curve do we fit to them to recover their orbits? Prior to Newton, Kepler had found that elliptical orbits could be fitted to the observed positions of the planets. This result came to be known later as "Kepler's second law." It is called that, for example, in Maxwell's *Matter and Motion* (1894, 110). From it, one can infer that each planet is attracted to the Sun by a force that varies inversely with the square of distance from the Sun as the planet moves through its orbit. That an elliptical motion is associated with this inverse square law is an early result proved by Newton in Book I of *Principia* ([1726] 1962, Proposition XI, Problem VI). Maxwell uses this result to infer from the elliptical motions of the planets to the inverse square law of gravity:

> Hence the acceleration of the planet is in the direction of the sun, and is inversely as the square of the distance from the sun. This, therefore, is the law according to which the attraction of the sun on a planet varies as the planet moves in its orbit and alters its distance from the sun. (112)

That is, we have the following inference:

Observed positions of the planets.

(warrant) $H_{ellipses}$: The planets move in their specific elliptical orbits.

____(deduce)

$H_{inv.\ square}$: The planets are attracted to the Sun by a force that varies with the inverse square of distance.

Newton himself, however, was more circumspect. This relation of support is straightforward only insofar as we assume that the fit of an ellipse to the observed motions is exact. Newton knew that it is not exact, so he did not offer Maxwell's inference in his *Principia*. That an elliptical motion is governed by an inverse square law of force is merely reported as a theorem of mathematics.

In its place, Newton ([1726] 1962) offered an inverted relation of support. The pertinent discussion comes later in Book III in his Proposition XIII, Theorem XIII. At this stage in the development, Newton had already inferred the inverse square law of gravity from other phenomena. He would now infer from the inverse square law to the elliptical motions of the planets. Noting the inversion explicitly, he wrote that

> Now that we know the principles on which they [the motions of the planets] depend, from these principles we deduce the motions of the heavens *a priori*. Because the weights of the planets towards the sun are inversely as the squares of their distances from the sun's centre, if the sun were at rest, and the other planets did not act one upon another, their orbits would be ellipses, having the sun in their common focus. . . . (420–21)

Newton offered here a relation of support that inverts the one given above by Maxwell:

Observed positions of the planets.

(warrant) $H_{inv.\ square}$: The planets are attracted to the Sun by a force that varies with the inverse square of distance.

____(deduce)

$H_{ellipses}$: The planets move in their specific elliptical orbits.

The observed positions of the planets are still needed as a premise in the inference since an inverse square law of attraction from the Sun is also compatible with parabolic and hyperbolic trajectories. They are ruled out by the periodic motion of the planets. Then specific positions of the planets at specific times are needed to recover the specific ellipse that is the orbit of each planet.

Newton's inference, however, is qualified by an idealization indicated in his remark above "if the sun were at rest, and the other planets did not act one upon another." The orbits of the planets are not exactly elliptical because of perturbations from the gravitational attraction of the other planets. These deviations are generally negligible at the level of accuracy of Newton's analysis. However, a noticeable perturbation was produced by the massive planet Jupiter acting on the motion of Saturn.[1] It is greatest when the two planets are nearest each other: that is, when they are in conjunction. "And hence arises," Newton concluded, "a perturbation of the orbit of Saturn in every conjunction of this planet so sensible, that astronomers are puzzled with it" ([1726] 1962, 421).

5. The Exactness of the Inverse Square Law

Newton ([1726] 1962) did not explicitly incorporate the inference from the elliptical orbits of the planets to the inverse square law in the carefully developed sequence of propositions in Book III of *Principia*. However, an important step in that sequence was something close to this inference. It concerned the inverse square law of gravity. How did Newton know that the correct law is exactly an inverse square law? Might a similar law work as well or even better? Does gravity conform to the inverse square law only as an approximation? Perhaps the force varies with distance r according to $1/r^{2+\delta}$, where δ is some small number close to zero?

In one of the most brilliant analyses of his *Principia*, Newton showed that we have strong evidence for the force of attraction conforming exactly with the inverse square law. Under such a law, Newton had shown, the unperturbed planets move along elliptical paths fixed in space. The aphelion of each

1 Less noticeable, Newton reported, are the perturbations in Jupiter's motion because of the attraction of Saturn. He reported other perturbations as "yet far less" ([1726] 1962, 422). The exception was the sensible disturbance to the orbit of the Earth because of the Moon.

planet — the point of greatest distance from the Sun — will be fixed in space, and the planet will return to it after a complete circuit of 360° around the Sun. The major axis of the ellipse, the line of the apsides connecting aphelion and perihelion, will be correspondingly fixed.

This fixity would be lost, Newton now showed, if the law differed from an inverse XLV square law. In Proposition XLV, Corollary 1, of Book I, he considered the case of bodies orbiting in near-circular orbits. He showed that, if the law of attraction differed from an inverse square law, then a planet would not return to its aphelion after a circuit of 360° around the Sun. It would need to complete more or less of the circuit according to how much the force deviated from an inverse square law. That is, for a $1/r^{2+\delta}$ force law, the planet would return to its aphelion after passing $360°/\sqrt{(1 - \delta)}$. The result was remarkably robust, holding even when the deviation from the inverse square law δ was not small.

Since the planets do move in near-circular orbits, Newton could apply his result to the motions of the planets. If we set aside known perturbations, then the planets do trace fixed elliptical orbits, returning to their aphelia after a 360° circuit around the Sun. Newton could conclude with satisfaction in Book III, Proposition II, Theorem II that

> [The inverse square law] is, with great accuracy, demonstrable from the quiescence of the aphelion points; for a very small aberration from the proportion according to the inverse square0 law of the distances would (by Cor. 1, Prop. XLV, Book I) produce a motion of the apsides sensible enough in every single revolution, and in many of them enormously great. ([1726] 1962, 406)

In summary form, this argument is a version of Maxwell's argument since it infers from a property of the elliptical orbits of the planets to the exact inverse square law of gravity:

Observed positions of the planets.

(warrant) H_{ellipses}: The planets move in their specific elliptical orbits.

Newton's Proposition XLV, Corollary 1, Book I.

____(deduce)

$H_{\text{inv. square}}$: The planets are attracted to the Sun by a force that varies with the inverse square of distance.

The overall structure of the relations of support displayed here is of the two hypotheses accruing support from the observed positions of the planets over time. Although the two component inferences are deductive, the combined relations of support are inductive and can be summarized as

Observed positions of the planets.

____(induction)

H_{ellipses}: The planets move in their specific elliptical orbits.

$H_{\text{inv. square}}$: The planets are attracted to the Sun by a force that varies with the inverse square of distance.

In broad strokes, the relations of support recounted here in Sections 4 and 5 are between the two hypotheses $H_{\text{inv. square}}$ and H_{ellipses}. They enter into the mutual relations of support pictured in the arch analogy of Figure 8.2.

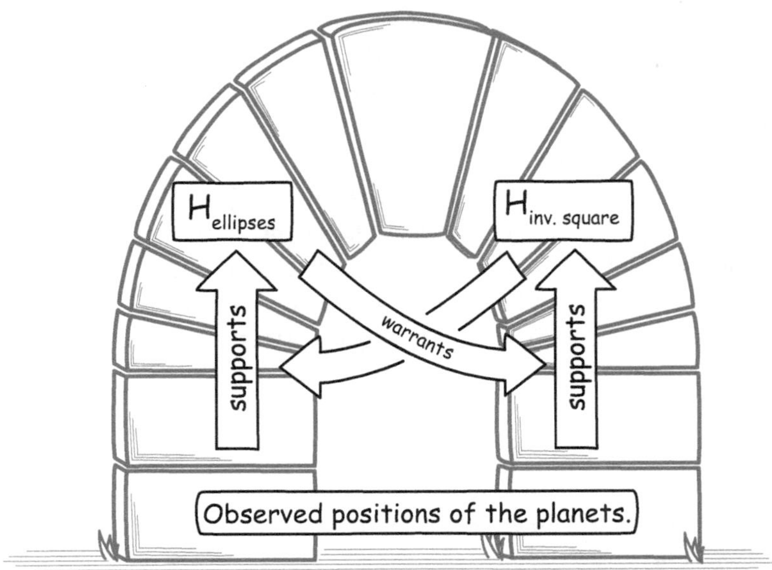

Figure 8.2. Elliptical orbits and the inverse square law

6. Conclusion

We have seen here two pairs of propositions in Newton's *Principia* ([1726] 1962) that mutually support one another. A close reading of his text is likely to reveal more. A natural candidate is Kepler's harmonic rule that relates the period and mean radii of planetary and lunar orbits: $(\text{period})^2$ is directly proportional to $(\text{radius})^3$. Newton infers from this harmonic rule to his inverse square law. We now routinely invert the inference and infer from the inverse square law to the harmonic law.

Such inversions are encouraged by a development common in maturing theories. We are inclined initially to infer from the elliptical orbits of the planets to the inverse square law of attraction, for the elliptical orbits are closer to observations. As the theory matures, we find multiple supports for the inverse square law. We also recognize that Newton's fully elaborated system corrects the simple statement that the planets move in ellipses, for in some cases the perturbing effects of other celestial bodies move them away from their ellipses. Then it becomes more natural to invert the relation of support and see the inverse square law as supporting a corrected version of the original observations of elliptical orbits.

Another example of this inversion is found in the role of atomic spectra in the foundation of quantum theory, as related in Chapter 9, "Mutually Supporting Evidence in Atomic Spectra." Ritz's combination principle supports the discrete energy levels of Bohr's 1913 theory of the atom and thus the quantum theory that developed from it. The developed quantum theory, however, entails a version of the Ritz principle, corrected by selection rules. This complication indicates the inverted relation of support.

REFERENCES

Maxwell, James Clerk. 1894. *Matter and Motion*. London: Society for Promoting Christian Knowledge.

Newton, Isaac. (1726) 1962. *Mathematical Principles of Natural Philosophy*. 3rd ed. Translated by Andrew Motte, revised by Florian Cajori. Berkeley: University of California Press.

———. 1888. *A Catalogue of the Portsmouth Collection of Books and Papers Written by or Belonging to Sir Isaac Newton*. Cambridge, UK: Cambridge University Press.

Norton, John D. 2021. *The Material Theory of Induction*. BSPS Open Series. Calgary: University of Calgary Press.

9

Mutually Supporting Evidence in Atomic Spectra

1. Introduction

Gases and vaporized metals, when heated or energized by electric discharges, emit light or electromagnetic radiation in the invisible parts of the spectrum. In the nineteenth century, spectroscopists began detailed measurements of the frequencies emitted by various substances. The most striking result was that, commonly, the emitted spectra did not consist of a continuous range of frequencies but only specific frequencies organized regularly in series. Identifying which frequencies were emitted by each substance under which circumstances proved to be a challenge that occupied spectroscopists for decades. Their efforts required many ingenious approaches. What resulted was a complicated network of relations of evidential support that is the subject of this chapter. In it, we will see mutual relations of support crossing over each other and at two levels.

We will look only at the simplest of the emission spectra, that of hydrogen, for itis already sufficient to display this multiplicity of relations of mutual support. We will take as the simplest item of evidence the proposition that excited hydrogen produces electromagnetic radiation at such and such frequency or wavelength. One such item asserts the fact that a prominent line in the hydrogen spectrum, the first "H_α" line of the Balmer series, is at wavelength 656.2 Angstroms. Once a spectroscopist has identified some lines in the spectrum of a substance, it is possible to identify others by means of a device introduced in 1908 by Walther Ritz, his "combination principle."

It asserted that adding or subtracting the frequencies of certain[1] known lines in a spectrum will yield more lines.

If there are two lines with the frequencies[2] ν_{12} and ν_{23} of the right type, then there is a third line at the frequency $\nu_{13} = \nu_{12} + \nu_{23}$. These additions are easily inverted. If we have lines at the frequencies ν_{12} and ν_{13}, then there is a third line at the frequency $\nu_{23} = \nu_{13} - \nu_{12}$. And, if we have lines at the frequencies ν_{23} and ν_{13}, then there is a third line at the frequency $\nu_{12} = \nu_{13} - \nu_{23}$. Each of these applications of the Ritz combination principle expresses a relation of support. There are three, and they cross over one another in relations of mutual support:

Lines at ν_{12} and ν_{23} support a line at ν_{13}.

Lines at ν_{12} and ν_{13} support a line at ν_{23}.

Lines at ν_{23} and ν_{13} support a line at ν_{12}.

There are more than just a few of these sets of mutually supporting items of evidence. Since the emission spectrum of hydrogen contains infinitely many lines, there are infinitely many of them.

In Sections 2 and 3, I recall the discovery of the various series of lines of the hydrogen spectrum and their systematization by Ritz through his combination principle. In Section 4, I explore how the principle allows a dense network of relations of mutual support among the lines. If the Ritz combination principle is taken as a premise, then these relations of support are expressed by deductive inferences. They combine to produce a totality in which the observed lines of the hydrogen spectrum provide inductive support for the series of infinitely many lines.

In Section 5, I ask a further evidential question. What supports the Ritz combination principle? Is it merely to be supported as a generalization about observed lines in the spectrum? Which fact warrants it? The decisive theoretical development came in 1913 when Niels Bohr proposed an atomic mechanism capable of producing precisely the spectra observed. It became one of the foundations upon which modern quantum theory was built. Bohr's theory, to be outlined in Section 6, proposed that the lines arise when an

1 The word *certain*, meaning "some carefully chosen," indicates an important restriction. The principle does not work for all pairs of lines.

2 The two indices arise from the simple two parameter formulae (1)–(6) below, found empirically to systematize the frequencies of the lines present.

excited electron drops or jumps down from a higher to a lower energy state. Each jump leads to an emission of radiant energy with a frequency proportional to the energy emitted. This mechanism provided a direct explanation of the Ritz combination principle. The two frequencies ν_{12} and ν_{23} corresponded to two emissions in a two-step jump. If the jump is taken in a single step, then the frequency $\nu_{13} = \nu_{12} + \nu_{23}$ comes directly from the requirement that the two-step jump or the single-step jump liberate the same quantity of energy.

The Ritz combination principle provides another instance of the crossing over of relations of support but at a more elevated level of the theory. On the one hand, as described in Section 7, the combination principle, taken as a datum from observational spectroscopy, provides evidential support for the Bohr theory, and it was reported as such. Using a few notions from his theory, the principle translates directly in the emission mechanism that Bohr proposed. On the other hand, as reported in Section 8, the converse relation of support also holds. Once quantum theory is established, it entails the Ritz combination principle. The converse relation of support is important, for what quantum theory eventually provides is a corrected version of the principle. Some of the lines that the original Ritz principle predicts are "forbidden": that is, they correspond to electron jumps precluded by quantum theory. What results is an embellished Ritz combination principle, supplemented by "selection rules" that indicate which lines are forbidden.

2. The Discovery of Regularities in Emission Spectra

The emission spectrum of hydrogen contains lines at many frequencies. They are called "lines" since the early methods of spectroscopy captured the different frequencies present in the light as lines on a photographic plate. The frequency or wavelength of the light was recovered from distance measurements on the plate. An example from Fowler (1922, 8) is shown in Figure 9.1.

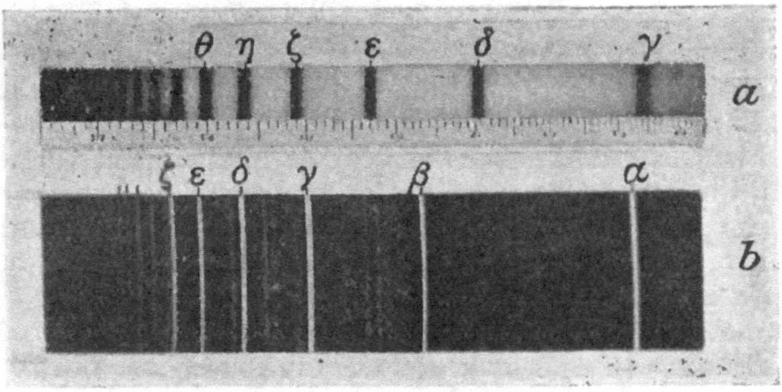

Figure 9.1. A spectrograph of the spectrum of hydrogen

The first formula to embrace some of these lines with enduring success was posited by Balmer ([1885] 1967) for strong lines in the visible spectrum. In modernized form, his formula for the frequencies of lines in the "Balmer series" was

$$\nu(2, m) = R(1/2^2 - 1/m^2) \qquad (\text{"Balmer"}) \qquad (2)$$

where R is a constant. The different values of $m = 3, 4, 5, \ldots$ gave specific lines in the spectrum shown in Figure 9.1.

$$H_\alpha: \nu(2, 3) = R(1/2^2 - 1/3^2)$$

$$H_\beta: \nu(2, 4) = R(1/2^2 - 1/4^2)$$

$$H_\gamma: \nu(2, 5) = R(1/2^2 - 1/5^2)$$

$$H_\delta: \nu(2, 6) = R(1/2^2 - 1/6^2)$$

. . .

In the following decades, similar formulae were found for other lines in the hydrogen spectrum:

$$\nu(1, m) = R(1/1^2 - 1/m^2) \qquad m = 2, 3, 4, \ldots \qquad (\text{"Lyman"}) \qquad (1)$$

$$\nu(3, m) = R(1/3^2 - 1/m^2) \qquad m = 4, 5, 6, \ldots \qquad (\text{"Paschen"}) \qquad (3)$$

$$\nu(4, m) = R(1/4^2 - 1/m^2) \qquad m = 5, 6, 7, \ldots \qquad (\text{"Brackett"}) \qquad (4)$$

$$\nu(5, m) = R(1/5^2 - 1/m^2) \qquad m = 6, 7, 8, \ldots \qquad (\text{"Pfund"}) \qquad (5)$$

Each series is named after the spectroscopist responsible for its identification.

This quick recitation of the various formulae masks the magnitude of the problems faced by the spectroscopists. Decades separated the recovery of these series. Although Balmer's formula was reported in 1885, the terms of the Paschen series began to be verified around 1908, as announced by Ritz (1908). Lyman (1914) reported his ultraviolet spectrum in a letter to *Nature* that year. Brackett (1922) reported more lines in the Paschen spectrum and the first two members of the newly discovered Brackett series. Pfund (1924) reported the first line of the Pfund series.

There were multiple problems to be overcome. The first four lines of the Balmer spectrum, H_α to H_δ, are easiest to find since they are in the visible spectrum. The Lyman series lies in the ultraviolet spectrum, and the remaining series are in the infrared spectrum. These different ranges require different instrumentation to separate the frequencies and register them. Controlled conditions, such as low pressures, are needed to manifest sharp lines. Then some of the lines reported have celestial origins in spectrographs taken of stars. Since we have no independent samples of the matter of the stars, how do we know just which excited matter produced them? How are they to be matched up with spectra produced by excited matter on Earth? The spectrograph in Figure 9.1 shows such a case. The upper set of lines arises in light from the star Sirius. The lower set comes from light emitted by excited hydrogen in a terrestrial laboratory. Fowler (1922, 7) suggests that the celestial lines can be identified as an extension of those in a spectrum found terrestrially if they fall near enough on a definite curve. Figure 9.2 shows such a curve from Fowler (14). The vertical axis plots the *m* of (1), (2), and (3), and the horizontal axis plots frequency.

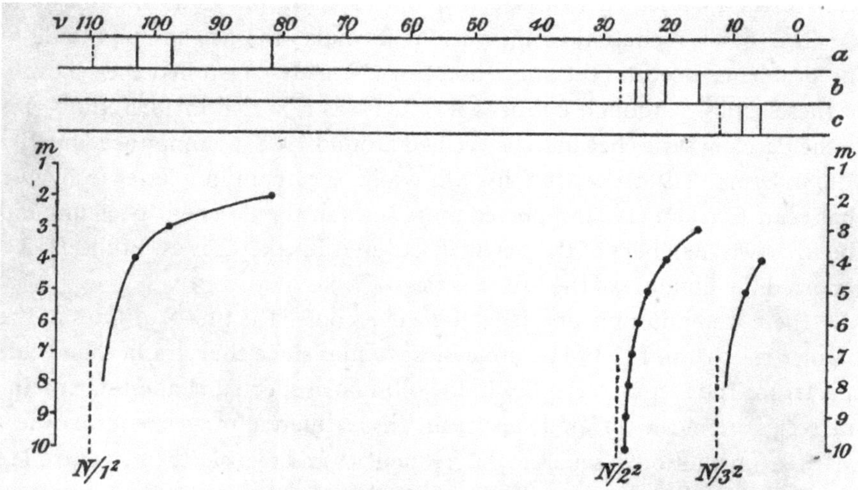

FIG. 2.—THE THREE SERIES OF HYDROGEN.
(*a*)=Lyman series; (*b*)=Balmer series; (*c*)=Ritz-Paschen series.

Figure 9.2. Frequencies of series form definite curves

A trace of these different sources appears in Bohr's (1913a) celebrated paper on his theory of the atom. Bohr notes that then only nine lines of the Balmer series had been observed terrestrially in vacuum tubes, whereas thirty-three had been observed in celestial spectra (9).

Finally, even when definite series are identified in some spectrum, it is not always clear that all of the series derive from the same substance. Sommerfeld (1923, 207–08) reports two series originally attributed to hydrogen because of the similarity to the Balmer formula (2) for hydrogen. They are

$\nu = R(1/1.5^2 - 1/m^2)$ m = 2, 3, 4, . . . ("Principal series")

$\nu = R(1/2^2 - 1/(m+0.5)^2)$ m = 2, 3, 4, . . . ("Second subsidiary series of hydrogen")

One outcome of Bohr's atomic theory of 1913 was that these series would result from an atom with a nuclear charge twice that of hydrogen, so that the constant *R* in these formulae is four times greater than that for hydrogen. That is, they derive from helium and not hydrogen. This conversion is easily accomplished by multiplying the above formulae by 4/4. We now have

$\nu = (4R)\,(1/3^2 - 1/m^2)$ m = 4, 5, 6, . . .   ("Principal series")

$\nu = (4R)\,(1/4^2 - 1/m^2)$ m = 5, 6, 7, . . .   ("Second subsidiary series of hydrogen [?]")

The attribution of the spectra to helium was already made immediately by Bohr (1913b) in a letter to *Nature.*

3. The Ritz Combination Principle

Given the variety and difficulty of the problems facing the spectroscopists in locating and grouping spectral lines, any assistance in the heuristics would be useful. Such was offered by Ritz (1908). Rydberg (1890, 331) had noted that formulae for spectral lines could be simplified if they were written in terms of wave number, the inverse of wavelength.[3] Then the formulae could be expressed as a difference of two terms, as done in (1) to (5) above. This fact enabled Ritz (1908, 523) to propose what he called his "principle of combination" (*Kombinationsprinzip*). Its value, as Ritz noted in the first sentence of his paper,[4] is that one could use known spectral series to discover new ones. He applied it to a range of spectra, including those of hydrogen, helium, and the alkali and alkaline earth metals.

A good statement of the principle is provided by Ritz himself in a note found posthumously in his papers and published as an appendix to Ritz (1908) in his *Gesammelte Werke* (*Collected Works*) (1911, 162). Sommerfeld (1923, 205) quotes Ritz as giving this formulation:[5]

> By additive or subtractive combination, whether of the series formulae themselves, or of the constants that occur in them,

---

3 The spectroscopists preferred to report wavelengths since they were more directly measurable than frequency. To convert wavelengths to frequencies required multiplication by the speed of light: frequency = (speed of light) / wavelength. Using inverse wavelength as a surrogate for frequency avoids systematic errors introduced by errors in the value of the speed of light employed.

4 "In the following, it will be shown that, from known spectral series of an element, one can derive new series without introducing any new constants. Through this especially, almost all the series and lines recently discovered in the alkalis by Lenard, Konen and Hagenbach, Saunders, Moll, Bergman etc. come to be represented exactly" (Ritz, 1908, 523).

5 Sommerfeld's (1923) report is abridged. In place of "certain newly discovered lines from those known earlier," Ritz's (1908, 523) text specifies lines of alkalis then recently discovered by Lenard and others, as well as new elements, in particular helium.

formulae are formed that allow us to calculate certain newly discovered lines from those known earlier.

The principle is incomplete since it does not specify which additions and subtractions are those that yield new lines. The necessary supplement is provided in each application by a formula that represents the line frequency as a difference of two terms. Its application to hydrogen assumed that the series of hydrogen conforms to a general formula

$$\nu(n, m) = R(1/n^2 - 1/m^2) \qquad n = 1, 2, 3, \ldots \quad m = 2, 3, 4, \ldots \tag{6}$$

in which we always have $m > n$. It follows that a new line in the spectrum can be identified by taking the difference in the frequencies of two known lines, as long as the expression (6) for each shares a common term. For example, the lines Hα and Hβ can be subtracted in this way since they share a $1/2^2$ term eliminated by the subtraction:

$$\text{H}\beta\text{:}\ \nu(2, 4) = R(1/2^2 - 1/4^2)$$

$$\text{H}\alpha\text{:}\ \nu(2, 3) = R(1/2^2 - 1/3^2)$$

__subtract______________

$$\nu(3, 4) = R(1/3^2 - 1/4^2)$$

What results is the first line $\nu(3, 4)$ of the Paschen series (3), not an established series in 1908. It led to an immediate affirmation of the correctness of Ritz's proposal. In his paper, Ritz (1908, 522) reported with obvious satisfaction that Paschen had informed him by letter ("Nach einer brieflichen Mitteilung") that he had observed just this line in the infrared spectrum.

4. Mutually Supporting Evidence

For Ritz, the combination principle was valuable as a means of discovering new lines. At the same time, it was the warrant for an inference from the existence of some lines to others. The evidence of the lines H_α and H_β of the Balmer series supports the line $\nu(3, 4)$ of the Paschen series. This subtraction can be reversed into an addition that supplies a different relation of support:

$$H_\alpha: -\nu(2, 3) = R(1/2^2 - 1/3^2)$$

$$\nu(3, 4) = R(1/3^2 - 1/4^2)$$

__add____________________

$$H_\beta: \quad \nu(2, 4) = R(1/2^2 - 1/4^2)$$

That is, the frequencies of the H_α line and the $\nu(3, 4)$ line can be added to recover the H_β line. In this addition, the common $1/3^2$ terms cancel. That is, the H_α line and the $\nu(3, 4)$ line support the H_β line.

These two relations show the crossing over of relations of support. In the first, the H_β provides support for the $\nu(3, 4)$ line. In the second, the $\nu(3, 4)$ line provides support for the H_β line. Since the full range of series covered by relations (6) has infinitely many lines, there will be infinitely many of these relations of support, crossing over in many ways.

These relations can be captured in infinite sets. For example, the Ritz combination principle can be applied to the infinitely many lines of the Balmer series (2) to support the Paschen (3), Bracket (4), and Pfund (5) series. For the first, lines in the Balmer series can be subtracted to cover the entire Paschen series:

$$\nu(2, m) = R(1/2^2 - 1/m^2) \quad (m > 4) \text{ Balmer}$$

$$H_\alpha: \nu(2, 3) = R(1/2^2 - 1/3^2)$$

__subtract________________________________

$$\nu(3, m) = R(1/3^2 - 1/m^2) \quad (\text{m} > 4) \text{ Paschen}$$

Additional lines are needed as supplementary evidence if series in the sequence of (1), (2), (3), (4), and (5) are to support those earlier in the sequence. For example, we take as an extra datum $\nu(1, 2)$, the first line of the Lyman series (1), and then the entire Lyman series is recovered by addition from the Balmer series:

$$\nu(1, 2) = R(1/1^2 - 1/2^2)$$

$$\nu(2, m) = R(1/2^2 - 1/m^2) \quad (m > 2) \text{ Balmer}$$

__add________________________________

$$\nu(1, m) = R(1/1^2 - 1/m^2) \quad (\text{m} > 1) \text{ Lyman}$$

If we take as an extra datum the H_α line of the Balmer series, then the Paschen series supports the Balmer series.

$$\nu(3, m) = R(1/3^2 - 1/m^2) \quad (m > 3) \quad \text{Paschen}$$

$$H_\alpha\text{: } \nu(2, 3) = R(1/2^2 - 1/3^2)$$

__add____

$$\nu(2, m) = R(1/2^2 - 1/m^2) \quad (m > 3) \quad \text{Balmer}$$

Two of these relations of support cross over one another and can be represented more compactly as

Ritz combination principle
Balmer series
____deduce______________
Paschen series

Ritz combination principle
H_α line
Paschen series
____deduce______________
Balmer series

Similar computations realize many more like-structured relations of mutual support that cross over each other, including

the Paschen series supports the Bracket and Pfund series;

the Bracket and Pfund series support the Paschen series;

the Bracket series supports the Pfund series;

the Pfund series supports the Bracket series;

et cetera.

It is noteworthy that all of the individual relations of support just described are implemented by deductive inferences. We can infer deductively from some subset of lines, via the Ritz combination principle, to the larger portions and even the entire set in (6). Nonetheless, accepting the entirety of the series does involve inductive risks. Those risks enter in accepting the premises that figure in the individual deductions. We take a small inductive risk in accepting the correctness of the report of the existence of each line. Most notably, we take considerable inductive risk in accepting the combination principle, since it has infinite scope. That the risk is considerable is seen most easily from the fact that later investigations introduced a small "fine structure" splitting of the lines in the series described above.[6] More significantly, as we shall see below, the Ritz combination principle itself needed to be modified by selection rules that prohibit certain lines when we move beyond the hydrogen spectrum.

Although combining deductive relations to yield inductive support overall might appear at first to be paradoxical, it is not so. All that has happened is that the inductive risks taken in accepting the premises of the deductions are the only inductive risks that we need to take. Once they are taken, we can proceed with maximally secure deductive arguments. This type of support is inductively more secure than combining inductive relations of support in a similar way. No further inductive risk is taken in accepting these component deductive inferences, whereas further inductive risk would be taken if they were replaced by inductive inferences. In Chapter 2, "Large-Scale Structure: Four Claims," I reflected on other examples of deductive relations of support combining to provide overall inductive support.

The massively entangled network of relations of mutual support goes well beyond the heuristic guidance of Ritz's original purpose. For that narrower purpose, the most useful are the inferences from readily available lines to those not yet discovered. My concern here, however, is not so narrow. It is to discern the full structure of the relations of inductive support.

5. Supporting the Ritz Combination Principle

The inferences reported in the previous section all employ the Ritz combination principle as a premise. None of the inferences in that section provides support directly for the Ritz combination principle. Rather, they all merely use it. With the qualification noted below, the principle is a standard part of atomic spectroscopy.

What evidence supports the Ritz combination principle? One might be tempted to answer that we have many instances of the general formula (6) and no counterexamples. So we can inductively infer to (6) and from it deduce the Ritz combination principle for the hydrogen spectrum. The trouble is that a generalization — any generalization — requires a warranting fact. So far, it has been unclear what that fact is.

---

6 The splitting, reported by Sommerfeld in 1916, resulted from relativistic corrections to Bohr's atomic theory. Sommerfeld found that differences in the eccentricities of the elliptical electron orbits of the theory led to slight differences in their energies (1915, 1923, 474).

We might be tempted to say that, when a general formula this simple fits all of the cases at hand, we have a license to infer to it. I developed the familiar difficulty at length in Chapter 6 of *The Material Theory of Induction* (Norton 2021). We lack both a notion of simplicity precise enough to warrant these inferences, and we lack a factual basis for the inductive powers of such a notion.

As an intermediate attempt to warrant the generalization, we might suggest that a formula as simple as (6) can be as successful as it is only if it is part of a larger regularity whose precise character is not currently known to us. Something like this is plausible. However, it rests on the supposition of further facts not so far produced. That, at least, was the situation in 1908 when Ritz proposed his principle. In 1913, circumstances would change. Then Bohr proposed his novel theory of the atom. That theory used Ritz's principle and the formula (6) as evidential support. Soon the relation of support would become mutual when the quantum theory that emerged from Bohr's theory provided support for a modified version of Ritz's principle.

6. Bohr's Theory of the Atom

Bohr's celebrated theory of the atom was based on Rutherford's nuclear account of the atom. According to it, a hydrogen atom consists of a massive, positively charged nucleus with a light, negatively charged electron orbiting it. To this, Bohr added two ideas. Classical electrodynamics requires that this orbiting electron must radiate its energy electromagnetically and thus be pulled rapidly into the nucleus. Bohr simply posited otherwise:

I. There are stable orbits for the electron.

The energies of these orbits were to be computed by standard electrostatics. Bohr further supposed that electrons could jump between these stable orbits. Another idea connects these jumps to emission spectra:

II. When an electron drops down from a more energetic stable orbit to a less energetic one, closer to the nucleus, the energy E it loses reappears as electromagnetic radiation with a frequency ν, according to $E = h\nu$, where h is Planck's constant.

We can denote the (negative) energies of two stable orbits as W_1 and W_2, with $W_1 > W_2$. When an electron drops from the first to the second orbit, it emits electromagnetic radiation of frequency ν_{12} whose value, according to II, is

$$\nu_{12} = (W_1 - W_2)/h \tag{7}$$

Comparison with the general spectral formula (6) then allows us to identify the (negative) energies of the stable orbits as

$$\Omega(n) = R/n^2 \qquad n = 1, 2, 3, \ldots \tag{8}$$

The striking outcome here is that, from the spectral formula (6), we infer that the energies of the stable orbits do not form a continuous set. Rather, they form a discrete set whose members are indexed by *n*. Bohr's posits I and II do not presume discreteness. It is inferred from the evidence of the spectra.

The Bohr theory clarified Ritz's combination principle. In its original form, the principle was the recognition of a bare numerical regularity. It was a kind of scientifically useful numerology. Bohr's theory gave it a physical basis. Consider the case shown in Figure 9.3. An electron in an excited hydrogen atom drops to a lower energy orbit, emitting radiation of frequency ν_{12} with energy $E_{12} = h\nu_{12}$. In a second jump, it drops to a still lower energy orbit, emitting radiation of frequency ν_{23} with energy $E_{23} = h\nu_{23}$. Had the electron jumped directly from the first orbit to the final orbit, it would have emitted radiation of frequency ν_{13} with energy $E_{13} = h\nu_{13}$.

We have two cases, one with two successive jumps and the other with a single jump. They are between the same initial and final orbits. Thus, the energy radiated in each must be the same:

$$E_{13} = E_{12} + E_{23}$$

Applying $E = h\nu$ to each of these three energies, we recover

$$\nu_{13} = \nu_{12} + \nu_{23}$$

This last sum is the Ritz combination principle applied to the hydrogen spectrum. Its physical foundation is now displayed.

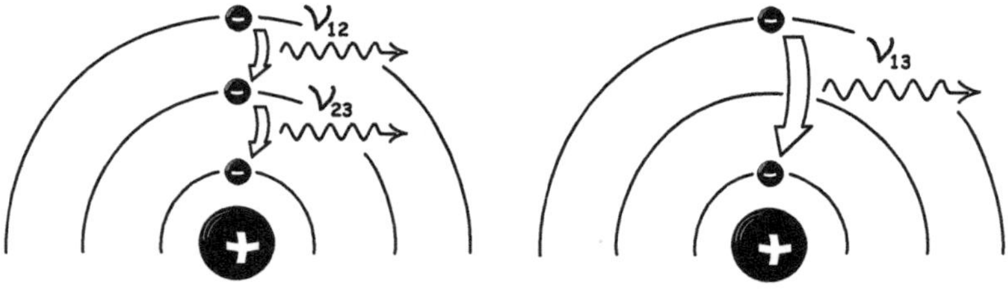

Figure 9.3. Physical basis of the Ritz combination principle

There is much more in Bohr's theory, and these details have been elaborated frequently elsewhere.[7] Using further conditions, Bohr concluded that the constant R in (6) and (8) is given by $R = 2\pi^2 me^4/h^3$, where m is the mass of an electron and e its charge. The value of R, computed from this formula using the best-known values of m, e, and h, Bohr reported, was 3.1×10^{15}. It matches closely enough to the value that he reported from spectral observations, 3.290×10^{15}.

Bohr also showed that the stable orbits of (8) coincided with the orbital angular momentum of the electron taking on integer values in units of $h/2\pi$. This formulation of the discreteness of the stable states of (8) became increasingly important as Bohr's theory evolved. In more elaborate versions of his theory, the "old quantum theory," this result was the simplest case of the quantization of action. In the "new quantum theory" that emerged in the mid-1920s, this result coincided with the fact that stable electron orbitals are eigenstates of the angular momentum operator.

7. The Ritz Combination Principle Supports Quantum Theory

The first half of the mutual relations of support is that the newly emerging quantum theory was supported by the Ritz combination principle. This support has been evident from the start. In a much-quoted remark, reported by Bohr's assistant and confidant, Léon Rosenfeld, Bohr remarked that, "as soon as I saw Balmer's formula, the whole thing was immediately clear to me" (quoted in Duncan and Janssen 2019, 14).

7 For an early authoritative textbook account, see Sommerfeld (1923, 211–18). Norton (2000) develops these details with special focus on the evidential relations.

The evidential support that the Ritz principle gave to Bohr's emerging theory was widely recognized. Max Born (1935, 85) was forthright about it: "A direct confirmation of this [Bohr's 1913a] theory can be seen in the following fact. . . ." He proceeded to explain in detail and with a figure similar to Figure 9.3 how Ritz's combination principle (identified by this name) is a consequence of the cascade of emissions described in the preceding section.

Sommerfeld (1923), in his early, authoritative volume on the old quantum theory, was similarly forthright. He introduced the Ritz combination principle by name, along with the quotation given above, and then explained its application in detail. He then characterized its significance:

> The principle of combination has maintained itself in the whole region of spectroscopy from infra-red to X-ray spectra as an *exact physical law* with the degree of accuracy that characterises spectroscopic measurement. It constitutes the foundation on which Bohr's theory of spectra rests, and is, in essence, identical with Bohr's law . . . [equation (7) above], which likewise taught us to regard the frequency of a spectral emission as the difference between two energy-levels. (205–06; Sommerfeld's emphasis)

If we approach the support relations materially, then we can be more precise in just what Ritz's combination principle provides to Bohr's theory. Bohr's posit II above associates spectral lines with electron jumps between stable orbits of different energy. Using the posit as a warranting fact, we infer *from* each spectral line *to* the existence of an electron jump in the hydrogen atom between stable energy states.

The Ritz combination principle adds something very important to this last inference. All that this inference gives us is the energy *differences* between the energies of the stable orbits. Posit II does not specify how these stable energy states are related. It might just be that each line derives from its own unique set of energy states and that no other line derives from electron jumps to or from them. The principle assures us that it is possible to find a single set of energies of stable orbits such that all of these orbits are accessible to the electron in a hydrogen atom. More precisely, it follows from the spectral formula (6) that such a set of energies is given by the relation (8) of the Bohr theory,[8] $\Omega(n) = R/n^2$.

If we follow Sommerfeld (1923) and take the Ritz combination principle as holding universally for all spectral lines, including those not observed, then the relation of support is deductive. All of the inductive risk is taken in accepting posit II provisionally as a hypothesis.

8. Quantum Theory Confirms the Ritz Combination Principle

In the early years of quantum theory, it was natural to focus on the support that the Ritz combination principle provided for the developing theory. The principle itself was rightly judged to be more securely supported by spectroscopic evidence. The developing quantum theory was speculative and even required physicists to overlook a glaring contradiction with classical electrodynamics. As quantum theory developed, became more established, and evolved into the later "new quantum theory," this orientation reversed. A more secure quantum theory provided support for the Ritz combination principle directly. The principle is a deductive consequence of the account given by quantum theory of the origin of the spectra. It also became more congenial to see support for the Ritz combination principle in quantum theory, for that support derived from a definite physical ontology and replaced what I called "numerology" above.

What further strengthens the inverted relation of support is that the development of quantum theory showed that the full Ritz combination principle, when applied to spectra beyond those of hydrogen, needed corrections.[9] It turned out that not all of the lines predicted by the principle occurred. Some transitions turned out to be "forbidden," and the determination of which ones are allowed was governed by "selection rules." The jumps allowed in Bohr's

---

8 The formula is determined only up to an additive constant that plays no role in energies and frequencies of the radiation emitted. The inference is only to the possibility of the single set of energies described. It does not preclude a more complicated set that simulates the behavior of the simpler set, even though in practice this complication would be dismissed as contrived.

9 Another example of this type of correction is seen in the previous chapter on Newton and the inverse square law of gravity. Kepler's elliptical orbits of the planets support the inverse square law. Yet that law, when developed systematically by Newton, leads to corrections to the elliptical orbits due to perturbations from other celestial bodies.

original theory were constrained only by energy conservation. Such jumps must also conform to the conservation of angular momentum. The emitting electron must lose just the angular momentum carried off by the emitted radiation.[10] In the case of the simple hydrogen spectrum, the additional condition does not further limit the spectra beyond the limitations of energy conservation. However, that is only a special case. Spectra of other elements do have forbidden lines.

Through these considerations, the reverse direction of support, from quantum theory to Ritz's combination principle, becomes more secure. The observation of spectra can give us direct evidence of only a finite subset of the infinity of lines possible. That finite evidence can support Ritz's combination principle among the lines observed. When we move past hydrogen spectra, this finite evidence can indicate when the principle fails. If, however, we derive the principle from a fully developed quantum theory, then we recover the principle in its most general form as it applies to the infinity of lines in some spectrum. We also recover a way of determining when certain lines are forbidden and a principled physical account of why they are forbidden.

This inversion had already occurred under the old quantum theory. As the theory developed, new quantum numbers were added, beyond the single quantum number n of Bohr's theory of 1913. Sommerfeld (1923) introduced the "azimuthal quantum number" among other numbers. His authoritative treatment of the old quantum theory included an extensive account of a selection rule for atomic spectra. He states it as follows:

> **The principle of selection** states: *the azimuthal quantum number can at the most alter by one unit at a time in changes of configuration of the atom.* (266; Sommerfeld's emphasis)

This selection rule was carried over[11] and vindicated by the wave mechanics of the new quantum theory. It was rapidly absorbed into textbook expositions, such as Pauling and Bright (1935, Section 40f).

In his article for *Review of Modern Physics*, Gibbs (1932, 307) reflects on the need to qualify the Ritz combination principle:

---

10 In the full quantum electrodynamic analysis, an emitted photon carries off $h/2\pi$ of angular momentum. It follows that the emitting electron can only jump to a state whose angular momentum differs from its starting state by $h/2\pi$.

11 Sommerfeld's "at most one unit" is replaced by exactly one unit for the quantum numbers l and m in the case of hydrogen. See Slater (1960, 183).

> The later development of the quantum theory has shown that for certain types of radiation some of the Ritz combination lines are "forbidden" or perhaps better are extremely improbable under ordinary circumstances. The degree of probability for these "forbidden" lines varies widely for different combinations and accordingly under certain conditions of pressure, electric field, and mode of excitation some of the more probable of these improbable or "forbidden" lines are observed.

After reporting the discovery of the Brackett and Pfund series, Gibbs recorded what amounts to the inversion of the relations of support:

> These series, both of which lie well out in the infrared, were discovered sometime after the theoretical basis for the combination principle had been completely changed and elaborated by the introduction of the quantum theory. Indeed the theoretical arguments advanced by Ritz in proposing this principle were quite unsound even in terms of the older classical theory. It is an excellent example of how a fundamentally correct idea is envisioned through false reasoning, to be later explained on an entirely new basis, the theoretical development of which was encouraged and assisted to some extent by the very idea itself. (307–08)

9. Conclusion

The investigation of atomic spectra and their relation to quantum theory illustrates the nonhierarchical structure of relations of evidential support. There is a massively entangled set of relations of support among the infinitely many propositions that assert the existence of specific spectral lines. The fact that warrants these relations of support is the Ritz combination principle. It too enters into nonhierarchical relations of support, for initially the principle provided important evidential support for the newly emerging quantum theory. As that theory developed and became better established, this relation of support was inverted. The quantum theory was seen as providing evidential support for the Ritz combination principle. This inversion is appropriate since quantum theory indicated that the Ritz combination principle had to be supplemented or corrected to accommodate "forbidden lines." The quantum theory could provide both a systematic means of identifying these forbidden lines and a physical basis for forbidding them.

REFERENCES

Balmer, J.J. (1885) 1967. "A Note on the Spectral Lines of Hydrogen." In *Atomic Spectra*, edited by W.R. Hindmarch, 101–07. Oxford: Pergamon Press.

Bohr, Niels. 1913a. "On the Constitution of Atoms and Molecules." *Philosophical Magazine*, Series 6, 26: 1–25.

———. 1913b. "The Spectra of Helium and Hydrogen." *Nature* 92: 231–32.

Born, Max. 1935. *Atomic Physics*. 8th ed. London: Blackie and Son.

Brackett, Frederick S. 1922. "Visible and Infrared Radiation of Hydrogen." *Astrophysical Journal* 56: 154–61.

Duncan, Anthony, and Michel Janssen. 2019. *Constructing Quantum Mechanics, Volume 1: The Scaffold 1900–1923*. Oxford: Oxford University Press.

Fowler, Alfred. 1922. *Report on Series in Line Spectra*. London: Fleetway Press.

Gibbs, R.C. 1932. "Line Spectra of the Elements, Part 1: Bibliography 1." *Reviews of Modern Physics* 4: 279–470.

Lyman, Theodore. 1914. "An Extension of the Spectrum in the Extreme Ultra-Violet." *Nature* 93: 241.

Norton, John D. 2000. "How We Know about Electrons." In *After Popper, Kuhn and Feyerabend: Recent Issues in Theories of Scientific Method*, edited by Robert Nola and Howard Sankey, 67– 97. Dordrecht: Kluwer.

———. 2021. *The Material Theory of Induction*.BSPS Open Series. Calgary: University of Calgary Press.

Pauling, Linus, and E. Bright Wilson. 1935. *Introduction to Quantum Mechanics*. New York: McGraw-Hill.

Pfund, August. 1924. "The Emission of Nitrogen and Hydrogen in the Infrared." *Journal of the Optical Society of America* 9: 193–96.

Ritz, Walther. 1908. "Üeber ein neues Gesetz der Serienspektren." *Physikalische Zeitschrift* 9: 523–29.

———. 1911. *Gesammelte Werke*. Paris: Gauthier-Villars.

Rydberg, Johannes R. 1890. "On the Structure of the Line-Spectra of the Chemical Elements." *Philosophical Magazine*, Series 5, 29: 331–37.

Slater, John C. 1960. *Quantum Theory of Atomic Structure*. Vol. 1. New York: McGraw-Hill.

Sommerfeld, Arnold. 1915. "Die Feinstruktur der Wasserstoff und der Wasserstoff-ähnlichen Linien." In *Sitzungsberichte der mathematisch-physikalischen Klasse der K. B. Akademie der Wissenschaften zu München*, 1915, 459–500 (presented January 8, 1916). Translated as "The Fine Structure of Hydrogen and Hydrogen-Like Lines." *European Physical Journal H* 39 (2014): 179–204.

———. 1923. *Atomic Structure and Spectral Lines*. Translated by H.L. Brose. London: Methuen.

10

Mutually Supporting Evidence in Radiocarbon Dating

1. Introduction

Consider two ways that we can date artifacts and samples. First, traditional methods of historical analysis and archaeology enable us to date artifacts, and the counting of tree rings enables us to date wood from ancient trees. Second, radiocarbon dating provides another means of dating these samples. What results are two sets of propositions concerning the age of specific artifacts. In Section 4, the first are called *H* (historical), and the second are called *R* (radiocarbon).

Each type of dating can provide evidence for the other type. That is, relations of support between these two sets of propositions proceed in both directions, analogously to the relations of support between the stones on either side of an arch.

The second type *R* can support the first type *H*: if we are interested in checking the historical dating of some artifact, then we can send a sample to a radiocarbon laboratory for dating.

The first type *H* can support the second type *R*: radiocarbon dating itself requires empirical calibration to correct for many confounding variables, such as changes in levels of atmospheric carbon 14. Historically dated artifacts and wood dated by tree ring counting can be used in this calibration process. In it, the evidence of these other methods of dating provides support for the recalibrated radiocarbon dating of the samples.

When the two methods agree for some sample, we have support relations passing in both directions. However, the circumstances of the sample might incline us to emphasize only one direction.

In Section 2, I will review briefly how radiocarbon dating works, and in Section 3 I will describe the need for and methods of independent calibration of radiocarbon dating. Finally, in Section 4, I will review how relations of evidential support cross over among the type *H* and type *R* propositions, using the example of the dating of the shroud of Turin and associated control samples.

To speak of just two mutually supporting methods oversimplifies greatly in the interests of brevity. An appreciation of the richness of the interactions of many lines of evidence employed in radiocarbon dating has been provided by Alison Wylie in several works, including Wylie (2016). For a related analysis of radiometric dating in geology, see Alisa Bokulich (2020).

2. How Radiocarbon Dating Works

Consider some ancient artifact such as a scrap of linen from an Egyptian mummy's wrapping or a thread from a medieval cloak. How are we to know its age? In the 1940s, William Libby hit upon a method so ingenious and important that it earned him the Nobel Prize in chemistry in 1960.[1] These artifacts are all derived from carbon-based plants. These plants derived their carbon from the CO_2 in the atmosphere. Virtually all of the atmospheric carbon is the stable isotope ^{12}C, "carbon 12." However, a tiny portion is a radioactively unstable ^{14}C. This tiny portion is decaying exponentially, with clocklike regularity, with a half-life of about 5,730 years. That means that, after 5,730 years, only half of the original amount of ^{14}C remains; after 2 x 5,730 = 11,460 years, only a quarter remains; and so on. Wait long enough and nearly none remains. Coal, formed from living plants several hundred million years ago, contains virtually no ^{14}C. By these simple calculations, we can determine the age of an artifact from two numbers: the amount of ^{14}C in the artifact at its formation and the amount of ^{14}C in the artifact now.

The second of these numbers can be determined by laboratory analysis. The first, however, presents a greater challenge. The amount of ^{14}C in the artifact at the time of its formation is fixed by the level of ^{14}C in the atmosphere at

1 An early mention of the method appears in the journal literature in brief closing remarks in Anderson et al. (1947).

that time. The isotope ^{14}C occurs in atmospheric carbon in roughly the ratio of 1 atom of ^{14}C to 10^{12} atoms of ^{12}C.[2] Although atmospheric ^{14}C is decaying with the half-life of 5,730 years, the atmospheric levels are maintained at roughly constant levels through a process that creates new ^{14}C atoms. Cosmic rays strike nitrogen atoms in the atmosphere and convert them to ^{14}C atoms. Since the rate of replenishment rises and falls with the intensity of the cosmic rays impinging on the atmosphere, there is a corresponding movement in the levels of ^{14}C. The ratio of 1 to 10^{12} is a rough estimate of a ratio that varies over time. Many other processes affect this ratio. Some have a large effect. The ratio dropped significantly after 1880 because of the large amounts of carbon-based fossil fuels burned in the industrial revolution. The ^{14}C in the atmosphere was diluted by essentially ^{14}C free carbon from the fossil fuels. This and other factors have sufficiently disrupted the rate of replenishment that radiocarbon dating of artifacts is practicable only to artifacts older than 300 years.[3]

3. The Need for Calibration

For artifacts older than 300 years, the variability in the atmospheric ^{14}C levels and other factors lead to incorrect dating, commonly an underestimate of the age of the artifact. In the early years of radiocarbon dating, when there were fewer means available to check it, a thorough analysis of the errors was not possible. Arnold and Libby (1951, 111) collected eighteen months of radiocarbon dating in a report presented as "an overall-check of the method which was the main purpose of the research." As a part of these efforts, they presented the historically known and radiocarbon ages of samples from ancient Egypt (wooden beams from tombs, wood from a funerary ship, wood from a mummiform coffin, ancient wheat and barley grains). They reported the radiocarbon ages of samples from many other locations but generally without historically determined ages.

By the 1960s, discrepancies between the radiocarbon and true dates of historical artifacts were becoming apparent. Stuiver and Suess (1966) reported on the accumulation of evidence of the discrepancies. The relationship

2 As cited by Key (2001, 2338).

3 These other effects include seventeenth-century rapid changes in solar magnetic intensity and the artificial production of ^{14}C as a result of atmospheric testing in the twentieth century. For more details and more general background, see Taylor (1997, 69).

between the two ages, they stressed, depends on so many potentially variable factors that it requires an approach other than the theoretical analysis that then gave radiocarbon ages:

> This relationship cannot be determined theoretically, but can be derived empirically by determination of the radiocarbon contents of samples of known age. (534)

They reported the existence of samples of known age from old wood whose age could be determined by the counting of tree rings. They expressed high hopes for samples that would soon be available of bristlecone pine wood more that 6,000 years old. These samples did meet their expectations and now play a central role in determining the relationship that they sought.

The corrections needed came to be summarized in calibration curves that map the radiocarbon age of a sample against that sample's true calendar age. The term "radiocarbon age" is precisely defined in the radiocarbon dating literature. It designates the age indicated by depletion of ^{14}C in the artifact if we make a series of convenient but false stipulations. They include the assumption of the constancy of reservoir ^{14}C levels, an incorrect but formerly used half-life of 5,568 years, the counting of time from 1950 AD as the zero point, and more.[4] Recent calibration data and curves have been provided by Reimer et al. (2013). Figure 10.1 is a calibration curve plotted from their data for samples created in the northern hemisphere.

4 For more details, see Taylor (1997, 67–68).

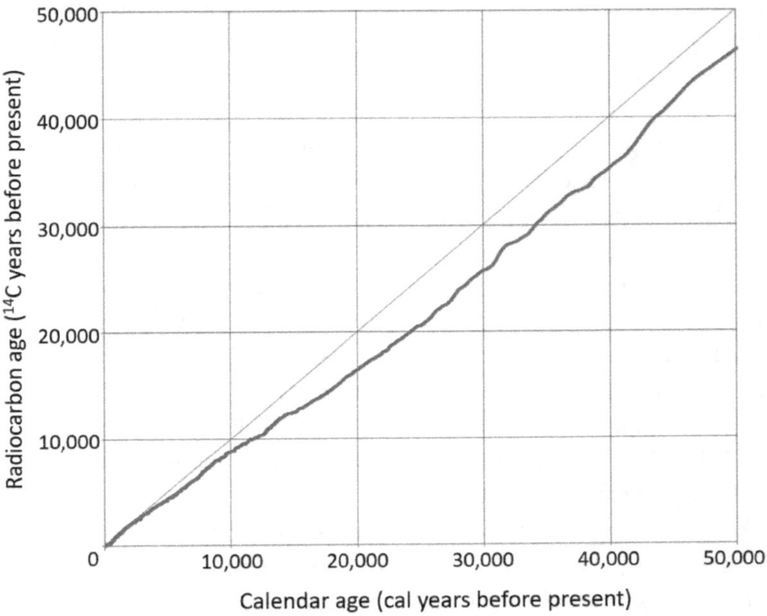

Figure 10.1. Northern hemisphere calibration curve, IntCal13; from data in Reimer et al. (2013) and reproduced in conformity with a Creative Commons CC BY-SA 3.0 license granted by the copyright holder at https://en.wikipedia.org/wiki/File:Intcal_13_calibration_curve.png

The curve shows that radiocarbon age might underestimate the true calendar age by as much as 20%. Once the curve has been used to correct the radiocarbon age, I call the new age the "recalibrated radiocarbon age."

4. Relations of Evidential Support

The relations of evidential support to be considered here are between two types of propositions:

> *H*: The historically determined age of a designated sample is the true age.
>
> *R*: The recalibrated radiocarbon age of a designated sample is the true age.

Here "historically determined" indicates that dating was carried out by the traditional methods of history, archaeology, and dendrochronology (tree ring dating), excluding radiocarbon methods.

So far, we have seen that propositions of type *H* are used to give evidential support to propositions of type *R*. Indeed, propositions of type *H* are used to construct the calibration curves that recalibrate the propositions of type *R*. Thus, they provide the evidential support for the correctness of the recalibrated ages.

However, the relations of evidential support can be reversed. Propositions of type *R* can support those of type *H*. We might become uncertain about the dating ascribed to some sample in a proposition of type *H*. Perhaps we might become unsure of the archaeological dating of 4,650 +/– 75 years of the acacia wood beam from the tomb of Zoser at Sakkara, listed in Arnold and Libby (1951, 111). We can use the recalibrated radiocarbon dating of samples from it to reaffirm its archaeological dating.

An interesting, concrete example of the crossing over of relations of support between the two types of propositions is provided by the radiocarbon dating of the shroud of Turin. As most people know, the shroud bears front and rear impressions of someone with injuries compatible with crucifixion. It is purported to be the burial shroud of Jesus. However, it did not appear on public display until the 1350s. In a careful series of tests reported in Damon et al. (1989), samples of the shroud were sent to three laboratories. In a failed effort to blind the tests, three control samples were also sent to each laboratory. The results showed agreement among the three laboratories for dating of all the samples. They concluded with 95% confidence that the linen of the shroud was created from flax grown sometime between 1260 and 1390 AD.

The crossing over of relations of inductive support arose in the context of the three control samples:

> *Sample 2*. Linen from a tomb excavated at Qasr Ibrîm. Dated by embroidery pattern and Christian ink inscription to the eleventh and twelfth centuries.

Sample 3. Linen from an early-second-century AD mummy of Cleopatra from Thebes. Radiocarbon dated to 110 BC–75 AD at 68% confidence.[5]

Sample 4. Threads from the cope of St. Louis d'Anjou. Dated by stylistic and historical evidence to 1290–1310 AD.

These three samples are dated by *H*-type propositions and then by *R*-type propositions from the three independent laboratories. Since the dating of all of the samples agrees in both types of propositions, we can read the relations of support in each case as passing in both directions.

The intended direction of the calibration of the laboratories is that the *H*-proposition dating of the samples provides evidential support for the laboratories' *R*-proposition dating. However, we can choose equally to read the evidential support as proceeding in the opposite direction: if there was any doubt about the dating of the three control samples, then their radiocarbon dating by the three independent laboratories affirms their correctness. That is, the *R* propositions provide evidential support for the *H* propositions.

REFERENCES

Anderson, Ernest C., et al. 1947. "Radiocarbon from Cosmic Radiation." *Science*, May 30, 576–77.

Arnold, James R., and Willard F. Libby. 1951. "Radiocarbon Dates." *Science* 113: 111–20.

Bokulich, Alisa. 2020. "Calibration, Coherence, and Consilience in Radiometric Measures of Geologic Time." *Philosophy of Science* 87, no. 3: 425–56. https://doi.org/10.1086/708690.

Damon, Paul E., et al. 1989. "Radiocarbon Dating of the Shroud of Turin." *Nature* 337, no. 6208: 611–15.

Key, Robert. 2001. "Radiocarbon." In *Encyclopedia of Ocean Sciences*, edited by John H. Steele, S.A. Thorpe, and Karl K. Turekian, 2338–53. Cambridge, MA: Academic Press.

5 Damon et al. (1989) indicate only a radiocarbon dating for the sample. However, the description of the artifact at the British Museum also indicates a historical dating. The mummified body is of "Cleopatra, daughter of Candace, a member of the family of Cornelius Pollius, Archon of Thebes" (https://www.britishmuseum.org/collection/object/Y_EA6707) and is elsewhere identified as "…daughter of Candace, a member of the Cornelius Pollius family, the Archon of Thebes, under the rule of Emperor Trajan" (https://egypt-museum.com/mummy-of-cleopatra/). Trajan's reign from 53 to 98 AD overlaps with the interval provided by radiocarbon dating.

Reimer, Paul J., et al. 2013. “IntCal13 and Marine13 Radiocarbon Age Calibration Curves 0–50,000 Years Cal BP.” *Radiocarbon* 55: 1869–87.

Stuiver, Minze, and Hans E. Suess. 1966. “On the Relationship between Radiocarbon Dates and True Sample Ages.” *Radiocarbon* 8: 534–40.

Taylor, Royal E. 1997. “Radiocarbon Dating.” In *Chronometric Dating in Archaeology*, edited by Royal E. Taylor and Martin J. Aitken, 65–96. New York: Springer.

Wylie, Alison. 2016. “How Archaeological Evidence Bites Back: Strategies for Putting Old Data to Work in New Ways.” *Science, Technology, and Human Values* 42: 203–25.

11

The Determination of Atomic Weights

1. Introduction

A table of the weights of the atoms of the elements of chemistry is commonly on display in high school science classrooms. Figure 11.1 shows an early example of the table, drawn from the work of Dmitri Mendeleev, the chemist most associated with the introduction of the table. We read familiar facts from it. A hydrogen atom has a weight of 1, near enough. An atom of carbon has a weight of 12. An atom of oxygen has a weight of 16. And so on. We then easily compute the weight of a molecule of water, whose composition is specified by the familiar formula H_2O. A water molecule has two atoms of hydrogen and one atom of oxygen. Its weight is 2 x 1 + 16 = 18.

Familiar as these facts are now, they did not spring into our textbooks the moment that Dalton (1808) proposed that ordinary matter consists of atoms of the elements hydrogen, carbon, oxygen, and so on. Rather, these were details that Dalton's theory failed to specify adequately. The omission was no oversight. The evidence that Dalton marshaled for his theory was too weak to pin down the relative weights of his atoms and the molecular formulae of simple substances such as water. These facts were hidden behind an evidential circle. Dalton could not know the correct molecular formulae until he had determined the correct atomic weights. But he could not determine the correct atomic weights until he had found the correct molecular formulae. He had no means adequate to break the evidential circle.

The determination of the weights of his atoms proved to be a recalcitrant problem whose solution required half a century of concerted efforts by chemists. That half century provides us with an illuminating study of a tangle of mutual relations of inductive support. Because of the great complexity of the facts of chemistry with its many elements, we shall see that these relations of support are far more complicated than, in the architectural analogy, two sides of an arch supporting each other. They are closer to the multiplicity of mutual support relations of an intricate vaulted ceiling, such as displayed in Chapter 2. We shall also see that higher-level hypotheses proved to be essential in the efforts to break the circularity that defeated Dalton. The most familiar of them is Avogadro's hypothesis. Its content is now taught to high school students, who memorize it as they did the lines of nursery rhymes. It already merited only a perfunctory statement in the 1911 eleventh edition of the *Encyclopaedia Britannica*, buried in the short entry for "Amedeo Avogadro": ". . . [U]nder the same conditions of temperature and pressure equal volumes of all gases contain the same number of smallest particles or molecules. . . ." In its time, however, it was an adventurous speculation, indulged only cautiously since it allowed chemists to determine atomic weights and molecular formulae. Adopting hypotheses such as Avogadro's incurred an evidential debt. We shall see that this debt was discharged through still more entangled relations of mutual inductive support at the corresponding higher levels of generalization.

Series	Zero Group	Group I	Group II	Group III	Group IV	Group V	Group VI	Group VII	Group VIII
0	x								
1	y	Hydrogen H=1·008							
2	Helium He=4·0	Lithium Li=7·03	Beryllium Be=9·1	Boron B=11·0	Carbon C=12·0	Nitrogen N=14·04	Oxygen O=16·00	Fluorine F=19·0	
3	Neon Ne=19·9	Sodium Na=23·05	Magnesium Mg=24·1	Aluminium Al=27·0	Silicon Si=28·4	Phosphorus P=31·0	Sulphur S=32·06	Chlorine Cl=35·45	
4	Argon Ar=38	Potassium K=39·1	Calcium Ca=40·1	Scandium Sc=44·1	Titanium Ti=48·1	Vanadium V=51·4	Chromium Cr=52·1	Manganese Mn=55·0	Iron Fe=55·9 Cobalt Co=59 Nickel Ni=59 (Cu)
5		Copper Cu=63·6	Zinc Zn=65·4	Gallium Ga=70·0	Germanium Ge=72·3	Arsenic As=75·0	Selenium Se=79	Bromine Br=79·95	
6	Krypton Kr=81·8	Rubidium Rb=85·4	Strontium Sr=87·6	Yttrium Y=89·0	Zirconium Zr=90·6	Niobium Nb=94·0	Molybdenum Mo=96·0	—	Ruthenium Ru=101·7 Rhodium Rh=103·0 Palladium Pd=106·5 (Ag)
7		Silver Ag=107·9	Cadmium Cd=112·4	Indium In=114·0	Tin Sn=119·0	Antimony Sb=120·0	Tellurium Te=127	Iodine I=127	
8	Xenon Xe=128	Cesium Cs=132·9	Barium Ba=137·4	Lanthanum La=139	Cerium Ce=140	—	—		— — — (—)
9		—	—		—	—	—	—	
10	—	—	—	Ytterbium Yb=173	—	Tantalum Ta=183	Tungsten W=184	—	Osmium Os=191 Iridium Ir=193 Platinum Pt=194·9 (Au)
11		Gold Au=197·2	Mercury Hg=200·0	Thallium Tl=204·1	Lead Pb=206·9	Bismuth Bi=208	—	—	
12	—	—	Radium Rd=224	—	Thorium Th=232	—	Uranium U=239		

Figure 11.1. A periodic table of the elements; from Mendeleev (1904, 26)

In Sections 2 and 3 of this chapter, I will review Dalton's "new system" of 1808 and how it is troubled by an evidential circularity in atomic weights and molecular formulae. Such circularities can be broken by an aptly chosen hypothesis. In Section 4, I review Dalton's failed attempt, guided by notions of simplicity, to select such a hypothesis. In Section 5, I review three hypotheses that came to guide work on atomic weights and molecular formulae over the next half century: Avogadro's hypothesis, Dulong and Petit's law of specific heats, and Mitscherlich's law of isomorphism. The ensuing analysis culminated in a celebrated synthesis of the chemical evidence and the support relations among them by Stanislao Cannizzaro (1858). In Sections 6–8, I review the evidential case presented by Cannizzaro. It emphasizes the interconnectedness of the relations at multiple levels. In Section 9, I review another relation of mutual support, this time between two sciences. For the chemists, Avogadro's hypothesis was supported by the equipartition theorem of the new physics of the kinetic theory of gases. For the physicists, the direction of the support was reversed. Finally, in Section 10, I record the transition of Avogadro's hypothesis from a useful speculation to an established rule. Dulong and Petit's law of specific heats was similarly established but with a crucial amendment that quantum effects lead it to fail at low temperatures.

2. Dalton's Atomic Theory

The atomic theory of matter has a venerable history, extending back to antiquity. Although it is easy to praise the early atomists as far-sighted visionaries, struggling to free themselves from the prejudices of their eras, a better assessment is less celebratory. Alan Chalmers (2009) has documented thoroughly how, for most of its life, the atomic theory was highly speculative. It had little empirical grounding and was thus rightly regarded with reserve or suspicion by those who practiced empirical science.

The turning point came in the early nineteenth century with Dalton's (1808) new proposal of a specific atomic constitution for matter in his *New System of Chemical Philosophy*. Curiously, though, his proposal was not the decisive factor in turning atomism from potentially fertile speculation to successful empirical science. The success of his proposal depended essentially on Antoine Lavoisier's work in chemistry a few decades earlier. Before it, just which were the elements of chemistry was unsettled. Was it to be the ancient choice of earth, air, fire, or water? Or was it the *tria prima* of the three principles of mercury, sulfur, and salt of Paracelsus? Or should we follow Boyle and discard the notion of elements entirely? Lavoisier had settled the matter when he collected his table of elements, as presented in his *Elements of Chemistry* ([1789] 1965). There he gave a subset of the familiar modern table of elements (175). It included hydrogen, oxygen, "azote" (nitrogen), sulfur, phosphorus, charcoal (carbon), and many more. Air and water, it was now found, are not elements after all. Air is a mixture of oxygen and azote. Water is a compound of oxygen and hydrogen. Combustion is not the release of phlogiston but the consumption of oxygen.[1]

Prior to Lavoisier's discoveries, an atomic theory had little hope of bridging the gap between specific properties attributed to atoms and the chemical properties of matter seen in the laboratory. One could speculate *ad nauseam* about the properties and behaviors of the most fundamental atoms or (breakable) corpuscles of matter. However, as long as these were atoms or corpuscles of air, water, fire, or earth, recovering the rich repertoire of chemical change then known to the chemists was precluded.

After Lavoisier, the prospects were quite different. Speculate that the simple bodies of Lavoisier's system are constituted of atoms peculiar to each and the pieces fall rapidly and easily into place. Dalton's good fortune was that his was the first prominent attempt at this speculation. Dalton associated a definite atom with each of Lavoisier's elements. The theory of chemical composition then became beautifully simple. The elements form compounds when their atoms combine in simple ratios. One carbon atom combines with one oxygen atom to make "carbonic oxide" (modern carbon monoxide CO). One carbon atom combines with two atoms of oxygen to make "carbonic acid" (modern carbon dioxide CO_2) (Dalton 1808, 215). We now take this simple idea for granted. However, its use with Lavoisier's table of elements is profound; the constancy of proportions in chemical composition is now explained at the atomic level.

Dalton was dependent on Lavoisier's proclamations of which are the elements. This dependence is shown by Dalton's retention of Lavoisier's identification of heat as a material substance. For Dalton, gases, liquids, and solids

1 There are also a few unexpected entries in Lavoisier's ([1789] 1965) table of "simple substances." It includes light and caloric, in which caloric is a material substance comprising heat. A "gas" for Lavoisier is defined as a body fully saturated with caloric (50). The oxygen that he prepared in his laboratory was for him really "oxygen gas" (52), elemental oxygen saturated with caloric.

were all quiescent at the atomic level. He had no kinetic conception of heat as atomic or molecular motion. Rather, the fundamental particles of matter were surrounded by atmospheres of heat. The expansion and contraction of matter with heating and cooling were explained by the addition to or subtraction of the substance of heat in these atmospheres, which would then enlarge or diminish.

3. A Circularity: Atomic Weights and Molecular Formulae

I now turn to the awkwardness that will govern the discussion to follow. Dalton's theory required atoms to combine in simple ratios when forming compounds: 1:1, 1:2, et cetera. However, Dalton had real difficulty in determining just which of those ratios should be for specific compounds. Famously, he decided that water is formed from one atom of hydrogen and one atom of oxygen, so that we would now write its molecular formula as HO rather than the familiar H_2O. This was just one of many molecular formulae that would require subsequent correction. Ammonia, for example, is NH in his account, not the modern NH_3.

I should note, as a matter of historical fidelity, that neither the term "molecular formula" nor the notation "HO" is Dalton's. I use them here for descriptive continuity with later work. Dalton drew circles representing each element and its compounds. The representation from his *New System* shown in Figure 11.2 is much reproduced and has near-iconic status. In it, hydrogen is "simple" 1 and drawn as a circle with a dot. Oxygen is "simple" 4 and drawn as a plain circle. The first "binary" (compound) 21 is water and represented by the two circles, one each for hydrogen and oxygen, side by side.

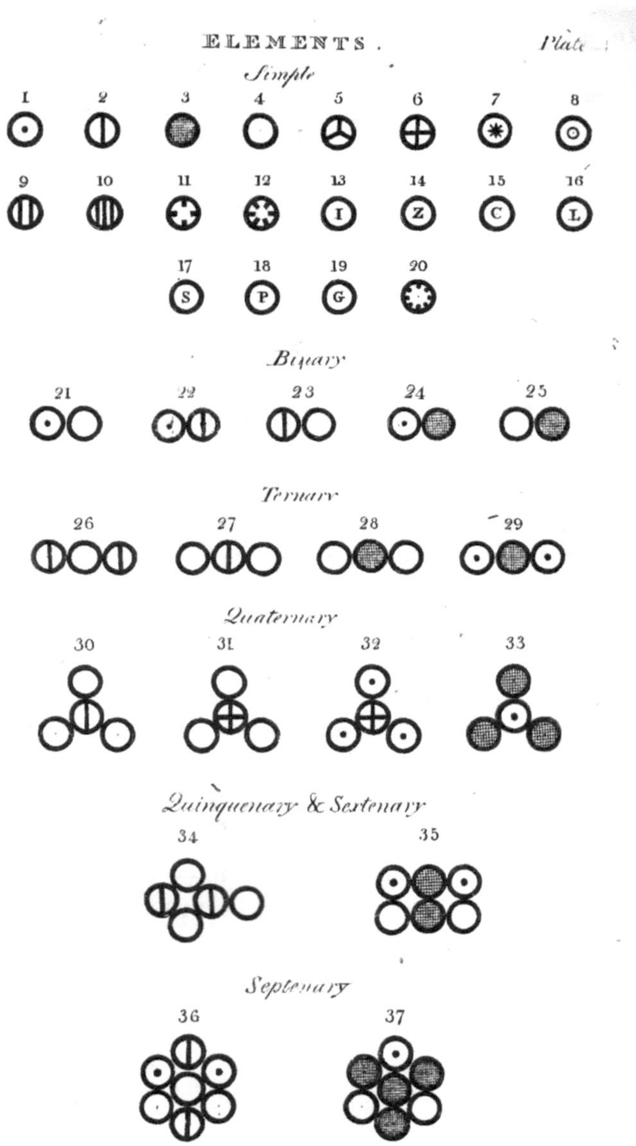

Figure 11.2. Dalton's illustration of the atomic elements and their compounds; from Dalton (1808, Plate 4, near 219)

The misidentification of the molecular formula of water and other compounds lay in no oversight or inattentiveness by Dalton. It lay in a serious incompleteness in his theory. One might know that 1g of hydrogen combines with exactly 8g of oxygen to produce water.[2] But how is one to know that this reaction involves two hydrogen atoms for each oxygen atom? That is, how can one know the correct molecular formula for water from the ratios of weights of the elements in it?

The problem would be solved by knowledge of the ratio of the weights of individual atoms. If we set the atomic weight of a hydrogen atom as the unit, then what would result if an oxygen atom has atomic weight 8? From the fact that 1g of hydrogen combines with 8g of oxygen to make water, we might propose that one atom of hydrogen has combined with one atom of oxygen to make water. That is, we find that water is HO.

However, what if the atomic weight of oxygen is really 16? Then, from the fact that 1g of hydrogen combines with 8g of oxygen to make water, we might propose that water forms by combining two atoms of hydrogen with one atom of oxygen. That is, water is H_2O. These possibilities can be multiplied indefinitely, and Table 11.1 shows some of them.

Table 11.1. Underdetermination of molecular formulae by combining weights

Combining weights to make water	Atomic weights	Molecular formula for water*
1g hydrogen : 8g oxygen	hydrogen = 1; oxygen = 1	HO_8
1g hydrogen : 8g oxygen	hydrogen = 1; oxygen = 2	HO_4
1g hydrogen : 8g oxygen	hydrogen = 1; oxygen = 4	HO_2
1g hydrogen : 8g oxygen	hydrogen = 1; oxygen = 8	HO
1g hydrogen : 8g oxygen	hydrogen = 1; oxygen = 16	H_2O
1g hydrogen : 8g oxygen	hydrogen = 1; oxygen = 32	H_4O
1g hydrogen : 8g oxygen	hydrogen = 1; oxygen = 64	H_8O

* More generally, each of these formulae belongs to an infinite class with the same ratio of atoms. If hydrogen has atomic weight 1 and oxygen has atomic weight 8, then the compound molecule could be HO, H_2O_2, H_3O_3, H_4O_4, et cetera.

2 This is the modern figure. Dalton (1808, 215) reports the ratio as "1:7, nearly."

The molecular formula for water is left underdetermined by the observed combining weights. Rather, these weights merely give us an infinite set of possible pairings of component atomic weights and molecular formulae. If we knew one member of the pair, then we would know the other. If we knew the atomic weights, then we would know the molecular formulae. If we knew the molecular formulae, then we would know the atomic weights. There is a tight circularity in these pairings. To know one, we need to know the other. But we cannot know the second unless we already know the first. Because of this circularity, the molecular formula for water and the atomic weights of its constituent atoms remain underdetermined.

4. A Failed Hypothesis of Simplicity

This circularity can be broken by an aptly chosen hypothesis. We shall soon investigate cases of hypotheses that were introduced speculatively and eventually found solid inductive support. They are the success stories. Hypotheses do not always fare well. A clear instance is the hypothesis that Dalton himself introduced to solve the problem of determining "the number of simple elementary particles which constitute one compound particle" (1808, 213) or the correct molecular formulae (to use the more modern expression). He defined compounds as binary, ternary, and so on by equations (213).

1 atom of A + 1 atom of B = 1 atom of C, binary.

1 atom of A + 2 atoms of B = 1 atom of D, ternary.

2 atoms of A + 1 atom of B = 1 atom of E, ternary.

1 atom of A + 3 atoms of B = 1 atom of F, quaternary.

3 atoms of A + 1 atom of B = 1 atom of G, quaternary.

&c., &c.

With these terms in place, Dalton made the elaborate, multi-part hypothesis that would enable him to determine molecular formulae independently of the relative atomic weights.

> The following general rules may be adopted as guides in all our investigations respecting chemical synthesis.

> 1st. When only one combination of two bodies can be obtained, it must be presumed to be a *binary* one, unless some cause appear to the contrary.
>
> 2nd. When two combinations are observed, they must be presumed to be a *binary* and a *ternary.*
>
> 3rd. When three combinations are obtained, we may expect one to be a *binary*, and the other two *ternary.*
>
> 4th. When four combinations are observed, we should expect one *binary*, two *ternary*, and one *quaternary*, &c.
>
> 5th. A *binary* compound should always be specifically heavier than the mere mixture of its two ingredients.
>
> 6th. A *ternary* compound should be specifically heavier than the mixture of a binary and a simple, which would, if combined, constitute it; &c.
>
> 7th. The above rules and observations equally apply, when two bodies, such as C and D, D and E, &c., are combined. (214; Dalton's emphasis)

In the briefest terms, this compound hypothesis amounted to the assertion that one should choose the simplest molecular formula or formulae available. These rules were not entirely arbitrary. They fit comfortably with the mechanical picture that Dalton had developed of how compounds form. (It would take us too far afield for me to explain how.)

For my purposes here, it was a hypothesis nonetheless and introduced provisionally. To remain in chemistry, eventually it had to accrue inductive support. This is a story of failure, not success. It did not find this support. The hypothesis led Dalton to incorrect molecular formulae, such as that water is HO. Thus, it proved to be incompatible with the other hypotheses introduced to determine the molecular formulae. These other hypotheses mutually supported one another and survived into standard chemistry. Dalton's hypothesis did not find support and was discarded.

5. Breaking the Circularity

Dalton was trapped in a circularity. To know the correct molecular formulae, he needed to know the correct, relative atomic weights. Yet to know the correct, relative atomic weights, he needed to know the correct molecular formulae. This circularity presented a serious challenge to chemists in the first half of the nineteenth century. It was broken and decisively so by the efforts of some of the greatest chemists of the era. They found other means for ascertaining molecular formulae or atomic weights. No one of them was decisive, but their accumulated import was.

Following are three of the most important.[3]

5.1. Avogadro's Hypothesis

When compounds form from elements, their weights combine in fixed ratios. One gram of hydrogen combines with exactly 8g of oxygen to produce water. This fact is explained elegantly in Dalton's atomic theory by his supposition that compounds form when elemental atoms combine in simple, whole number ratios.

Gay-Lussac had remarked in a memoir read in 1808 on a second fixed ratio that proved to be just as important. When gaseous elements combine, they also do so in fixed volume ratios.[4] Two volumes of hydrogen (under the same conditions of temperature and pressure) always combine with just one volume of oxygen to make water. An appealing explanation of this fixity of volume ratios is that each of the volumes contains the same number of atoms. We could then read directly from the 2:1 ratio of volumes that water forms when two atoms of hydrogen combine with one atom of oxygen to make water. The circularity is broken. Water is H_2O and not HO.

There is an initial plausibility to the idea. Although atoms of different elements might have different weights, we would be merely supposing that each atom occupies the same space.[5] It is natural to extend the hypothesis

3 They are selected since they play major roles in standard accounts of the determination of atomic weights written around the end of the nineteenth century: Meyer (1888, Part I, 1892), Pattison Muir (1890), and Wurtz (1881).

4 For a convenient compendium of Gay-Lussac's, Dalton's, and Avogadro's writings on the topic, see Dalton, Gay-Lussac, and Avogadro (1893).

5 At this time, prior to the kinetic theory of gases, the discussion proceeded with Dalton's model of gases as quiescent piles of atoms. Each atom was surrounded by a halo of caloric or heat.

to molecules compounded of atoms: a fixed volume of gas or vapor holds the same number of free atoms (if atomic) or molecules (if a molecular compound). However, the hypothesis then runs immediately into serious difficulties. Using modern notation not then in use, we represent the formation of water as

$$2H + O \rightarrow H_2O$$

$$\text{2 vol. hydrogen} + \text{1 vol. oxygen} \rightarrow \text{1 vol. water vapor}$$

This contradicts laboratory observations. Two volumes of hydrogen combine with one volume of oxygen to make *two* volumes of water vapor.

The solution to the puzzle was given by Avogadro (1811).[6] One had to give up the assumption that hydrogen gas and oxygen gas consist simply of free atoms of hydrogen and oxygen. Rather, both gases consist of molecules that, in this case, contain two atoms of hydrogen and two atoms of oxygen.[7] Using modern notation, the formation of water is represented by

$$2H_2 + O_2 \rightarrow 2H_2O$$

$$\text{2 vol. hydrogen} + \text{1 vol. oxygen} \rightarrow \text{2 vol. water vapor}$$

What resulted was a powerful new principle for the determination of molecular formulae. It is given a complete and canonical formulation by Cannizzaro (1858, 1):

> I believe that the progress of science made in these last years has confirmed the hypothesis of Avogadro, of Ampere, and of Dumas on the similar constitution of substances in the gaseous state; that is, that equal volumes of these substances, whether

Heating the gas increased the size of the halo, and that explained why heating a gas leads it to expand.

6 Translated as "Essay on a Manner of Determining the Relative Masses of the Elementary Molecules of Bodies, and the Proportions in Which They Enter into These Compounds" in Dalton, Gay-Lussac, and Avogadro (1893). An editor, "J. W.," remarks in the preface that "the English version of the French original will probably be found more faithful than elegant, especially so in the case of Avogadro's paper, where the French is always clumsy and occasionally obscure."

7 Avogadro's use of the term "molecule" in 1811 did not match modern usage. Avogadro used the term for what we would now label either an atom or a molecule. What we now distinguish as an atom he labeled *molecule élémentaire* ("elementary molecule").

simple or compound, contain an equal number of molecules: not however an equal number of atoms, since the molecules of the different substances, or those of the same substance in its different states, may contain a different number of atoms, whether of the same or of diverse nature.

Powerful as this hypothesis would prove to be, its early history was troubled. It did not gain ready acceptance for decades. Dalton himself had come out early against the hypothesis. An appendix to *Part II* of his *New System* contained a survey of some experiments on the combining volumes of gases. He found the results to contradict Gay-Lussac's claim that gas volumes combine chemically in simple, whole number ratios. Dalton concluded that

> The truth is, I believe, that gases do not unite in equal or exact measures in any one instance; when they appear to do so, it is owing to the inaccuracy of our experiments. . . . (1810, 559)

If Gay-Lussac's claim fails, then so must the stronger hypothesis of Avogadro.

5.2. Dulong and Petit's Law of Specific Heats

Avogadro's hypothesis provided independent access to atomic and molecular weights of gaseous substances. It also indirectly opened access to the atomic weights of nongaseous elements as long as they enter into compounds with elements that elsewhere take the gaseous state. However, the scope of this indirect access is limited.

Dulong and Petit (1819) reported quite a different method of determining the atomic weights of solid elements. In his atomic theory, Dalton had represented solid elements as consisting of quiescent atoms surrounded by halos of caloric (heat). Dulong and Petit reported that Dalton had supposed that the quantity of heat associated with each atom was the same, no matter the element. It would then follow that the atomic heat capacity — the amount of heat needed to raise each atom by one degree of temperature — would be the same for all elements. However, Dulong and Petit continued to note that the results that Dalton had derived from this hypothesis were "so inconsistent with experiment that it is impossible for us not to reject the principle upon which such determinations are founded" (190). They attributed the difficulty to the inaccuracy in data then available to Dalton. They proceeded to show that more careful measurements led to vindication of the law. It is asserted as

The atoms of all simple bodies have exactly the same capacity for heat.

In other words, the atomic heat capacity is the same for all elements.

The expression of the law in measurable quantities was not so simple. We cannot measure the atomic heat capacity directly. What we can measure is the specific heat. It is the heat needed to raise a unit weight (1g) of a body by one degree of temperature. It must be multiplied by the true atomic weight, expressed as grams per atom, to recover the atomic heat capacity.

(specific heat) x (true atomic weight) = (atomic heat capacity)

However, we do not know the atomic weights in grams per atom. All that we know is the relative atomic weights, taking some atom as an arbitrary unit. That is, we have

(relative atomic weight) = (unknown conversion factor) x (true atomic weight)

So the best quantitative expression for the law is that

(specific heat) x (relative atomic weight) = constant

where the constant must come out the same for all elements. Using the best values that they could find for both specific heats and relative atomic weights, Dulong and Petit (1819) proceeded to show that this relation returns the same constant for a list of elements. Table 11.2 shows the data that they reported.

Table 11.2. Dulong and Petit's data

	Specific heats	Relative weights of the atoms*	Products of the weight of each atom by the corresponding capacity
bismuth	0.0288	13.30	0.3830
lead	0.0293	12.95	0.3794
gold	0.0298	12.43	0.3704
platinum	0.0314	11.16	0.3740
tin	0.0514	7.35	0.3779
silver	0.0557	6.75	0.3759
zinc	0.0927	4.03	0.3736
tellurium	0.0912	4.03	0.3675
copper	0.0949	3.957	0.3755
nickel	0.1035	3.69	0.3819
iron	0.1100	3.392	0.3731
cobalt	0.1498	2.46	0.3685
sulfur	0.1880	2.011	0.3780

* The weights are relative to the atomic weight of oxygen. Multiplying them by 16 gives roughly the modern values, except for tellurium and cobalt.

The near constancy of the product in the final column indicates that the relative atomic weights are correct, at least relative to the elements in the table.

This constant is the atomic heat capacity of all atoms but expressed in some arbitrary system of units dependent on the unknown conversion factor mentioned above.

5.3. Mitscherlich's Law of Isomorphism

These two methods seem to have been the most important in breaking the circularity of atomic weights and molecular formulae. Other methods were also brought to bear. Mitscherlich's 1821 "law of isomorphism" is routinely mentioned in contemporary accounts (Meyer 1888, Part I, Section IV; Pattison Muir 1890, 345–47; Wurtz 1881, 55–60). In Mitscherlich's formulation, it asserts that

> Equal numbers of atoms similarly combined exhibit the same crystalline form; identity of crystalline form is independent of

> the chemical nature of the atoms, and is conditioned only by the number and configuration of the atoms. (quoted in Pattison Muir 1890, 345)

The law connects crystalline form with molecular formula so that a similarity of crystalline form suggests a similarity of molecular formula. A celebrated case — mentioned in both Pattison Muir (1890, 346) and Ramsay (1900, 17–18) — is gallium alum. So-called alums are sulfates of two metals. Potassium alum or potash alum, otherwise common alum, is a sulfate of potassium and aluminum. Gallium also forms an alum-like compound of sulfates of gallium and potassium and has a crystalline form similar to common alum. By invoking Mitscherlich's law of isomorphism, one can assume that the gallium merely replaced the potassium in the crystalline structure, and one can then determine gallium's atomic weight.

Despite its virtues, accounts of Mitscherlich's law are notable for their qualifications and warnings of the law's limited scope and fragility. Cannizzaro (1858) did not use it, as far as I can see.

6. The Vaulted Inductive Structure of Atomic Weights and Molecular Formulae

The methods just described are powerful and enable a complete determination of the atomic weights of the elements and thus the correct molecular formulae. Nevertheless, half a century after Dalton proposed his atomic theory, there was still a chaos of competing proposals. The Karlsruhe Congress of 1860 gathered about 140 of the leading chemists of Europe with the purpose of resolving the problem. The events of the congress have become a matter of legend in the history of chemistry.[8] Two years earlier Stanislao Cannizzaro had already published a solution to the problem. Relying heavily on Avogadro's hypothesis, he had successfully pieced together all of the parts of the puzzle and found a consistent set of atomic weights and molecular formulae. He had reported his success to *Il nuovo cimento* (Cannizzaro 1858), in which he sketched how he had led his students through his solution.

8 See Hartley (1966) and Ihde (1961) for accounts.

That set Cannizzaro outside the mainstream of work in chemistry, which remained skeptical of Avogadro's hypothesis.[9] He needed to mount a sustained defense of Avogadro's hypothesis even in 1860 at the Karlsruhe Congress. In spite of his efforts and the earlier publication of his solution, no agreement was reached at the congress. Rather, the decisive moment came at its close when Angelo Pavesi distributed copies of Cannizzaro's paper. When key participants, including Lothar Meyer and Dmitri Mendeleev, later studied Cannizzaro's paper, they were convinced, and his system was established as the standard.

This, at least, is the standard history. Chalmers (2009, Chapter 10) has argued that Cannizzaro's achievement is overrated. What is not acknowledged is his debt to the successes in prior work by organic chemists able to arrive at structural formulae for organic substances. Cannizzaro's methods, however, could yield atomic weights and molecular formulae but not the structural formulae.

My concern here, however, is narrower. It is the inductive structure of the case that Cannizzaro laid out for his values of atomic weights and molecular formulae and its later development. In short, that case exemplifies the massively complex interconnections suggested by the analogy with a vaulted ceiling. In the sections that follow, we shall see just a small portion of these interconnections.

- Section 7 will review relations of mutual support at the level of finest detail: that is, interrelations among the atomic weights and molecular formulae of specific substances.
- Section 8 will review relations of mutual support among the methods used. Specifically, there are relations of mutual support between Avogadro's hypothesis and the law of Dulong and Petit.
- Section 9 will review relations of support at the level of theory. That is, Avogadro's hypothesis in chemistry lends

9 Thorpe (1910, 64–65) recalls the situation:

> By the middle of the nineteenth century the hypothesis of Avogadro was practically forgotten and the law of volumes ignored. The atomic weights of the elements and the system of notation universally employed in England and Germany were based wholly upon equivalents.

support to an analogous hypothesis in statistical physics and conversely.

7. Mutual Support of Atomic Weights and Molecular Formulae

Cannizzaro's (1858) analysis depends heavily on Avogadro's hypothesis and the associated notion that elemental gases have molecular compositions, such as H_2, O_2, et cetera. The hypothesis requires that equal volumes of gases contain the same number of molecules. As a result, the mass density of a gas is directly proportional to the molecular weight of its constituent molecules. This observation provided the starting point for Cannizzaro's analysis. Cannizzaro prepared a large table of the densities of many gases of both elements and compounds. Table 11.3 lists just some of the densities from his large table (9). The units for mass density are selected so that molecular hydrogen gas has a density of 2.

The third column of the table includes further information of great importance. It divides the gas densities of compounds in proportion to the mass ratios of the constituent elements. For example, hydrochloric acid — hydrogen chloride HCl — forms from chlorine and hydrogen in the mass ratio of 35.5:1. Thus, the gas density of 36.5 for hydrochloric acid is broken up as deriving from a density of 35.5 of chlorine and 1 of hydrogen.

Table 11.3. Some of Cannizzaro's gas density data

Substance	Density	Component densities
hydrogen (H_2)	2	2 hydrogen
oxygen (O_2)	32	32 oxygen
chlorine (Cl_2)	71	71 chlorine
bromine (Br_2)	160	160 bromine
iodine (I_2)	254	254 iodine
mercury (Hg)	200	200 mercury
hydrochloric acid (HCl)	36.6	35.5 chlorine + 1 hydrogen
hydrobromic acid (HBr)	81	80 bromine + 1 hydrogen
hydroiodic acid (HI)	128	127 iodine + 1 hydrogen
water (H_2O)	18	16 oxygen + 2 hydrogen
calomel (mercurous chloride HgCl)*	235.5	35.5 chlorine + 200 mercury
corrosive sublimate (mercuric chloride $HgCl_2$)	271	70 chlorine + 200 mercury

* The modern formula is Hg_2Cl_2. However, above 400°C, calomel yields a vapor with the density that Cannizzaro indicated, now understood to result from a mixture of Hg and $HgCl_2$. See Selwood and Preckel (1940).

The table (unlike Cannizzaro's) includes the resulting molecular formulae for ease of reference. It is straightforward to arrive at them. A brief inspection of the table shows that the atomic weights of the elements present are overdetermined as the values of Table 11.4.

Table 11.4. Atomic weights inferred

Element	Atomic weight
hydrogen	1
oxygen	8
chlorine	35.5
bromine	80
iodine	127
mercury	200

To recapitulate Cannizzaro's analysis, recall that Avogadro's hypothesis tells us that the gas density is a surrogate for the molecular weight. Cannizzaro had conveniently chosen the unit for the gas density so that the gas density numerically equals the molecular weight. All that remains now is to find the combination of molecular formulae and atomic weights that returns the gas densities of Table 11.3.

Cannizzaro arrived at these combinations by noting how the component density of each element always appears as a multiple of some smallest unit. This smallest unit is the atomic weight. The simplest case is hydrogen, whose component densities are all multiples of 1. So we infer that the atomic weight of hydrogen is 1. We now read directly from the densities of Table 11.3 that the molecular formulae for hydrochloric, hydrobromic, and hydroiodic acids each have just one hydrogen atom. So their molecular formulae are HCl_x, HBr_y, and HI_z, where *x*, *y*, and *z* are unknown whole numbers. We also see that gaseous hydrogen is composed of molecules of two atoms, H_2. Water also has two atoms of hydrogen, so it is H_2O_w where *w* is some unknown whole number.

Proceeding in this way for the remaining elements completes the entries in Table 11.4 for the atomic weights and justifies the molecular formulae added to Table 11.3. Chlorine's component densities are multiples of 35.5, so that is its atomic weight. Chlorine gas is diatomic, Cl_2, and hydrochloric acid is HCl. Oxygen's component densities are multiples of 16, so that is its atomic weight. And so on.

For my purposes here, the important point is that the results are overdetermined. That means that only a portion of the data is needed to arrive at the full results. For example, the results for the remaining elements would remain the same if we dropped iodine and its compounds from the analysis. It would then follow that, if we reintroduce the data for iodine, the resulting assessment must agree with the earlier results. The atomic weight of hydrogen in iodine compounds must be the same as in water and hydrochloric and hydrobromic acids.

This overdetermination leads to multiple relations of mutual support. It means that we can take some subset of the results and find that it supports other parts of the results, and there is support in the converse direction.[10] For example, take the propositions that hydrogen gas and the halogen gases are diatomic: H_2, Cl_2, Br_2, and I_2. Using Avogadro's hypothesis and the gas density data, we now infer the atomic weights of these elements and from them that the hydro-halogenic acids have monovalent formulae, HCl, HBr, and HI. Or

10 An analogy to the overdetermination of two agreeing eyewitness accounts of some event might make this clearer. Each account provides support for the veracity of the other.

we can reverse the inference. From the monovalent formulae for the acids, we arrive at the diatomic molecular formulae of hydrogen and the halogens. These inferences can be represented as follows.

Hydrogen and the halogens are diatomic.		Hydro-halogenic acids are monovalent.	
gas density data ――――― Hydro-halogenic acids are monovalent.	(Avogadro's hypothesis)	gas density data ――――― Hydrogen and the halogens are diatomic.	(Avogadro's hypothesis)

As before, we can depict these relations of support as an arch (shown in Figure 11.3).

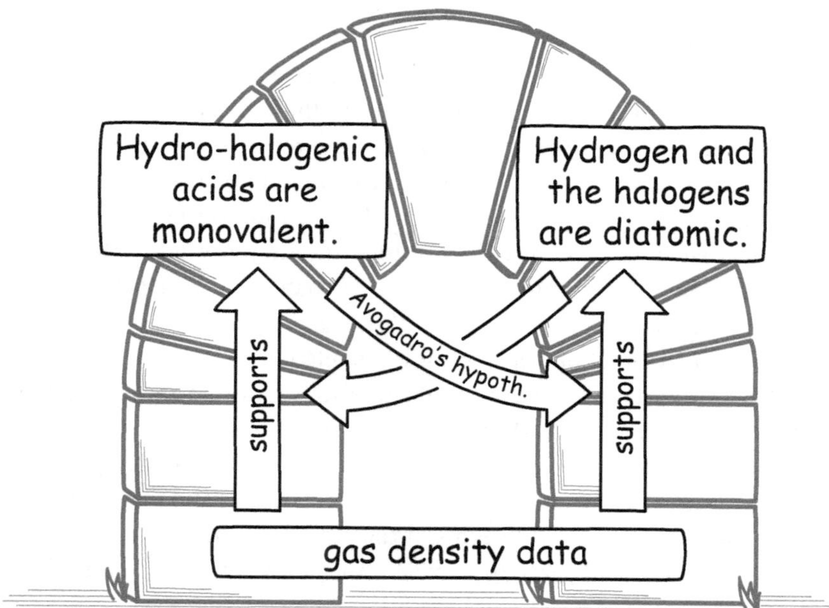

Figure 11.3. Mutual support of molecular formulae

The examples of mutual support are readily multiplied. For example, the diatomic composition of hydrogen and oxygen supports the molecular formula H_2O for water, and that formula supports the diatomic composition of hydrogen and oxygen. That is, we have the following inferences.

Hydrogen and oxygen are diatomic.

gas density data
———————— (Avogadro's hypothesis)
Water is H_2O.

Water is H_2O.

gas density data
———————— (Avogadro's hypothesis)
Hydrogen and oxygen are diatomic.

These further relations of mutual support, and many more of greater complexity, combine to form a vaulted structure of many entangled relations of support.

These two sets of inferences illustrate how hypotheses function at this fine-grained level. Avogadro assumed that hydrogen gas is diatomic as a provisional hypothesis while he pursued his main hypothesis concerning gas density. It followed that water is H_2O. However, the diatomic hypotheses need further support from elsewhere before their provisional status can be discharged. That is now provided by the other inferences concerning the hydro-halogenic acids.

This support for the diatomic hypothesis was already included in Avogadro's original essay. There Avogadro[11] noted the essential fact that hydrochloric acid gas (then still called "muriatic acid gas") is formed by combining unit volumes of hydrogen and chlorine to form *two* volumes of hydrochloric acid gas. This is incompatible with a monatomic constitution for hydrogen and chlorine, for then we have

$$H + Cl \rightarrow HCl$$

1 vol. hydrogen + 1 vol. chlorine → 1 vol. hydrochloric acid gas

11 Avogadro (1811), as translated in Dalton, Gay-Lussac, and Avogadro (1893, 45).

If both hydrogen and chlorine are diatomic, however, then compatibility with the observed volumes is restored:

$$H_2 + Cl_2 \rightarrow 2HCl$$

1 vol. hydrogen + 1 vol. chlorine → 2 vol. hydrochloric acid gas

Hydrogen enters into many more compounds. As the molecular formulae of these further compounds are found, the original hypothesis of the diatomic character of hydrogen receives correspondingly more support. What was a provisional hypothesis initially becomes a fixed part of a much larger network of relations of mutual support. Eventually, the diatomic hypothesis cannot be discarded without also having to discard the full set of atomic weights and molecular formulae developed in modern chemistry.

The density of the relations of mutual support is greater than can be seen in the above analysis. Table 11.3 reports only some of Cannizzaro's density data. His full set is larger, and as a result the number of compounds is still larger,[12] which in turn provides many more relations of mutual support.

8. Mutual Support of Avogadro's Hypothesis and the Law of Dulong and Petit

The inferences of the previous section depend on Avogadro's hypothesis. It is the material fact that warrants them. What grounds do we have for Avogadro's hypothesis? When it was introduced, its support in background theory was meager. His original suggestion was dependent on rather fragile suppositions about the nature of Daltonian atoms: the hypothesis follows from the assumption that the volume of caloric associated with each atom is independent of the type of element.

Cannizzaro had urged much more convincingly that the very success of the inferences of the previous section is already strong support for the hypotheses:

> Now, since all chemical reactions take place between equal volumes, or integral multiples of them, it is possible to express

12 Crudely, if one has n elements, then the number of binary pairings of elements increases as n^2. Although not all pairing will produce a new compound, the possibilities are still growing faster than n.

all chemical reactions by means of the same numerical values and integral coefficients. The law enunciated in the form just indicated is a direct deduction from the facts: but who is not led to assume from this same law that the weights of equal volumes represent the molecular weights, although other proofs are wanting? I thus prefer to substitute in the expression of the law the word molecule instead of volume. (1858, 13)

However, other proofs were not wanting. They could be found both in other parts of Cannizzaro's sketch (as we shall see in this section) and in relations to physical theories of gases (as we shall see in the next section).

His earlier analysis had suggested an atomic weight of 200 for mercury. However, Cannizzaro reported (1858, 22) that an incorrect atomic weight of 100 had been supposed elsewhere. To show the error, he now turned to a second method of determining atomic weights, by means of their elemental specific heats. The method is that of Dulong and Petit (1819), although they are not mentioned by name. To begin, Cannizzaro showed that the atomic weights found earlier for mercury, bromine, and iodine yield the constant atomic heat capacity required by Dulong and Petit. His data and computation are shown in Table 11.5.

Table 11.5. Cannizzaro's specific heat calculations for elements

Substance	Atomic weight	Specific heat	Atomic heat capacity*
solid bromine	80	0.08432	6.74560
iodine	127	0.05412	6.87324
solid mercury	200	0.03241	6.48200

* This atomic heat capacity of roughly 6.8 differs from that of Dulong and Petit (1819) of roughly 0.38 since Cannizzaro's atomic weights are taken in units in which the atomic weight of hydrogen is 1, whereas Dulong and Petit's Table 2 takes the atomic weight of oxygen to be 1. They both measure specific heat with the same units, however.

Cannizzaro (1858, 22–24) then extended the method to compounds. He supposed that the heat capacity of each atom remained the same, even when the atom is in a compound. That meant that the atomic heat capacity of each atom in some molecule was to be calculated by the new formula

$$\frac{\text{specific heat of compound} \times \text{compound molecular weight}}{\text{number of atoms per molecule}} = \text{constant}$$

where the constant was once again the atomic heat capacity in the same system of units as used in Table 11.5.

Using that assumption, Cannizzaro (1858) sought the atomic weight of mercury from the measured heat capacities of four halides of mercury: HgCl, $HgCl_2$, HgI, and HgI_2. Assuming that these were the correct molecular formulae and using the atomic weights already determined, Cannizzaro arrived at the results presented in Table 11.6.

Table 11.6. Cannizzaro's specific heat computation for some mercury halides

Formula	Molecular weight	Specific heat	Number of atoms per molecules	Atomic heat capacity
HgCl	235.5	0.05205	2	6.128872
HgI	327	0.03949	2	6.45661
$HgCl_2$	271	0.06889	3	6.22306
HgI_2	454	0.04197	3	6.35146

Once again the computed atomic heat capacities of the elements in the compounds come out to be almost the same constant. They are also not too distant from the atomic heat capacity for the elements computed in Table 11.5. This affirms the correctness of the formula and atomic weights of Tables 11.5 and 11.6.

For my purposes here, the important point is that the two principal methods employed — Avogadro's hypothesis and the constancy of atomic heat capacity — agree in the atomic weights and molecular formulae that they deliver for the subset of the substances to which they both apply.

Atomic and molecular weights and molecular formulae for mercury, chlorine, and mercury chlorides determined by Avogadro's hypothesis	=	Atomic and molecular weights and molecular formulae for mercury, chlorine, and mercury chlorides determined by atomic specific heats

This agreement is another manifestation of the overdetermination of Cannizzaro's results. However, as before, it can be expressed in terms of relations of mutual support. The correctness of the results delivered by atomic heat capacities for mercury chlorides is supported by the results of applying Avogadro's hypotheses to the same substances. The converse relation of support holds as well. These mutual relations of support can be represented in the arch analogy shown in Figure 11.4.

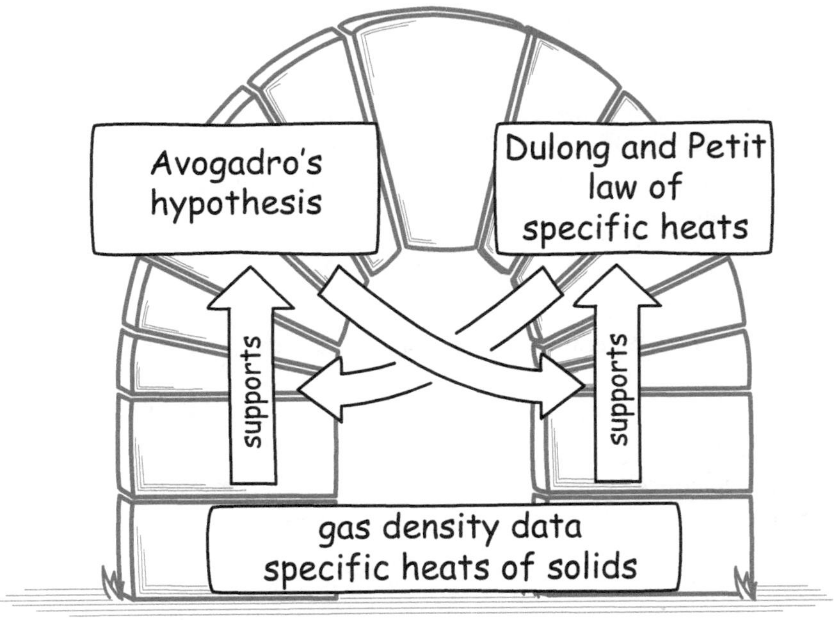

Figure 11.4. Mutual relations of support among Avogadro's hypothesis and Dulong and Petit's law

9. Mutual Support of Avogadro's Hypothesis in Chemistry and the Kinetic Theory of Gases

At the same time as Cannizzaro was using Avogadro's hypothesis to determine the correct atomic weights, a new science was emerging that would provide support for Avogadro's hypothesis. This was the kinetic theory of gases. It was advancing rapidly in the mid-1850s through the work of Krönig (1856), Clausius (1857), and Maxwell (1860). The theory sought to recover the mechanical properties of gases from the assumption that a gas consists of many molecules in rapid motion. In that theory, the pressure exerted by a gas on the walls of a containing vessel results from many collisions of the gas molecules with the wall. The heat energy of the gas corresponds to the kinetic energy of its molecules, and its temperature is proportional to the kinetic energy of each of its molecules.

An early and important achievement of kinetic theory was the recovery of the ideal gas law. According to it, the pressure P exerted by a volume V of gas at temperature T is given by

$$PV = n_m RT = nkT$$

The gas consists of n_m moles, that is, $n = n_m N$ molecules, where N is Avogadro's number, R is the ideal gas constant, k is Boltzmann's constant, and $R = Nk$.

This law already contains Avogadro's hypothesis. To see this, we merely rewrite the law as

$$n = PV/kT$$

It follows immediately that, if two samples of a gas have the same pressure P, volume V, and temperature T, then they contain the same number of molecules n.

It is possible, following Maxwell's later (1871, 295–96) development,[13] to isolate the assumptions used to arrive at Avogadro's hypothesis. First is a purely mechanical result about the pressure P exerted by n molecules of weight m:

$$(2/3)\ P = (1/2)\ nmv_{\text{rms}}^{\ 2}$$

where v_{rms} is the square root of the mean of the squared molecular velocities (rms = "root-mean-square"). Second is a result that Maxwell sought to prove in 1860: if two gases are at thermal equilibrium — that is, at the same temperature — then the mean kinetic energy of their molecules is the same. That is, they agree in the quantity $(1/2)\ mv_{\text{rms}}^{\ 2}$.

These two results are now applied to two volumes of gases of the same pressure, volume, and temperature. Respectively, they consist of n_1 and n_2 molecules, of molecular weight m_1 and m_2, and have rms velocities v_{rms1} and v_{rms2}. The condition of sameness of pressure entails

13 Curiously, Maxwell misattributes the hypothesis as the "Law of Gay-Lussac."

(pressure) $$(1/2)\, n_1 m_1 v_{rms1}^{\ 2} = (1/2)\, n_2 m_2 v_{rms2}^{\ 2} \quad (1)$$

The condition of thermal equilibrium entails that their kinetic energies are equal:

(thermal equilibrium) $$(1/2)\, m_1 v_{rms1}^{\ 2} = (1/2)\, m_2 v_{rms2}^{\ 2} \quad (2)$$

It follows immediately from (1) and (2) that

(Avogadro's hypothesis) $$n_1 = n_2 \quad (3)$$

asserts that the two volumes of gases hold the same number of molecules. I have labeled the three equations so that we can summarize this last inference as

(pressure)

(thermal equilibrium)

(Avogadro's hypothesis)

Needless to say, chemists such as Cannizzaro were delighted with this affirmation of a core assumption of their analysis by physicists, especially given the doubts still prevailing about Avogadro's hypothesis. Cannizzaro (1858, 4) mentions confirmation by Clausius (1857). He was far more buoyant, however, about the significance of this independent support for Avogadro's hypothesis when he gave the Faraday Lecture at the Chemical Society on May 30, 1872:

> . . . [A]t the same time physicists, by considering the constitution of gases under a new point of view, have been brought, independently of chemical considerations, to the supposition of equal numbers of molecules in equal volumes of perfect gases, to which Avogadro and Ampère had previously been led by different modes of interpreting physical phenomena.
>
> Who can fail to see in this long and unconscious march of the science, around and towards a fixed point, the decisive proof of the theory of Avogadro and Ampère? A theory to which we have been led by setting out from different and even opposite points — a theory which has enabled us to forsee several facts which experience has confirmed, must be something more than a mere scientific fiction. It must indeed be either the

actual truth, or the image of that truth, seen through media interposed between our intelligence and the reality. (947–48)

Lothar Meyer was one of the chemists who turned to Cannizzaro's views after the congress in 1860. He also reported with enthusiasm that the physicists had found independent support for Avogadro's hypothesis. In his more popular *Outlines of Theoretical Chemistry* (1892, 32–33), he noted that "this idea of Avogadro has received decisive confirmation as a result of the new development of the mechanical theory of heat." After a qualitative review of how the confirmation arises, he concluded that "this is one of the most powerful arguments in support of Avogadro's hypothesis. Its truth is now no longer disputed."[14]

The chemists were eager to show that Avogadro's hypothesis gains support from the kinetic theory of gases. The physicists, however, were happy to display the relation of support proceeding in the other direction: that is, *from* the chemists' establishment of Avogadro's hypothesis by chemical means *to* key results in the kinetic theory. Since Avogadro's hypothesis in physics had neither the central role nor the controversial history that it had in chemistry, the display of this reverse inference was less prominent in physics. However, it was present.

In its simplest form, it is as follows. The chemists were eager to report that (1) ("pressure") and (2) ("thermal equilibrium") entailed (3) (Avogadro's hypothesis). However, a quick inspection of the algebra relating (1), (2), and (3) shows that (2) could be inferred from (1) and (3). That is,

(pressure)

(Avogadro's hypothesis)

(thermal equilibrium)

This inversion of the chemists' inference was actually the one first reported by Clausius in his paper in 1857 on the kinetic theory of gases. Clausius first

14 Meyer's more technical text (1888, 23) gives more details of the reasoning sketched in equations (1)–(3) and concludes that ". . . Avogadro's hypothesis attains the same degree of probability which the kinetic theory of gases has obtained."

reported Krönig's (1856) derivation of the pressure formula (1) and then continued that

> If we apply this [(1) (pressure)] to simple gases, and assume that, when pressure and temperature are the same, equal volumes contain the same number of atoms — a hypothesis which for other reasons is very probable, — it follows that, in reference to their translatory motion, the atoms of different gases must have the same *vis viva* [kinetic energy]. (Section 11)

One might wonder why Clausius wanted to proceed in this reverse direction. The reason is that the result (2) ("thermal equilibrium") is not easy to attain by purely dynamic arguments concerning the collisions of molecules. Maxwell's (1860) paper offered a demonstration of it in conjunction with his derivation of the Maxwell velocity distribution for the gas molecules.

However, even Maxwell was happy to claim independent support for the results of the kinetic theory of gases from the research of the chemists. In his *Encyclopaedia Britannica* article "Atom," Maxwell (1875, 455–56) reviewed briefly the inference to Avogadro's hypothesis (3) from the assumptions (1) (pressure) and (2) (thermal equilibrium). He then noted that the same hypothesis[15] had been recovered by the chemists in their investigations of chemical combinations. He continued that

> This kind of reasoning, when presented in a proper form and sustained by proper evidence, has a high degree of cogency. But it is purely chemical reasoning; it is not dynamical reasoning. It is founded on chemical experience, not on the laws of motion.
>
> Our definition of a molecule is purely dynamical. A molecule is that minute portion of a substance which moves about as a whole, so that its parts, if it has any, do not part company during the motion of agitation of the gas. The result of the kinetic theory, therefore, is to give us information about the relative masses of molecules considered as moving bodies. The consistency of this information with the deductions

15 Once again misattributed to Gay-Lussac.

> of chemists from the phenomena of combination, greatly strengthens the evidence in favour of the actual existence and motion of gaseous molecules. (456)

These relations of mutual support are made possible by the logical interdependence of the relations (1), (2), and (3). Hence, Andrew Meldrum (1904, 24), adopting a skeptical stance, could review the logic of the demonstration of Avogadro's hypothesis in the kinetic theory and conclude that

> This puts the proof of Avogadro's hypothesis from the kinetic theory of gases in its true light. The hypothesis is but one out of two hypotheses which are contingent on one another. Either granted, the other can be proved.

10. Hypothesis No More

The appeal of Avogadro's hypothesis was that it provided an independent way to determine molecular weights and thereby defeat the circularity that had trapped Dalton. It was introduced provisionally in 1811 and faced what amounted to Dalton's claim of incompatibility with experiment. It languished for decades until Cannizzaro found it to be just the vehicle that he needed to determine the true molecular formulae and atomic weights.

At this point, Avogadro's hypothesis was being used as just the sort of provisional warrant for inference described in Chapter 2. It was indulged because of its great utility. Starting with the ratio of the densities of two gases, the hypothesis warranted an inference to the ratios of their molecular weights. It is the analog of the stone supported by scaffolding while the remaining stones of the arch are put in place.

The provisional status of the hypothesis had to be discharged, however, just as the scaffolding supporting the stones of an arch or vault eventually has to be removed. This burden was taken seriously. We have seen above how support for the hypothesis gradually accrued through the success of the overall project. Its results are overdetermined. That means that a part can become support for another part and conversely. Just this happened with the agreement of the results derived through Avogadro's hypothesis and through Dulong and Petit's law of specific heats. That allowed each to support the other. For Cannizzaro, the derivation of Avogadro's hypothesis from the kinetic theory of gases supplied what he called above the "decisive proof."

As the supports mounted, Avogadro's hypothesis lost its hypothetical character. It became a *rule*, a certainty of textbook chemistry. In his *Theoretical Chemistry from the Standpoint of Avogadro's Rule and Thermodynamics*,[16] Nernst (1904, 39–40) reported that ". . . Avogadro (1811) advanced a hypothesis which, after much opposition, has come to be recognized as an important foundation of molecular physics, as well as of all chemical investigations." Nernst proceeded to list four types of support. The hypothesis explains Gay-Lussac's result about combining volumes. It supplies molecular weights that agree with those derived from purely chemical investigations. It is derived independently from the kinetic theory of gases. It is able to deal successfully with a challenge from abnormal vapor densities.

In this chapter, I have traced the development and use of Avogadro's hypothesis as an illustration of how hypotheses are used in inductive inference in science. A second illustration could be provided by Dulong and Petit's law of specific heats. In brief, it warrants an inference from observed properties (specific heats of solids) to relative atomic weights. The law had a provisional status originally. One serious problem was that the constancy of the atomic heat capacity of the law was found to hold only in certain temperature ranges, notably failing for low temperatures. However, it gained support through its successful application. It also gained support from the new statistical physics that developed from the kinetic theory of gases. In brief, a simple model for a crystalline solid is a lattice of atoms held in place with spring-like forces. Statistical physics entails a constant molar heat capacity for such a system.[17] Perhaps the greatest triumph of the analysis came when Einstein (1907) explained the deviations from constancy of the molar heat capacity at low temperatures as deriving from the quantization of energy.

---

16 The German word is *Regel*: *Theoretische Chemie vom Standpunkte der Avogadro'schen Regel und der Thermodynamik*.

17 Each atom has three translational degrees of freedom and three position degrees of freedom associated with the conservative forces holding it in place in the lattice. The equipartition theorem assigns mean energy $kT/2$ to each degree of freedom so that there is a mean energy $6kT/2 = 3kT$ per molecule or $3RT$ per mole. It follows immediately that the atomic heat capacity is the constant $3k$ and that the molar heat capacity is the constant $3R$.

REFERENCES

Anonymous. 1911. "Amedeo Avogadro." In *Encyclopaedia Britannica*, 11th ed., vol. 3, 66. Cambridge, UK: Cambridge University Press.

Avogadro, Amedeo. 1811. "Essai d'une manière de déterminer les masses relatives des molécules élémentaires des corps, et les proportions selon lesquelles elles entrent dans ces combinaisons." *Journal de physique* 73: 58–76.

Cannizzaro, Stanislao. 1858. "Lettera del Prof. Stanislao Cannizzaro al Prof. S. de Luca; sunto di un corso di filosofia chimica, fatto nella R. Universita' di Genova." *Il nuovo cimento* 7: 321–66. Translated as *Sketch of a Course of Chemical Philosophy*. Alembic Cub Reprints No. 18. Edinburgh: Alembic Club; Chicago: University of Chicago Press, 1911.

———. 1872. "Considerations on Some Points of the Theoretical Teaching of Chemistry." *Journal of the Chemical Society* 20: 941–67.

Chalmers, Alan. 2009. *The Scientist's Atom and the Philosopher's Stone: How Science Succeeded and Philosophy Failed to Gain Knowledge of Atoms*. Dordrecht: Springer.

Clausius, Rudolf. 1857. "Über die Art der Bewegung, die wir Wärme nennen." *Annalen der Physik* 100: 353–79. Translated as "On the Kind of Motion We Call Heat." *Philosophical Magazine* 14 (1857): 108–27.

Dalton, John. 1808. *New System of Chemical Philosophy. Part I*. Manchester: R. Bickerstaff.

———. 1810. *New System of Chemical Philosophy. Part II*. Manchester: R. Bickerstaff.

Dalton, John, Joseph-Louis Gay-Lussac, and Amedeo Avogadro. 1893. *Foundations of the Molecular Theory: Comprising Papers and Extracts by John Dalton, Joseph-Louis Gay-Lussac and Amedeo Avogadro (1808–1811)*. Alembic Club Reprints No. 4. Edinburgh: William F. Clay.

Dulong, Pierre-Louis, and Alexis-Thérèse Petit. 1819. "Recherches sur quelques points importants de la théorie de la chaleur." *Annales de chimie et de physique* 10: 395–413. Translated as "Research on Some Important Aspects of the Theory of Heat." *Annals of Philosophy* 14 (1819): 189–98.

Einstein, Albert. 1907. "Die Plancksche Theorie der Strahlung und die Theorie der Spezifischen Wärme." *Annalen der Physik* 2: 180–90.

Hartley, Harold. 1966. "Stanislao Cannizzaro, F.R.S. (1826–1910), and the First International Chemical Conference at Karlsruhe." *Notes and Records of the Royal Society of London* 21: 56–63.

Ihde, Aaron J. 1961. "The Karlsruhe Congress: A Centennial Retrospective." *Journal of Chemical Education* 38: 83–86.

Krönig, August. 1856. "Grundzüge einer Theorie der Gase." *Annalen der Physik* 99: 315–22.

Lavoisier, Antoine. (1789) 1965. *Elements of Chemistry, in a New Systematic Order, Containing All the Modern Discoveries*. Translated by Robert Kerr. New York: Dover.

Maxwell, James Clerk. 1860. "Illustrations of the Dynamical Theory of Gases." *Philosophical Magazine* 19: 19–32; 20: 21–37.

———. 1871. *Theory of Heat*. London: Longmans, Green, and Company.

———. 1875. "Atom." In *Encyclopaedia Britannica*, 9th ed., vol. 3, 36–49. Reprinted in *The Scientific Papers of James Clerk Maxwell*, vol. 2, 445–84. Cambridge, UK: Cambridge University Press, 1890.

Meldrum, Andrew N. 1904. *Avogadro and Dalton: The Standing in Chemistry of Their Hypotheses*. Aberdeen: n.p.

Mendeleev, Dmitri. 1904. *An Attempt towards a Chemical Conception of the Ether*. Translated by George Kamensky. London: Longmans, Green, and Company.

Meyer, Lothar. 1888. *Modern Theories of Chemistry*. Translated by P. Phillips Bedson and W. Carleton Williams. London: Longmans, Green, and Company.

———. 1892. *Outlines of Theoretical Chemistry*. Translated by P. Phillips Bedson and W. Carleton Williams. London: Longmans, Green, and Company.

Nernst, Walter. 1904. *Theoretical Chemistry from the Standpoint of Avogadro's Rule and Thermodynamics*. 4th ed. Translated by R.A. Lehfeldt. London: Macmillan.

Pattison Muir, M.M. 1890. "Atomic and Molecular Weights." In *Watt's Dictionary of Chemistry*, 2nd ed. vol. 1, edited by M.M. Pattison Muir and H. Foster Morley, 336–60. London: Longmans, Green, and Company.

Ramsay, William. 1900. *Modern Chemistry: Theoretical*. London: J.M. Dent.

Selwood, P.W., and Ralph Preckel. 1940. "The Structure of Mercurous Chloride Vapor." *Journal of the American Chemical Society* 62: 3055–56.

Thorpe, Edward. 1910. *History of Chemistry. Vol. II. From 1850 to 1910*. New York: G.P. Putnam.

Wurtz, Charles Adolphe. 1881. *The Atomic Theory*. Translated by E. Cleminshaw. New York: D. Appleton and Company.

12

The Use of Hypotheses in Determining Distances in Our Planetary System

1. Introduction

How distant from us are our nearest neighbors in space: the Moon, the Sun, and the planets?[1] This basic problem of astronomy proved to be a most challenging one that exercised astronomers from antiquity to as late as the nineteenth century. It provides a revealing case study of how hypotheses are used to extend the otherwise limited inductive reach of evidence.

One might expect that these distances could be determined by simple measurement, much as a terrestrial surveyor can determine the location and height of an inaccessible mountain peak. However, distances even to our closest body, the Moon, are so great that they present formidable challenges. Accurate triangulation of great distances requires extremely accurate angular measurements that were mostly beyond ancient astronomers, except perhaps for the closest body, the Moon. Even then, the ancient astronomers needed to await the opportunities provided by solar and lunar eclipses to break otherwise fatal evidential circles. The difficulty of making precise enough measurements meant that these methods were able to estimate distances only to the Moon and, so some extent, the Sun. These early efforts are described in Sections 2, 3, and 4 below.

The introduction of telescopes to astronomy in the seventeenth century made possible more accurate angular measurements. However, measurements of distance by means of triangulation, or parallax, as it is called in the

1 I thank Bernard Goldstein for helpful comments on an earlier draft.

astronomical literature, were limited at best to our closest planets, Mars and Venus. In Section 5, I recount the seventeenth-century measurement of the parallax of Mars, and in Section 6 I recount the eighteenth-century observations of the transits of Venus across the face of the Sun.

We find from all of these efforts that triangulation by itself is unable to provide much. This remains true even with careful telescopic measurements and a willingness to sail to distant parts of the globe to make them. The eighteenth-century measurements of the transits of Venus, by themselves, gave only angular displacements. Something more was needed if they were to deliver the distances to Venus and the Sun.

That essential extra was provided by hypotheses about the configuration of these celestial bodies. These hypotheses could extend the inductive reach of the few measurements available, and determinations of the distances to all of the celestial bodies mentioned became possible. This approach had been used from the first moments of ancient Greek astronomy and remained the primary approach used to the end of the nineteenth century. In the following sections, I review three different types of hypotheses used: Pythagorean and Platonic harmonies (Section 8), Ptolemy's *Planetary Hypotheses* (Section 9), and Copernicus' hypothesis of a heliocentric planetary system (Sections 10 and 11).

Examination of these three different hypothetical supplements gives us an opportunity to see how the hypotheses were used and should be used. The use of a hypothetical supplement takes on an evidential debt that must be discharged by finding independent evidence for the hypotheses. Only then have the results of the investigation been given proper inductive support. The need to discharge this debt is underscored by the fact that each hypothetical supplement reviewed leads to a different system of distances. Further evidence for the harmonic and Ptolemaic hypotheses was not secured, and they were discarded. The Copernican hypothesis, however, accrued considerable support. The most important was Newton's discovery of a mechanics that gave a dynamical foundation for the motions hypothesized in heliocentric astronomy.

What resulted was the edifice of classical mechanics. It combined astronomy and celestial and terrestrial mechanics in a single system, in which each part provided evidential support for the others. This crossing over of relations of inductive support is illustrated in the particular case of Kepler's third law and the inverse square law of gravity. Each, as I show in Section 12, provides inductive support for the other.

This reliance on hypotheses to enable the determination of distances within the planetary system persisted up to the nineteenth century, the latest extent of the history reviewed here. With the twentieth century, direct measurements of distances to celestial bodies became possible through laser and radar ranging.

2. An Evidential Circle: The Distances and Sizes of the Moon and Sun

How distant from us is the Moon? To appreciate just how formidable a question it was for ancient astronomers to answer, consider the majestic splendor of a full Moon rising over the eastern horizon at sunset. It is easy to imagine that the Moon is small and rises from a nearby place just over the horizon. That misapprehension is soon dispelled.[2] A house on a distant hill becomes larger as we approach it. But the Moon does not. No matter how far east we might venture, we see the Moon of the same size rising. Our eastward travels, from horizon to horizon, do not perceptibly diminish the distance to the Moon. We then realize that it is much more distant than we first thought. That means that it must be much larger than we first thought. How much larger is it?

That question leads to the first evidential circle. The disk of the full Moon fills about half a degree in our visual field. If we knew the size of the Moon, then we could calculate its distance by simple geometry. But if it were two, three, or four times larger, then it must be two, three, or four times more distant. As Figure 12.1 shows, many pairs of distances and sizes yield the same angular size in our visual field of half a degree. We cannot know the distance to the Moon until we know its size. But we cannot know its size until we know its distance.

2 This misapprehension is compounded by the "Moon illusion," in which it appears larger when near the horizon, although its measurable angular size is unchanged.

Figure 12.1. Many size-distance pairs for the Moon yield half a degree of angular size

All that these simple observations tell us is that the distance to the Moon must be large but otherwise leave it undetermined.

What of the relative distances of the Sun and Moon? We observe that the Sun has about the same angular size as the Moon of about half a degree. This equality is most easily learned from eclipses of the Sun. Then the Moon aligns with the Sun and almost perfectly obstructs it. Sometimes the Moon blocks out the Sun completely. Sometimes there is an "annular" eclipse in which the Moon blocks out the Sun, except for a thin annular ring of the Sun's surface encircling the Moon.

That the Moon eclipses the Sun shows that the Moon must be closer to us than the Sun. Are they roughly the same distance from us? If they are the same size, then they must be roughly the same distance from us. But if the Sun is two, three, or four times larger than the Moon, then by simple geometry the Sun must be two, three, or four times more distant from us than the Moon. As Figure 12.2 shows, we cannot know which until we know the true size ratio of the Sun to the Moon. But we cannot know that ratio until we know the ratio of the distances. We are trapped once again in an evidential circle.

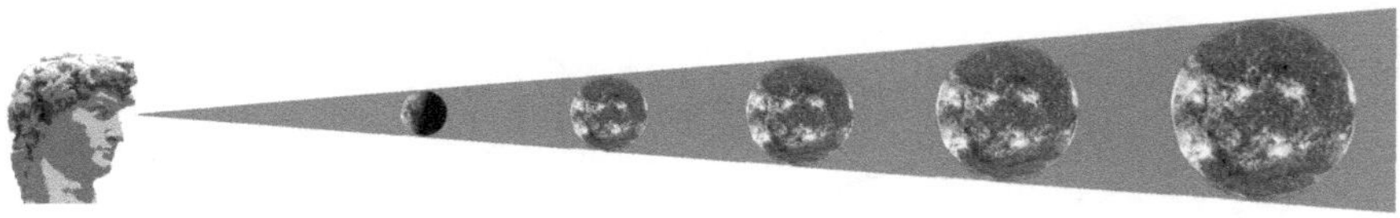

Figure 12.2. Many possible ratios of distances to the Sun and Moon

3. Aristarchus: Breaking the Evidential Circles

Both circles can be broken if we expand the evidence considered and are ingenious enough to do it in just the right way. This was the principal content of a remarkable document authored by Aristarchus of Samos, who lived roughly from 310 to 230 BCE. The work is presented in Greek and English translation in Heath (1913) under the title "Aristarchus on the Sizes and Distances of the Sun and Moon." Aristarchus breaks the evidential circle with two expansions of the evidence brought to bear. First, he introduces the angular positions of the Sun and Moon when the latter is precisely half illuminated: that is, at "dichotomy." Second, he introduces the behavior of the Moon during an eclipse of it, when it passes through the Earth's shadow.

When the Moon is exactly half full, the Sun, Earth, and Moon form a right-angled triangle, with the right angle at the Moon. The angle at the Earth is recoverable as the observable angular separation of the Sun and Moon. The shape of the triangle, shown in Figure 12.3, is thereby fixed, and the ratio of the Earth-Sun to Earth-Moon distance can be read from it.

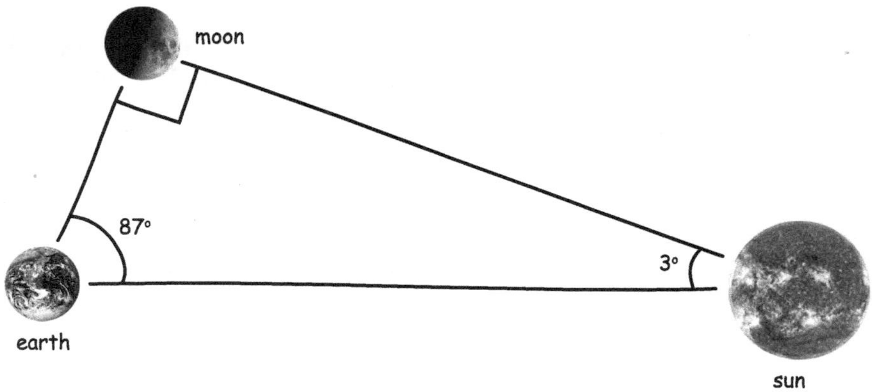

Figure 12.3. The Earth, Moon, and Sun at lunar dichotomy (not drawn to scale)

All that is needed is the angular separation of the Sun and Moon at dichotomy, as seen from the Earth. That is provided by the fourth of six hypotheses announced by Aristarchus (as given in Heath 1913, 353):

> That, when the moon appears to us halved, its distance from the sun is then less than a quadrant by one-thirtieth of a quadrant.

Since a quadrant is 90°, Aristarchus reports here that the angular separation of the Sun and Moon is 87°. After some analysis, he arrives at a ratio for the Earth-Sun to Earth-Moon distance that lies between 18:1 and 20:1.[3]

The method is ingenious and correct. However, it required the unattainable: an accurate measurement of the angular separation of the Sun and Moon at the moment of dichotomy. Aristarchus greatly underestimated the true ratio of 389:1.[4]

As far as the ratios of distances were concerned, Aristarchus had broken the evidential circle. He had established, he believed, the ratio of distances

---

3 Aristarchus did not have tables of tangents to consult, which now makes our computation trivial. The exact result is tan 87 = 19.08.

4 Dreyer (1953, 136) diagnoses the error as follows: "The method, though theoretically correct, is not practical, as the moment when the moon is half illuminated cannot be determined accurately. The angle of 'dichotomy' is in reality 89° 50' instead of 87°."

to the Sun and Moon. He could then infer directly to the ratio of the diameters of the Sun and Moon. It must be the same. It must also lie between 18:1 and 20:1.

Aristarchus then turned to determine not just the ratios of the distances to the Sun and Moon but also their individual values. They were expressed as ratios with the diameter of the Earth, whose value was then known well enough. Heath (1913, 399) presumes that Aristarchus did as Archimedes did and accepted Dicaearchus' estimate of a circumference of 300,000 stades. Eratosthenes' famous measurement of the Earth's size came later. Aristarchus realized that these individual distances could be recovered from phenomena observable at the time of an eclipse of the Moon. To determine these individual distances, he introduced a decisive new datum concerning an eclipse of the Moon (Heath 1913, 353):

> That the breadth of the [Earth's] shadow is [that] of two moons.

That is, as the Moon passes through the umbra, the conically shaped, full shadow of the Earth, the Moon's diameter is just half that of the umbra, as Figure 12.4 shows. What results is a complicated geometric figure that has been reproduced in many old manuscripts and modern treatises and is not drawn to scale in Figure 12.4. The lower figure depicts the essential geometry and is reproduced from Heath's (1913, 330) analysis.[5]

5 This work is in the public domain.

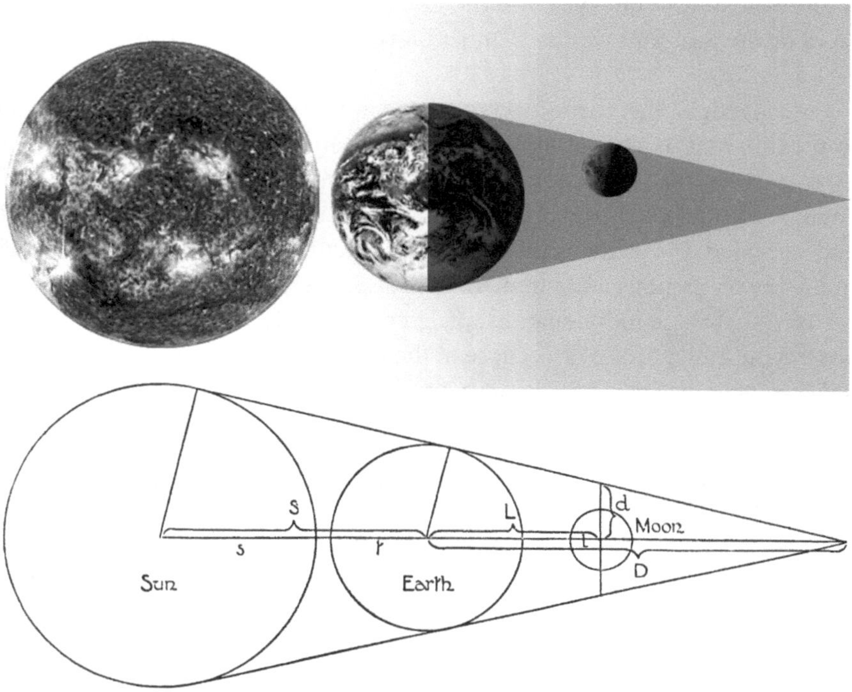

Figure 12.4. Aristarchus' figure for determining the distances to the Sun and Moon

Since the ratio of the diameters of the Sun *s* and Moon *l* are known, as are the ratios of the distances from the Earth to Sun *S* and Earth to Moon *L*, it turns out that the geometry of Aristarchus' figure is fixed. This might not be obvious from inspection of the figure, and it takes some calculations to determine it. Since they are tedious and not especially illuminating, I refer the reader to Heath's (1913, 330–31) reconstruction. Aristarchus arrived at a diameter for the Sun as a ratio of the Earth's diameter that lies between 19/3 and 43/6 and a diameter for the Moon as a ratio of the Earth's diameter that lies between 19/60 and 43/108. Once again, with the diameters of the Sun and Moon determined, it was a simple matter to determine the distances to the Sun and Moon from the known angular size of each as seen from the Earth.

The actual numbers reported by Aristarchus are quite far from the actual ratios in our Sun-Moon-Earth system.[6] His calculations depended on his earlier underestimate of the ratio *S/L* of the Earth-Sun and Moon-Sun distances. They were compounded by his taking erroneously that the angular size of the Moon is 2°, whereas Archimedes in the *Sand-Reckoner* had attributed the correct 1/2° to Aristarchus.[7]

Although Aristarchus' final numbers differ greatly from the actual values, his methods were correct and ingenious, marred only by the need for an impractical datum and a curious error in estimating the Moon's size. Van Helden (1985, 7) singles out Aristarchus' second Moon eclipse technique as

> . . . a method that, when fully developed by Hipparchus and Ptolemy, was to be the centerpiece of all determinations of absolute celestial distances until the seventeenth century.

4. Measurements of Parallax

The methods reviewed so far require that the disks of the Sun and Moon be discernible. As long as astronomers use only naked-eye methods, they cannot determine distances to the planets, for optical instruments are needed to resolve their disks. There is a general method that, in principle, is capable of determining the distance to any celestial object visible from the Earth. That is the measurement of its parallax. It is the difference in direction of some object as seen from different places on the Earth. Measuring it requires observations to be taken at two different places at the same time. For the case of a rotating Earth, parallax can also be measured from one position on the Earth when the rotation moves that position to another location in space.

Horizontal parallax uses the Earth's radius as the baseline for measurement.[8] Figure 12.5 shows an observer at *A* on the Earth's surface who finds the object at *P* to be at its zenith: that is, directly overhead. A second observer at *B*, located at a distance of one-quarter of the Earth's circumference, finds

6 Aristarchus' ratio for the Sun is 6.3 to 7.2, where the modern figure is 109. His ratio for the Moon is 0.31 to 0.40, where the modern figure is 0.27.

7 See Heath (1913, 311–14) for an analysis of this curious error.

8 It is distinguished from annual parallax, in which the radius of the Earth's annual orbit around the Sun is used as the baseline for measurement. It is applied when considering distances to stars.

the object to have just dipped below the horizon. If we draw *BC* parallel to *AP*, then the bearings of the object at *P* differ for these two observers by the angle of parallax, *CBP*. This angle is equal to the angle *BPA*, the angle subtended by the Earth's radius from *P*.

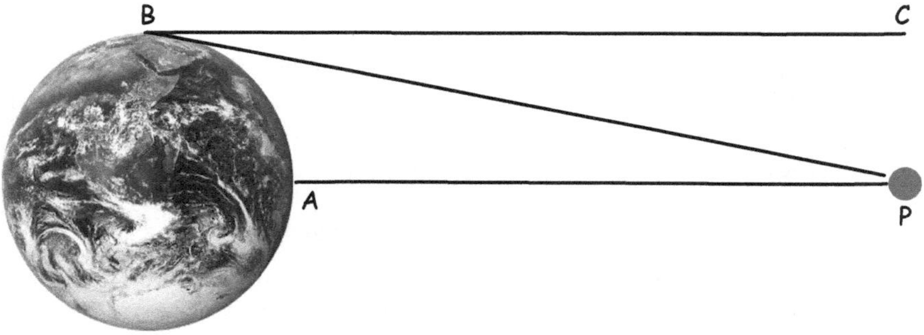

Figure 12.5. Horizontal parallax

This angle is called "horizontal parallax" since the name reflects *B*'s observing *P* on the horizon. For a distant object, the angle is small[9] and related inversely to the distance to the object by

distance = radius of the Earth / horizontal parallax in radians

In practice, horizontal parallax is not measured directly. A smaller displacement on the Earth's surface is used and horizontal parallax inferred from it.

Once again, eclipses provided opportunities for potentially informative measurements. An eclipse of the Sun will be total when seen from one part of the Earth's surface yet only partial when seen from another part. Encoded in this difference is a measure of the parallax of the Moon. Hipparchus and Ptolemy after him applied this approach to records of lunar eclipses to estimate lunar parallax.[10] Although the method is correct in principle, its successful application is difficult because of the need to measure angles precisely. The Moon's parallax of about 1° is the largest for celestial objects. Others are

9 The figure greatly exaggerates the angle. For the Moon, the horizontal parallax varies about roughly a degree. For the Sun, it is about 8.8 seconds of arc: that is, 2.4 thousandths of a degree.

10 For details, see Van Helden (1985, 10–19).

dauntingly smaller. Measurement of the tiny solar parallax of 8.8 seconds of arc was beyond the reach of the ancient astronomers.

5. The Parallax of Mars

The difficulty of measuring tiny parallactic angles was only overcome centuries later when telescopic observations were possible. Even then, the approach was indirect. The Earth-Sun distance was the most sought once heliocentric Copernican astronomy became established. It determined, as we shall see below, all of the other distances. However, direct measurement of the parallax of the Sun remained beyond the astronomers' reach, if only because the brilliance of the Sun precluded direct observations locating it against the stellar background. Instead, it proved to be feasible to determine the parallax of Mars and, using the known ratio of sizes of the orbits of Mars and the Earth, then compute the Earth-Sun distance.

The best known of these measurements of parallax from the seventeenth century is Cassini and Richer's measurement of the parallax of Mars in 1672 using simultaneous measurements of the position of Mars from France and Cayenne in South America. The opportunity for the measurements was an opposition of Mars to the Sun. That meant that Mars was making one of its closest approaches to the Earth and thus susceptible to the most accurate measurements. Their efforts yielded the parallax of Mars at this time in its orbit and thus its distance from the Earth. Using the then known ratio of the sizes of the orbits of the Earth and Mars, the crucial Earth-Sun distance could be estimated. Cassini and Richer arrived at an Earth-Sun distance of 87,000,000 miles, comfortably close to the modern value of about 93,000,000 miles.[11] Both Berry (1898, 205–09) and Van Helden (1985, Chapter 12) emphasize that the closeness of these numbers is less impressive once one recognizes the large margin of error associated with the Cassini and Richer result.

6. The Transits of Venus

The ancient astronomers had found solar eclipses to afford opportunities to determine the parallax of the Moon. These eclipses arise when the Moon passes exactly between the Earth and the Sun. An analogous circumstance

11 See Long (1742, 290, 292) for an early account.

arises when the planet Venus passes exactly between the Earth and the Sun. Since Venus is so much farther away from the Earth than the Moon, the effect is much less dramatic. Venus appears telescopically as a tiny dot migrating over the surface of the Sun. If this "transit of Venus" is observed from different locations on the Earth's surface, then Venus will be seen to transit across the disk of the Sun in different locations on the disk.

The path of Venus traces a chord across the circular disk of the sun. Determining the length of the chord fixes its location on the disk. The longest chords are diameters of the circle; the shorter the chord, the farther it is from a diameter. The most accurate way to estimate the difference in chord lengths was to time how long the transit took, when viewed from different locations. Since a transit requires about six hours, accurate times of transit were well within the grasp of measurement of early clocks. The transit times reflect directly the chord lengths and thus reveal differences of location of the transits against the Sun's disk.

Observing a transit of Venus from different places on the Earth enabled the parallax of Venus and the Sun to be determined. Of the expeditions to observe the transit of Venus, the best known, especially to Australians, is that of Captain Cook, who sailed to Tahiti for this purpose in 1769. The measurements of the Cook expedition were compared with those taken in other locations, notably Lapland. The resulting parallax of the Sun was determined to be in the range of eight to nine seconds of arc, in agreement with the modern value of about 8.8 seconds of arc.[12] Subsequent transits were observed in 1874, 1882, and more recently in 2004 and 2012.

The calculation of the parallax of Venus and the Sun from these observations must correct for many factors. The highly simplified analysis in Figure 12.6 brings out the element most important for my purposes here.

---

12 For accounts of the transits, associated measurements, and calculations, see Airy (1881, 144–60) and Newcomb (1892, 177–92). That these expeditions and measurements were of considerable popular interest in the nineteenth century is suggested by the publication of popular works such as Forbes (1874).

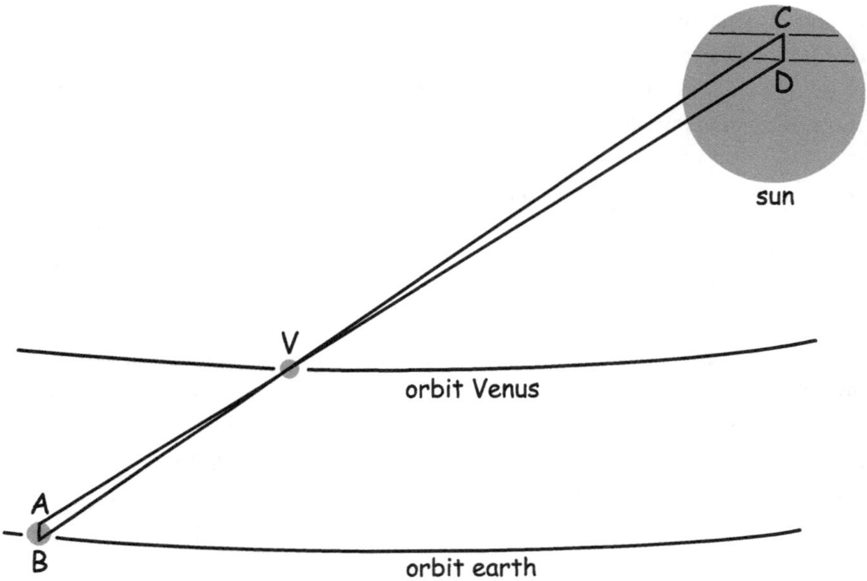

Figure 12.6. Transit of Venus; redrawn from Airy (1881, 153)

Points *A* and *B* are the locations of two observers on the Earth's surface. They are as widely separated as possible. *A* might be in the northern hemisphere. *B* might be in the southern hemisphere. The lines of sight *AVD* and *BVC* pass through Venus at *V* to different locations *D* and *C* on the Sun's disk. The distance *CD* is the separation between the two transit paths observed. If the absolute distance of *CD* can be determined, then it can be scaled up to give the absolute diameter of the Sun. Since the angular size of the Sun as seen from the Earth is readily measured, the distance to the Sun can be recovered.

Triangle *ABV* and *DCV* are similar. Thus, the distance sought, *CD*, can be found from the formula

$$CD = (DV / AV) \cdot AB$$

The distance *AB* is the known distance between the two observers on the Earth. The ratio *DV/AV* is determined by the ratios of the sizes of the planetary orbits. These last ratios were given by Copernican astronomy, as we shall see below.[13] Without knowledge of this ratio, we would be trapped once again in the familiar evidential circle. A small ratio *DV*/*AV* would lead to a small distance *CD* and a small Earth-Sun distance. A large ratio *DV*/*AV* would lead to a large distance *CD* and a large Earth-Sun distance. Some further datum, such as the absolute length *CD* itself, would be needed to break the circle.

7. The Need for Hypotheses

The efforts recounted above reveal the limits of simple geometric triangulation as a means of determining distances to bodies in our planetary system. This approach was able to arrive at a distance to the Moon and, when pressed to the extreme in the seventeenth century, a distance to Mars at its closest approach to the Earth. Even as late as the eighteenth and nineteenth centuries, these methods of triangulation had to be supplemented by further knowledge of the planetary system if their results were to be extended to a determination of the Earth-Sun distance. The seventeenth-century determination of the distance to Mars could be extended only to an estimate of the distance to the Sun by drawing from the known ratio of the sizes of the orbits of the Earth and Mars. The eighteenth-century and nineteenth-century observations of the transits of Venus were unable to return any absolute planetary distances until they were augmented by the known ratio of the sizes of the orbits of the Earth and Venus.

At the close of the nineteenth century, observations of the transit of Venus remained the best way to determine distances within the solar system. After a lengthy treatment of the transits of Venus, Simon Newcomb (1892, 192–99), then a leading authority in astronomy, added a discussion entitled "Other Methods of Determining the Sun's Distance, and Their Results." The promise of these "other methods" went unfulfilled. Newcomb could only say of them that ". . . at least two of which [methods] we may hope, ultimately, to attain a greater degree of accuracy than we can by measuring parallaxes" (192).[14]

---

13 Hipparchus' analogous determination of the parallax of the Moon at the time of a solar eclipse avoided the need for a corresponding ratio. Hipparchus could assume that the Sun was so much farther from us than the Moon that the Sun's rays arrived in parallel lines on the Earth.

14 How things change! Lasers, reflected off mirrors left on the Moon by manned and unmanned missions in the 1960s and early 1970s, now determine the distance to the Moon to within a few centimeters. We can now also use radar echoes to measure distances to the planets.

From the earliest times, the sort of supplement needed was already present as hypotheses of various types. Our histories of astronomy treat the early ones dismissively since most of these early supplements were in error. Since my concern here is not the correctness of the results but the appropriateness of the inductive strategies, we can arrive at a more favorable appraisal. Direct evidence, such as distance measurements by triangulation, can fail to give us the extent of the results sought, such as the distances to the Sun and distant planets. We can then conjecture or hypothesize those facts that would extend the inductive reach of the evidence available to us. This is an entirely responsible epistemic strategy as long as we remember that adopting a hypothesis takes on an inductive debt. It has to be discharged by further investigation that will provide independent inductive support for the hypothesis. Only then have we given the new results a solid foundation inductively in evidence.

8. Pythagorean and Platonic Harmonies

The Pythagorean and Platonic tradition in antiquity provided a rich if chaotic set of hypotheses concerning the distances to the celestial bodies. Their basis was a combination of ideas in musical harmony and simple arithmetic relations. In his creation myth, *Timaeus*, for example, Plato offers the following relative distances:

Moon	1
Sun	2
Venus	3
Mercury	4
Mars	8
Jupiter	9
Saturn	27

These ratios arise from interleaving the numbers of two geometric progressions: 1, 2, 4, 8 and 1, 3, 9, 27. They are just a small part of a rich collection of proposals.[15]

15 For a terse review, see Dreyer (1953, 62, 178–82).

If Plato's ratios are correct, then the inductive benefit is immediate. We need only determine the absolute distance to any one of these celestial bodies. Then the absolute distances to all of the others can be determined from the ratios. What would suffice is just one of the later determinations of the distance to the Moon by Aristarchus or Hipparchus.

It is easy for us now to dismiss these harmonic hypotheses as wild speculations.[16] They were indeed highly speculative. That they were likely incorrect would have been apparent to Aristarchus and Hipparchus. Plato located the Sun at twice the distance from the Earth as the Moon, but both Aristarchus and Hipparchus determined that the Sun is much more distant. That does not make Plato's conjectures epistemically irresponsible. Conjectures of some sort were the only way then available to advance the project of determining distances to celestial bodies beyond the distances accessible to measurement by triangulation. Might it just be that this particular implementation of the harmonic idea is flawed? Might further refinement produce a proposal that can survive independent scrutiny?

Johannes Kepler has unchallengeable credentials in astronomy. He believed that these harmonic ideas found their proper expression in the new heliocentric Copernican astronomy. His *Mysterium cosmographicum* of 1596 accounted for the number of planets and the ratios of planetary orbits by a celebrated geometric construction involving the five Platonic solids, nestled between spheres. The image from that work, shown here as Figure 12.7, has been widely reproduced.

16 Dreyer (1953, 181) writes that "in reality therefore we ought hardly to take the planetary intervals, as determined by the sphere-harmony, seriously; the whole doctrine is quite analogous to that of astrology, but is vastly more exalted in its conception than the latter, and it deserves honourable mention in the history of human progress."

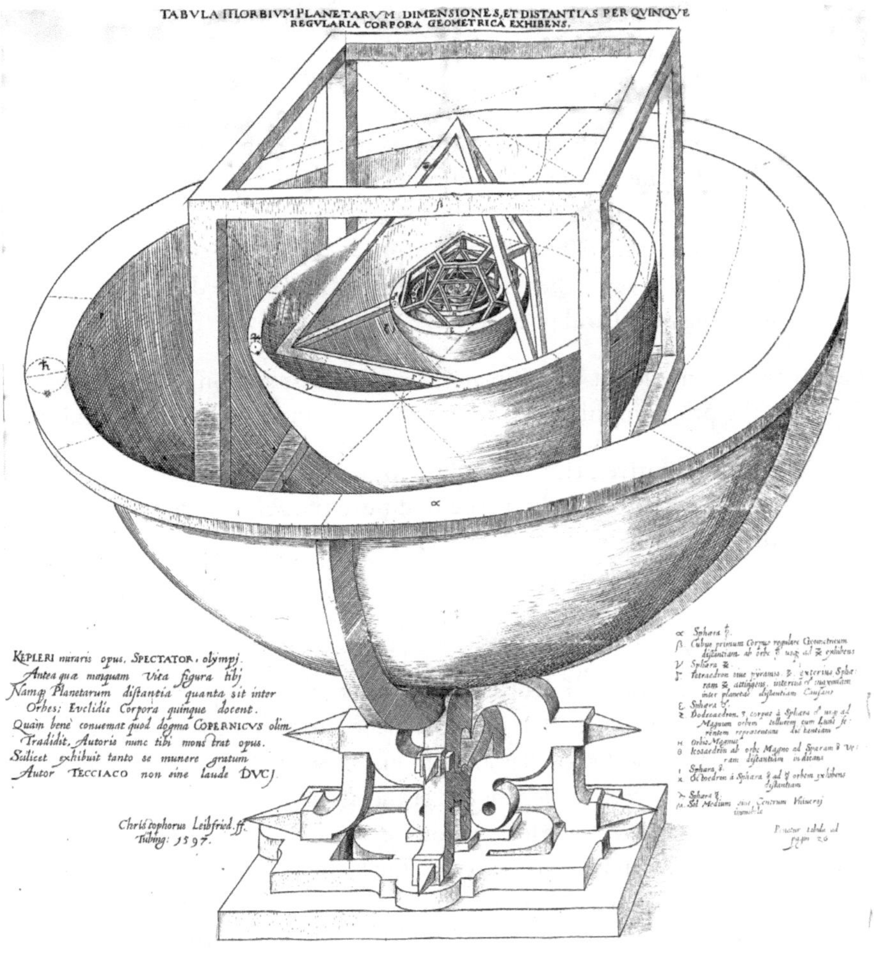

Figure 12.7. Kepler's construction

Kepler's *Harmonices mundi* of 1619 proceeded to find musical harmonies in planetary motions. Although we now dismiss these parts of Kepler's work as mistaken, they were part of a serious investigation. They were hypotheses that failed to find independent evidential support. Had they found such support, we would now be celebrating Kepler's prescience.

The tradition of seeking mathematical harmonies persists. In his Herbert Spencer lecture in 1933, an older Einstein revealed his conversion to a form of mathematical Platonism.[17] He wrote that

> Our experience hitherto justifies us in believing that nature is the realisation of the simplest conceivable mathematical ideas. I am convinced that we can discover by means of purely mathematical constructions the concepts and the laws connecting them with each other, which furnish the key to the understanding of natural phenomena. (1933, 274)

Lest there be any doubt that Einstein saw his formulation of these ideas as fulfilling the program initiated millennia ago in ancient Greece, he added

> But the creative principle resides in mathematics. In a certain sense, therefore, I hold it true that pure thought can grasp reality, as the ancients dreamed. (1933, 274)

9. Ptolemy's *Planetary Hypotheses*

The supreme expression of geocentric astronomy in antiquity was Ptolemy's second-century AD *Almagest*. It provides elaborate geometric constructions of the motions of the celestial bodies: the Moon, the Sun, and the planets. The constructions, however, were independent of the absolute size of the orbit of each body. Take, for example, the construction for Venus. This planet moves roughly with the Sun in its annual course around the heavens. But Venus is sometimes ahead of the Sun and sometimes behind it. This direct and retrograde motion was accounted for in Ptolemy's construction by attaching the planet to a rotating epicycle, as shown in Figure 12.8. The epicycle's center moves along a deferent circle such that this center remains aligned with the mean Sun. (The actual motion of the Sun deviates slightly from the mean motion.)

17 See Norton (2000) for an account of Einstein's conversion.

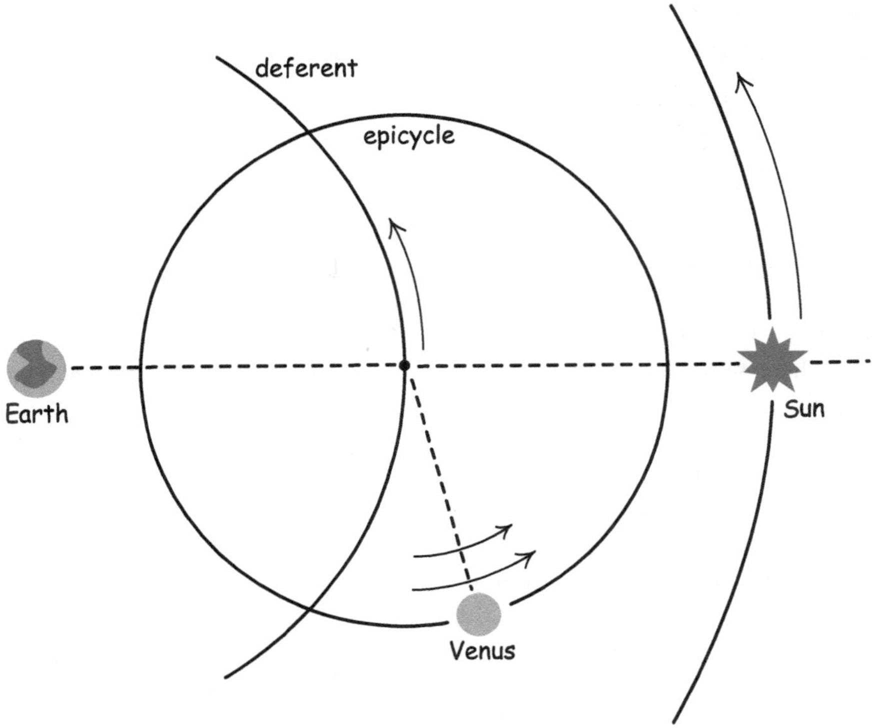

Figure 12.8. The Ptolemaic epicycle for Venus

The figure shows the motion of Venus drawn within that of the Sun. That is not needed to recover the retrograde motion of Venus. As long as the alignment of the center of the epicycle and the Sun is retained, the construction for Venus could be expanded so that its motion would be outside that of the Sun, Mars, Jupiter, or Saturn. The construction for each celestial body in Ptolemy's *Almagest* could be scaled up or down so that any order of the Sun and planets was possible.

The determination of the absolute sizes of these trajectories was taken up in a later work by Ptolemy, *Planetary Hypotheses*. The portions of it dealing with these distances have been lost in the extant Greek texts. Goldstein (1967) found them in a later Arabic translation, and his paper presents an English translation of the Arabic along with the original Arabic text.[18] In addition to the geocentric supposition, Ptolemy's analysis depended on two hypotheses:

Order. The celestial bodies increase in distance from the Earth in the order Moon, Mercury, Venus, Sun, Mars, Jupiter, Saturn.

Packing. The celestial bodies are packed together as closely as their geometrical constructions allow.

These hypotheses provided Ptolemy with the ratios of the distances to the celestial bodies. He could then combine these ratios with his estimate of the absolute distance to the Moon to recover the distances to all of the celestial bodies.

These hypotheses did not derive from considerations of musical and mathematical harmony. Rather, they rested on prosaic, physical considerations. To recover *Order*, we know that the Moon is closer to the Earth than the Sun and stars since the Moon eclipses them. The rest of the order was harder to pin down. The stars have the fastest motion in the Ptolemaic system, with Saturn, then Jupiter, and then Mars lagging successively more behind them. Assuming that proximity in speed reflects proximity in space, Ptolemy could conclude that Saturn is the closest to the stars; then comes Jupiter and then Mars. By this criterion, the Sun, Venus, and Mercury come next. However, the criterion could not give an order for them since their average motion against the stars was the same. Ptolemy settled on the order the Sun, then Venus closer to the Earth, and then Mercury closer still. He reasoned that the closeness of Mercury to the Moon was justified by the similarity of their eccentric motions and since the frequent retrograde motion of Mercury resembled the turbulent motions of the air above the Earth's surface. Similar reasoning placed Venus at the next distant position.

To establish the absolute distances to these celestial bodies, Ptolemy employed the fact that his constructions would take each body nearer to and farther from the Earth. The epicycle shown in Figure 12.8 does this, as does Ptolemy's use of eccentric circles: that is, circles whose centers are slightly displaced from the Earth. Ptolemy could determine from these constructions the ratio of the distances of closest approach to the Earth (perigee) and the farthest displacement (apogee). He now assumed (*Packing*) that all of the constructions were packed together as closely as the geometry allowed, without

18 For further analysis, see Van Helden (1985, 21–27).

the danger of any of the trajectories intersecting. That is, the apogee of the Moon will coincide with the perigee of Mercury, the apogee of Mercury will coincide with the perigee of Venus, and so on.

Ptolemy could only offer the physical plausibility of this packing assumption: "This arrangement," he wrote, "is most plausible, for it is not conceivable that there be in Nature a vacuum, or any meaningless and useless thing" (quoted in Goldstein, 1967, 8). He could not have been so certain of the assumption, for he proceeded to allow that, if there are empty spaces, then the distances cannot be smaller than those that he had determined.

Starting with his value of 64 Earth radii for the apogee of the Moon, Ptolemy used the ratios of perigee to apogee to determine stepwise the distances to all of the celestial bodies. The perigee of Mercury is then 64 Earth radii. The ratio of perigee to apogee for Mercury is 34:88, so its apogee is at 64 x (88/34) = 166 Earth radii. Continuing these calculations leads to the results summarized in Table 12.1.[19]

Table 12.1. Ptolemy's distances in units of Earth radii

	Perigee	Apogee	Ratio
Moon	33	64	33:64
Mercury	64	166	34:88
Venus	166	1,079	16:104
Sun	1,160	1,260	57.5:62.5
Mars	1,260	8,820	1:7
Jupiter	8,820	14,187	23:37
Saturn	14,187	19,865	5:7

Ptolemy encountered one discrepancy. His independent estimate of the perigee of the Sun is 1,160, which does not match the computed apogee of Venus of 1,079. He suggested that the discrepancy might derive merely from slight errors in the underlying observations. To continue, Ptolemy used the independently derived figure of 1,160 for the Sun's perigee.

19 Ptolemy's text delivers these results in a continuous narrative. This convenient tabular summary is provided by Van Helden (1985, 27). He notes that the value of the apogee of Jupiter of 14,187 is a small error of calculation and should be 14,189.

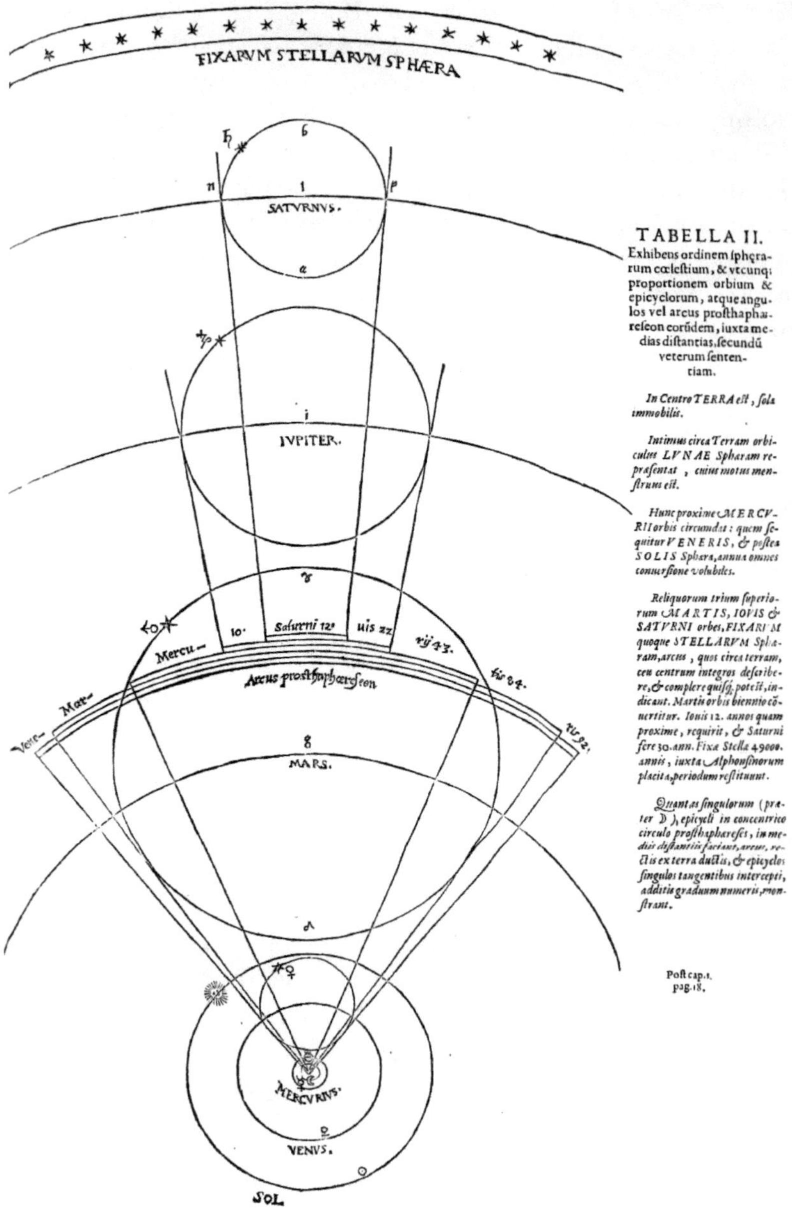

Figure 12.9. Kepler's drawing of the Ptolemaic system

Kepler's *Mysterium cosmographicum* of 1596 has a figure (Figure 12.9 here) that includes all of these celestial bodies in Ptolemy's system, with their epicycles, drawn approximately to the scale set by the distances of Table 12.1.

In assuming the geocentric configuration of celestial bodies and in making the assumptions of order and packing, Ptolemy took on an inductive debt. Until it was discharged — that is, until independent evidence for the assumptions was found — the evidential case for his distances was incomplete. Ptolemy counted as evidence for his packing hypothesis the closeness of the two estimates of the distance to the Sun's perigee: the packing-derived estimate of 1,079 and the independent estimate of 1,160. Although encouraging, that closeness was not enough to discharge the inductive debt. Further independent support was needed. Although Ptolemy's system remained the authoritative system for over a millennium, that further independent support never came. His system was abandoned in favor of another whose inductive debts were discharged and with spectacular success.[20]

10. The Copernican Hypothesis

Nicolaus Copernicus' *On the Revolutions of the Heavenly Spheres* of 1543 is somewhat tame in purely astronomical terms. In the simplest concept, it merely rearranges the circles of Ptolemy's geocentric system in a more apposite way. It is in another sense Earth moving. That rearrangement sets the Earth into twofold motion: spinning on its axis and orbiting the sun.

This basic supposition of Copernican heliocentric astronomy was routinely known as the "Copernican hypothesis" or "hypotheses" in the sixteenth and seventeenth centuries. Moxon's (1665) *Tutor* offered the reader on its title page *an Explanation of the Copernican Hypothesis and Spheres.* Hooke (1674) uses the expression liberally. In the sixteenth century, the term "hypothesis" was tainted by Osiander's surreptitious insertion of an anonymous preface into Copernicus' 1543 work. Osiander reduced Copernicus' proposal to a mere convenience of calculation that did not reveal true causes. He wrote that "these hypotheses need not be true nor even probable. On the contrary, if they provide a calculus consistent with the observations, that alone is enough" (Dobrzycki 1978, xvi).

---

20 For a survey of the persistence of Ptolemy's packing hypothesis through to the time of Kepler in the sixteenth century, see Goldstein and Hon (2018).

Copernicus himself made little use of the term but did not seem to be averse to it. Conveniently, he did equate the term in its usage by the Greeks with "principles and assumptions" (Dobrzycki 1978, 7).[21] As far as I can see, the term "hypothesis" does not appear in his earlier draft manuscript "Commentariolus." However, the main proposals of his heliocentric astronomy are called "assumptions" (Rosen 1971, 58). Rheticus uses the term "hypothesis" freely in his preliminary accounting of Copernicus' proposal, *Narratio prima*, written prior to 1543.[22] He goes to some pains to defend the truth of the hypotheses that he identifies in Copernicus' system. His defense foreshadows the present notion of hypothetico-deductive confirmation: it is a mark of truth if a hypothesis has true consequences. Rheticus puts it this way: "Aristotle says: 'That which causes derivative truths to be true is most true'" (Rosen 1971, 142).[23] In this context, then, common use of the term "hypothesis" referred to an adventurous proposal. Contrary to Osiander's pessimism, its truth could be secured through argument and evidence, and it was thus secured as we moved from the sixteenth century to the seventeenth century.

For my purposes here, what matters is that adoption of Copernicus' heliocentric system proved to be the key step in expanding astronomers' capacity to determine the distances to celestial bodies. Ptolemy needed to add hypotheses, *Order* and *Packing*, to his geocentric constructions in order to fix the ratios of these distances. Copernicus needed no such additions to determine the ratios of the orbital sizes. His heliocentric constructions already fixed them.

The recovery of these ratios followed from how Copernicus' system reduced the number of independent assumptions needed compared with those required by Ptolemy. Consider, for example, Ptolemy's construction for Venus as shown in Figure 12.8. Copernicus realized that two motions in Ptolemy's system were really just one. That is, the annual motion of the center of the epicycle of Venus along the deferent and the annual motion of the Sun were not real motions at all. Rather, there was just the single annual motion of the Earth around a central point near the Sun and then around the Sun itself in

21 Copernicus writes that astronomy's "principles and assumptions" were "called 'hypotheses' by the Greeks."

22 Reproduced in translation in Rosen (1971).

23 There is an extensive secondary literature on Copernicus' attitude to hypotheses; see Rosen (1971, 22–33).

later developments of heliocentrism, such as by Kepler. If an observer on the Earth was unaware of its motion, then it would appear that both the Sun and Venus were orbiting the Earth. These two circles were just apparent motions arising from displacing the true motion of the Earth to Venus and the Sun.

To accommodate this realization, Copernicus rearranged the circles in Figure 12.8 to recover those in Figure 12.10. As shown at the top of the latter figure, the two circles of the deferent of Venus and the Sun were collapsed into a single circle, and that circle was transposed to become the orbit of the Earth around the Sun. The epicycle of Venus now became its true orbit, centered on the Sun.

This new heliocentric construction for Venus no longer admitted the arbitrary rescaling of planetary distances that troubled Ptolemy's system. The maximum elongation of Venus — the maximum angular distance that it strayed from the Sun — was about 45°. That fact of observation immediately fixed the ratio of the sizes of the orbits of Venus and the Earth. The line *EV* in Figure 12.11 traces the line of sight to Venus at its maximum elongation. Since *EV* is tangent to the circle of the orbit of Venus, *EVS* is a right angle. If we take the simplest case of the angle *EVS* equal to 45°, then the triangle *EVS* is right-angled, with equal sides *EV* and *VS* adjacent to the right angle of triangle *EVS*. Using Pythagoras' theorem, it follows that the ratio of the size of the orbit of Venus to that of the orbit of the Earth, *SV* to *SE*, is 1 to $\sqrt{2}$: that is, 0.71 to 1.

This last calculation is simplified by assuming that the orbit of Venus is a perfect circle centered on the Sun. The deviations from this simplification complicate the determination only slightly.[24] A similar rearrangement gives the Copernican construction for Mercury and the determination of its orbital size.

The outer planets — Mars, Jupiter, and Saturn — required slightly different rearrangements. Their epicycles were not the representations of their true motions, merely the superposition of the Earth's motion onto their true motions. A similar analysis within the circles of the Copernican rearrangement gives the ratios of the sizes of these outer planetary orbits to that of the Earth. The analysis is a little more complicated. A greatly oversimplified version conveys the basic geometry of the analysis. Contrary to the reality, we assume that an outer planet is not moving. Then we can determine the ratio

24 For details, see Van Helden (1985, 43–44).

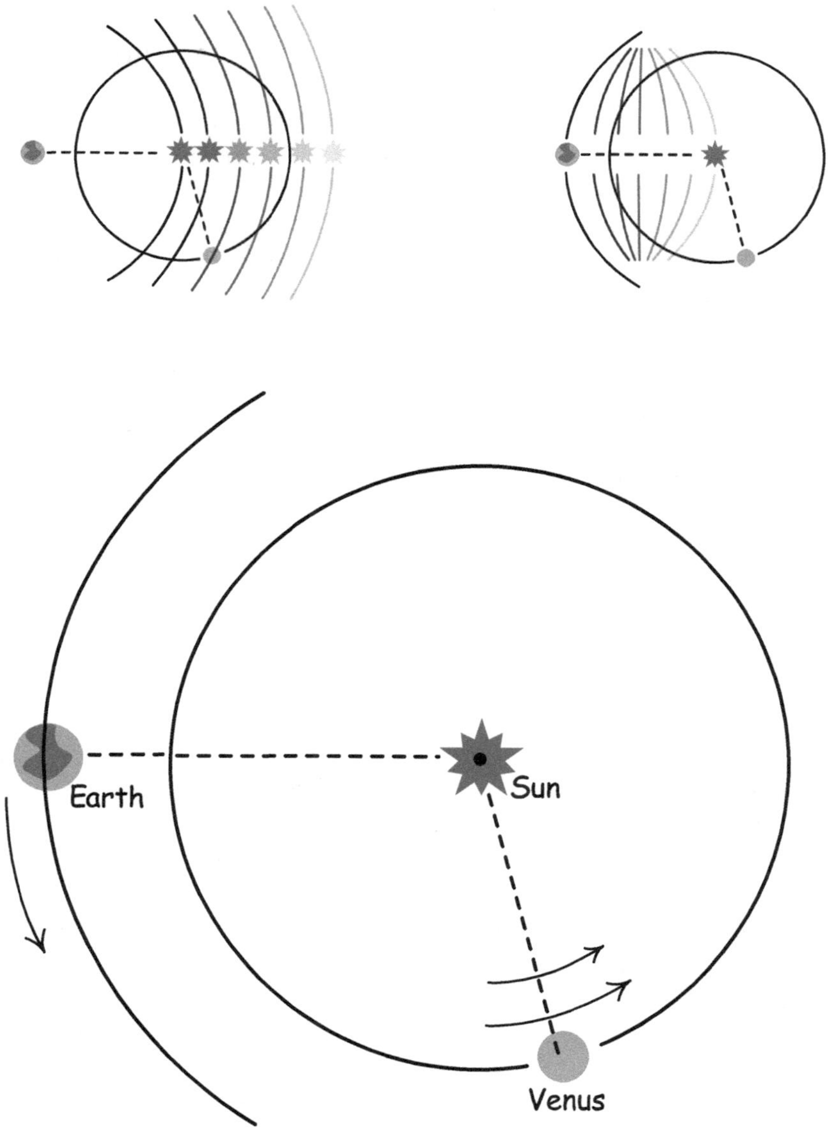

Figure 12.10. Venus in the Copernican system

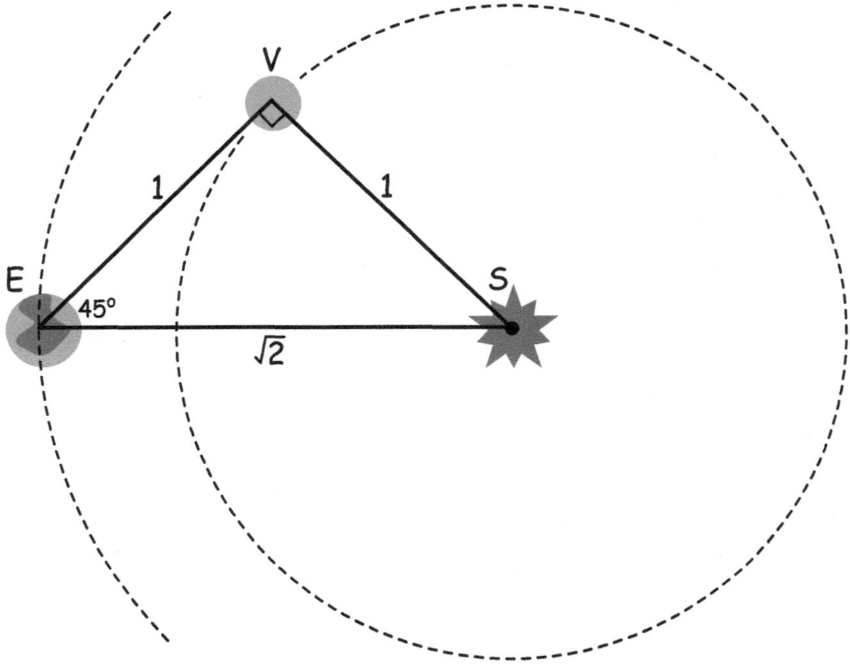

Figure 12.11. Fixing the size of the orbit of Venus

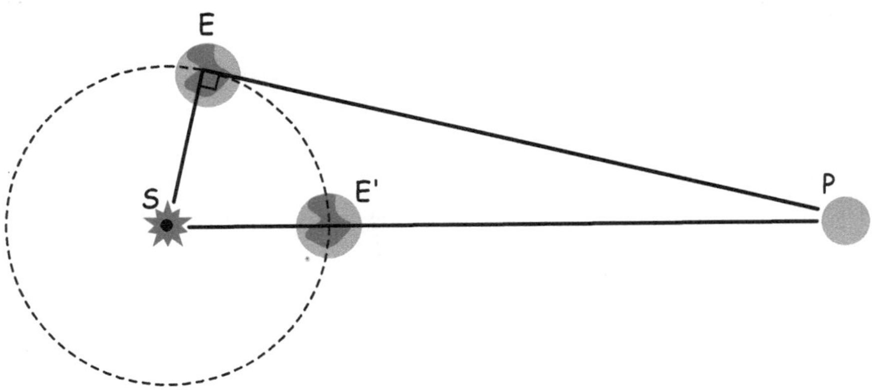

Figure 12.12. Distance to an outer planet

of the sizes of the orbits by checking how far the Earth progresses in its orbit between two orientations. First, the distant planet *P* is in direct opposition to the Sun *S*, indicated by the Earth at *E*' in Figure 12.12; second, the distant planet *P* is at quadrature — that is, at a right angle to the Earth-Sun distance — indicated by the Earth at *E* in Figure 12.12.

The angle *ESP* is known from how far the Earth has moved in its orbit. Observing the change in which stars are directly overhead at midnight would give the angle directly. Simple trigonometry on the right-angle triangle *SEP* tells us that the ratio of sizes *SP*/*SE* is 1/cos(*ESP*). This method is inapplicable in practice since the planet *P* will move during the time that the Earth progresses from *E*' to *E*. In the case of slow-moving Saturn, which has a period of 29.5 years, the movement will be slight. However, the analysis must correct for it. The correction is straightforward.[25]

11. Securing the Copernican Hypothesis

Copernican heliocentric astronomy and its later refinements proved to be key to the determination of planetary distances in the centuries that followed. It provided the ratios of the sizes of the orbits of the planets. All that astronomers needed was a single absolute measurement of one distance; then all of the rest could be recovered from the ratios. This was the procedure used after the seventeenth-century determination of the parallax of Mars and the eighteenth-century observations of the transits of Venus. This was the same strategy used by Ptolemy. His determination of the distance to the Moon triggered a cascade of computations that gave all of the distances. However, the difference was that independent evidence for Ptolemy's hypotheses never emerged. His inductive debt was never discharged. The Copernican hypothesis fared much better.

To begin, the Copernican system had an advantage over the Ptolemaic system in the practical challenges of securing evidential support. The Copernican system needed fewer independent hypotheses and thus fewer independent items of evidence. Ptolemy had to posit as an independent hypothesis that the centers of the epicycles of Mercury and Venus always aligned with the mean Sun, as shown in Figure 12.8. This alignment was automatic in the Copernican system since the center of the orbits of Mercury and Venus

25 For a simplified construction, see Crowe (2001, Chapter 6).

simply was the mean Sun. Similarly, Ptolemy had to posit that the epicycles of the outer planets — Mars, Jupiter, and Saturn — moved in perfect concert with the motion of the Sun, such that their retrograde motion coincided with their opposition to the Sun. Copernicus needed no such posits. These effects followed automatically from his recognition that these epicycles were merely the superposition of the Earth's annual motion on the true motions of the outer planets. Even just to recover an order for the planets in their distances from the Earth, Ptolemy had to posit additional hypotheses concerning their periods and motions. Copernicus needed no such additional posits. In his system, the relative distances of the planets from the Sun could be recovered from careful measurements of planetary positions.

As time passed, further evidence emerged. Galileo used his telescope to observe Venus in 1610, and he reported his results in his *Letters on Sunspots* of 1613. He saw Venus exhibiting a variety of Moon-like phases that could only be if its motion took it both closer to the Earth than the Sun and farther from the Earth than the Sun. This contradicted Ptolemy's system in which Venus is always closer to the Earth than the Sun but fit the Copernican hypothesis that Venus orbits the Sun.

It was Isaac Newton who made the decisive advance that fully discharged whatever residual inductive debt heliocentric astronomy might have carried. His *Principia* of 1687 provided a complete mechanics for the motions of the bodies in heliocentric astronomy. At the same time, celestial mechanics was combined with terrestrial mechanics in a single unified system. Any challenge to heliocentric cosmology would end up eventually having to challenge the entirety of this new physics.

12. Crossing of Relations of Support

The most useful relationship concerning the ratios of sizes of planetary orbits in the new astronomy is Kepler's so-called third law.[26] It asserts in its modern form that the square of the periods of a planet's orbit T^2 is proportional to the cube of the semi-major axis of its elliptical orbit R^3. Since the periods of two planets are accessible to measurement, the relationship provides a rapid determination of the ratios of their distances from the Sun. The relationship

26 Called thus, for example, by Maxwell (1894, 113).

between this law and Newton's mechanics provides a striking illustration of how relations of inductive support can cross over one another.

The distance-period relationship for the planets was first reported by Kepler for the mean distance from the Sun, among the many harmonies of his *Harmonices mundi* of 1619. In Book III of his *Principia*, Newton ([1726] 1962, 401–05) enumerated the phenomena from which his system of the world would be inferred. Phenomenon IV was his relation for the planets, asserted in terms of the mean distances. Phenomena I and II asserted the same relation for the moons of Jupiter and Saturn. Within Newton's mechanics, this relation could be translated almost immediately into a result central to his system: the acceleration due to the gravitational attraction of a body such as the Sun diminishes with the inverse square of distance. We can see how rapidly the result follows if we take the simple case of a planet or moon in a perfectly circular orbit of radius R with period T. It follows that the speed of the object is $V = 2\pi R/T$. Newton's mechanics sets the centrally directed acceleration A of such a motion equal to V/R^2. We can now combine these relations as

$$A = \frac{V^2}{R} = \frac{(2\pi)^2 R^2}{T^2} \cdot \frac{1}{R} = \frac{(2\pi)^2 R^3}{T^2} \cdot \frac{1}{R^2} = \text{constant} \cdot \frac{1}{R^2}$$

where Kepler's third law allows us to set R^3/T^2 to a constant.

Here we have the first relation of support:

from Kepler's third law *to* Newton's inverse square law of gravity.

It is possible to run the inferences in the above equalities in reverse and thereby infer Kepler's third law from the inverse square law:

$$A = \text{constant} \frac{1}{R^2} = \frac{V^2}{R} = \frac{(2\pi)^2 R^2}{T^2} \cdot \frac{1}{R} = \frac{(2\pi)^2 R^3}{T^2} \cdot \frac{1}{R^2}$$

We read from these equalities that R^3/T^2 must be a constant if we first assume the inverse square law. Thus, it is possible to have a relation of support that proceeds in the other direction:

from Newton's inverse square law of gravity *to* Kepler's third law.

Since the relation is a deduction, given the requisite background assumptions of Newton's mechanics, it is especially strong.

This second inference is commonly given in mechanics texts. Is it merely a formal derivation purely of mathematical interest? Or should we also conceive of it as a relation of evidential support proceeding in a direction opposite to that of Newton's original relation? That we can and should so conceive of it follows from a complication revealed by more careful analysis. The analysis above requires that the mass S of the central body, such as the Sun, should be considerably greater than the mass P of the orbiting body, such as a planet. When this assumption is relaxed, Maxwell (1894, 113–15) gives the correction that must be applied to the original form of Kepler's third law:

$$R^3 = \text{constant } (S+P)\ T^2$$

Deviations from the original law are small according to this formula as long as P is much smaller than S. However, for cases in which P becomes large in relation to S, the orbital periods will become smaller than predicted by the original relation from the orbital sizes. Maxwell proceeded to show that such deviations have been measured for the more massive planets Jupiter, Saturn, and Uranus.

Thus, Newton's mechanics does not merely recover Kepler's third law. Rather, it tells us the circumstances within which the law holds and gives a more general law that will hold when we deviate from those circumstances. In doing this, Newton's mechanics provides evidential support for Kepler's third law.

13. Conclusion

The determination of distances in our planetary system illustrates how hypotheses are used to extend the otherwise limited inductive reach of evidence. This is a procedure used widely in science. What makes the present case study revealing is that the investigations extended over millennia. That means that its stages are readily dissected. We can see in this slow development that evidence unaided by hypotheses was limited in its reach. Direct measurements of distances to celestial bodies by triangulation returned very little in spite of the most energetic and ingenious efforts. This reach was decisively furthered by various systems of hypotheses: harmonic, Ptolemaic, and Copernican. That each of the three considered here yielded different results underscores the provisional nature of the results. They are given a secure inductive foundation only when independent evidence is found for the hypotheses used and the inductive debt taken on in assuming them is discharged.

REFERENCES

Airy, George B. 1881. *Popular Astronomy: A Series of Lectures*. London: Macmillan.

Berry, Arthur. 1898. *A Short History of Astronomy*. London: John Murray.

Crowe, Michael J. 2001. *Theories of the World from Antiquity to the Copernican Revolution*. Mineola, NY: Dover.

Dobrzycki, Jerzy, ed. 1978. *Nicholas Copernicus On the Revolutions*. Translated by Edward Rosen. N.p.: Polish Scientific Publishers; n.p.: Macmillan.

Dreyer, J.L.E. 1953. *A History of Astronomy from Thales to Kepler*. New York: Dover.

Einstein, Albert. 1933. "On the Methods of Theoretical Physics." In *Ideas and Opinions*, 270–76. New York: Bonanza, 1954.

Forbes, George. 1874. *The Transit of Venus*. London: Macmillan.

Goldstein, Bernard R. 1967. "The Arabic Version of Ptolemy's *Planetary Hypotheses*." *Transactions of the American Philosophical Society* 57: 3–55.

Goldstein, Bernard R., and Giora Hon. 2018. "The Nesting Hypothesis for Planetary Distances and Its Persistence over the Centuries and across Cultures." In *Knowledge in Translation: Global Patterns of Scientific Exchange, 1000–1800 CE*, edited by Patrick Manning and Abigail Owen, 209–26, 343–47. Pittsburgh: University of Pittsburgh Press.

Heath, Thomas. 1913. *Aristarchus of Samos: The Ancient Copernicus*. Oxford: Clarendon.

Hooke, Robert. 1674. *An Attempt to Prove the Motion of the Earth from Observations*. London: John Martyn.

Long, Roger. 1742. *Astronomy*. Vol. 1. Cambridge, UK: Printed for the Author.

Maxwell, James Clerk. 1894. *Matter and Motion*. London: Society for Promoting Christian Knowledge.

Moxon, Joseph. 1665. *A Tutor to Astronomy and Geography: Or, The Use of the Copernican Spheres*. London: Printed for Joseph Moxon.

Newcomb, Simon. 1892. *Popular Astronomy*. 6th ed. New York: Harper and Brothers.

Newton, Isaac. (1726) 1962. *Mathematical Principles of Natural Philosophy*. 3rd ed. Translated by Andrew Motte. Revised by Florian Cajori. Berkeley: University of California Press.

Norton, John D. 2000. "'Nature in the Realization of the Simplest Conceivable Mathematical Ideas': Einstein and the Canon of Mathematical Simplicity." *Studies in the History and Philosophy of Modern Physics* 31: 135–70.

Rosen, Edward, ed. and trans. 1971. *Three Copernican Treatises*. 3rd ed. New York: Octagon.

Van Helden, Albert. 1985. *Measuring the Universe: Cosmic Dimensions from Aristarchus to Halley*. Chicago: University of Chicago Press.

13

Dowsing: The Instabilities of Evidential Competition

1. Introduction

In Chapter 4, "The Uniqueness of Domain-Specific Inductive Logics," I addressed the possibility that a single collection of empirical facts might evidentially support multiple sciences equally well. This circumstance would negate the power of evidence to determine a definite theory and compromise the uniqueness of our mature sciences. Worse, since these facts also determine the applicable inductive logic, we would then have multiple logics applicable in the same domain. Inductive anarchy would prevail.

In that earlier chapter, I argued that this possibility has not arisen in the case of mature sciences, well supported by an extensive body of empirical evidence. There is, for example, only one periodic table of the elements and only one chemistry derived from it. I also argued that the material theory of induction provides a mechanism that precludes the persistence of equal support for such multiple sciences. It is based on an instability in the competition among rival theories. Insofar as the differences between competing theories manifest in empirically decidable disagreements, evidence can point in favor of one over the other.[1] One theory then secures more facts than its rival. Since background facts so secured can then authorize more inductive inferences, that gain enhances its inductive reach while weakening that of its rival. The enhanced theory is then better placed to achieve more successes at

1 If the differences between them have no empirical manifestation, then we must ask whether the differences matter. Are they the same theories empirically but dressed up in different theoretical clothing? Do they contain elements superfluous to their empirical content?

the expense of its rival. A continuation of the process leads to the evidential dominance of one theory.

Where might we look to see this process within real sciences? The natural place is among the many fields of endeavor labeled as pseudosciences: astrology, parapsychology, telepathy, telekinesis, crystal healing, psychic surgery, and many more. These endeavors purport to offer bodies of knowledge in competition with established science. Each proposes facts radically at variance with standard science. If they are correct, then these facts would induce an inductive logic different from that of standard science.[2]

These endeavors are routinely disparaged by established science. The term "pseudoscience" is not intended to be flattering. In my view, these pseudosciences are properly disparaged, for the case has been made abundantly that they lack proper evidential support. The tradition of challenging the evidential credentials of these endeavors is as old as these endeavors themselves. Recently, a leading role among many in these efforts has been taken by CSICOP (Committee for the Scientific Investigation of Paranormal Claims). It was founded in 1976 and later renamed as CSI (Committee for Skeptical Inquiry). Its major vehicle of publication is the magazine *Skeptical Inquirer*, whose pages have offered evidential scrutiny of extraordinary claims since the magazine's inception in 1976 as *The Zetetic*.

My goal in this chapter is not once again to make the evidential case against these many pseudosciences. Rather, it is to see whether their evidential collapses resulted from the mechanism sketched earlier. It would be impractical and redundant to trace the collapse in many of these sciences. One will suffice as an illustration. The practice of dowsing is well suited to this analysis. The practice itself is narrowly defined: a dowser walks over a candidate area of land seeking underground water sources or, in the original tradition, metallic ores. The dowser employs some instrument as a detector. A forked hazel twig is traditionally preferred. The detection event is unambiguous: the detector moves, clearly and sometimes even violently, in response to the water or metal ores sought. Finally, success or failure

2 Another example of a variant logic is among conspiracy theorists. Many proceed under the assumption that nefarious hidden powers are systematically misleading the public for their own ends. The presumption of this fact leads the conspiracy theorists to an inverted inductive principle: strong evidence against their theory is actually evidence of the perfection of the deception by the hidden powers. Evidence "against" is really evidence "for."

is unambiguously determinable. Either there is water present there or not or the metal ore sought is there or not. There has been a long-standing debate over the effectiveness of dowsing. Its proponents are zealous in offering extraordinary tales of unlikely successes. Its critics are equally zealous in denouncing the practice as superstitious hokum.

The literature on dowsing is so massive that I make no effort to do it justice here. My goal is solely to investigate the competition between proponents and skeptics and to show that an instability in the competition leads to the collapse of the scientific credibility of dowsing and the evidential dominance of its scientific skeptics.

In Section 2, I briefly sketch the emergence of dowsing in the historical literature. In Sections 3 to 6, I recount the factual disputes surrounding dowsing: which physical theory if any governs the process (Section 3), how water sought by dowsers is distributed geologically (Section 4), whether there really is any effect in the first place (Section 5), and finally whether the effect can be merely unconscious self-deception (Section 6). In Section 7, I review how proponents and skeptics end up presuming different inductive logics because they differ in their presumptions of the prevailing facts. In Section 8, I conclude by displaying the instability that leads to the evidential dominance of the skeptics.

We shall see that the competition unfolded on two levels: that of theory and that of phenomena. At the level of theory, in the sixteenth century, proponents and skeptics had positions of comparable strength. The physical interaction between metallic ores and the dowser's rod fit well enough with the qualitative understanding of electric and magnetic effects. With continuing investigations in each field, theories of electricity and magnetism developed by the end of the nineteenth century into a quantitatively precise candidate theory of everything. This dominant theory supported the inference that there is no physical effect in nature corresponding to dowsing. The proponents of dowsing had nothing to match. They were reduced to speculating that the effect derived from some sort of qualitatively described psychic process.

At the level of phenomena, proponents and skeptics were once again in comparable positions in the sixteenth century. Proponents could point to a well-established and apparently successful practice of dowsing. Skeptics could point to the uncomfortable fact that dowsing did not work for everyone. The discovery of the ideo-motor principle in the nineteenth century allowed skeptics to block the inference from the motion of the dowser's rod to a real process of detection. The motion was the result of unconscious muscular actions by the dowser. The proponents could offer no comparable account of why dowsing failed for some. Proponents could infer from the success of the later tradition of water dowsing to the reality of a real process of water detection. The inference was warranted by the assumption that underground water was sparsely distributed and otherwise hard to locate. The inference lost its warrant with the recognition that underground water tables are widely dispersed and hard to miss in a random drilling. The failure of dowsers to detect their targets was established in the twentieth century for all but the most ardent believers by statistical analysis of well-crafted tests.

Finally, the successes of skeptics at the theoretical and phenomenological levels were mutually reinforcing. The theory deployed by skeptics left no niche for dowsers' physical processes of detection. Using this as a warranting fact, skeptics could infer from the failure of dowsers in tests to the conclusion that any apparent dowsing successes in the phenomena must be spurious. Conversely, the failure of dowsers in these tests supported the conclusion that skeptical theorists had not somehow overlooked a theoretical process that could underpin dowsing.

2. The Phenomenon Established

The modern tradition in dowsing seems to have started among the miners in Saxony and the Hartz mountains in what is now modern-day Germany. It was well established by the sixteenth century. From there, it spread over Europe and beyond. The process presumed to create the detection was direct physical interaction between underground metallic ores and the dowser's instrument. Since the interaction, apparently, was manifested routinely, it was reasonable to expect some general theoretical basis for it. That such an interaction was possible lay well within the contemporary state of physical theorizing. Barrett (1911, 169) suggested that a then common belief was that certain trees are attracted by metallic ores and droop over them. Agricola ([1556] 1912, 39), who gave the first extended account of dowsing, reported the belief; proponents of dowsing asserted "that movement of the twig is caused by the power of the veins and sometime[s] this is so great that branches of trees growing near a vein are deflected toward it." It was then only a small step to detach a twig from the tree and use its attraction to the metallic ores as a means of detection.

Such an attraction seems to be little different from the attractions then known in electrostatic phenomena and magnetism. Agricola ([1556] 1912, 39) likened the action to that of a magnet attracting iron. Proponents of dowsing, he reported, explained the failure of some people to succeed at dowsing through "some peculiarity of the individual, which hinders and impedes the power of the veins." His report revealed the rudimentary nature of the relevant science, for he likened this explanation to the supposed power of garlic to weaken a magnet: "For a magnet smeared with garlic juice cannot attract iron . . ." (39). Garlic has no such power, of course, and that is a fact easily recovered by a simple test.[3]

3. Disputes over the Theory of Dowsing Processes

At its inception, the effect of metallic ores on the dowser's twig was *likened* to the effects of electrical and magnetic attraction. It was rudimentary to see that the dowsing effect was not mediated by then known magnetic and electrical actions. Most ores sought by dowsing were not magnetic, and twigs were not susceptible to known magnetic action. Then known electrical actions persisted only if the systems were carefully insulated. The theoretical question, then, was whether dowsing had revealed a physical process to be added to the known processes of magnetism, electricity, and gravity. We shall see that, in the ensuing centuries, theories of electricity, magnetism, and gravity grew in strength. Yet accounts of the mechanism of dowsing languished. They lagged in their attempts to copy the latest developments in these other theories. By the end of the nineteenth century, there was no longer a theoretical niche in which dowsing processes could reside. There was no credible physical mechanism. We shall see that the most articulate of the proponents had to resort to clairvoyance and psychic processes as the foundation of dowsing.

3.1. The Effluvial Theory of Dowsing

Agricola ([1556] 1912) reported no theoretical foundation for the phenomenon beyond its similarity in some aspects to other processes, such as magnetic attraction. Here his level of reporting was comparable to that of Gilbert's *De magnete*, the influential treatise on electricity and magnetism published almost half a century later in 1600. His work was devoted to establishing the

3 For a brief history of this curious notion, see May (1979).

observed phenomena of magnetism and electricity and speculating about how the magnetism of the Earth might be associated with celestial processes. There was no detailed proposal for the mechanism of magnetic and electrical effects.[4]

Matters were soon to change. William Pryce's (1778) treatise on mining argued strongly in favor of the efficacy of dowsing. It included an extensive theory of the mechanism, formulated in terms of the effluvia proposed by the then popular corpuscular philosophy:

> It [the dowsing rod] was much talked of in France towards the end of the seventeenth century; and the corpuscular philosophy was called in to account for it. The corpuscles, it was said, that rise from the Minerals, entering the rod, determine it to bow down, in order to render it parallel to the vertical lines which the effluvia describe in their rise. In effect the Mineral particles seem to be emitted from the earth: now the Virgula [dowsing rod] being of a light porous wood, gives an easy passage to those particles, which are very fine and subtle; the effluvia then driven forwards by those that follow them, and pressed at the same time by the atmosphere incumbent on them, are forced to enter the little interstices between the fibres of the wood, and by that effort they oblige it to incline, or dip down perpendicularly, to become parallel with the little columns which those vapours form in their rise. (114)

Pryce turned from this report to an extended narrative aimed at establishing the plausibility of this theory of effluvia, drawing from the work of Robert Boyle (1673) though giving no citation to it. Perhaps he intended Boyle's energetic promotion and defense of effluvia. In any case, the effluvial theory described by Pryce bears a striking similarity to the effluvial theory of magnetism advocated by Descartes in his *Principles of Philosophy* ([1644] 1982, Part IV). Pryce concluded his defense of the effluvial theory with an analogy to magnetism. Effluvia from the Earth can magnetize iron as shown by

4 Contrary to some later reports (as given in Bynum, Browne, and Porter 1981, 111), the notion of *effluvia* seems to have no major role in *De magnete*. I found only one use of the word in the volume (Gilbert [1600] 1893, 78).

> ... the polarity and magnetism of an old Iron bar taken from a church window, where it has stood upright for many centuries, is proved to derive its virtue from the magnetick effluvia of the earth. (116)

We are encouraged to make the unspoken inference that effluvia from mineral ores can also act on dowsers' twigs.

We can assess the equivocal status of the theory in the mid-seventeenth century of Descartes and Boyle from the latter's own synoptic report on dowsing. Boyle (1669, 92) concluded his essay "Of Unsucceeding Experiments" with the lament that, "What to determine concerning the truth of this perplexing experiment, I confess not to know."

3.2. Resistance by Skeptics

At the same time as proponents of dowsing were advancing theories of its operation, there was a persistent tradition of theoretical skepticism. Agricola's ([1556] 1912) earliest account of dowsing is often reported by proponents of it. They regularly omit mention of his astute skepticism of the process. Agricola noted how unlike dowsing was from the well-established processes of electrical and magnetic attractions:

> But, in truth, all those objects which are endowed with the power of attraction do not twist things in circles, but attract them directly to themselves; for instance, the magnet does not turn the iron, but draws it directly to itself, and amber rubbed until it is warm does not bend straws about, but simply draws them to itself. If the power of the veins were of a similar nature to that of the magnet and the amber, the twig would not so much twist as move once only, in a semi-circle, and be drawn directly to the vein. . . . (41)

Dowsing, Agricola noted, was a theoretical anomaly in his time whose properties were unlike electricity and magnetism. That, of course, precluded it from having an electrical or magnetic nature.

Since Pryce's (1778) work was a practical manual for mining, we should not expect it to provide the most up-to-date science. The effluvial theory of dowsing that Pryce reported represented the level of theorizing from a century earlier. At the time of his writing, physical theorizing had changed.

Descartes' qualitative speculations about effluvia had been replaced by quantitative measures of forces. Newton's precise quantitative account of gravity in his *Principia* of 1687 had supplanted Gilbert's speculation about the role of magnetism in celestial motions and Descartes' cosmic vortices. In 1785, seven years after Pryce's work was published, Charles Coulomb presented seven memoires to the French Académie royale des sciences in which he reported his careful quantitative measurements of electrical forces.

These theoretical troubles for dowsing continued. As long as theories of electricity, magnetism, gravitation, and other forces remained qualitative, dowsers could speculate that their twigs were responding to some combination of these forces within the standard scientific repertoire or some additional but analogous force. Over the course of the next 100 years, theories of electricity and magnetism matured into the precise electrodynamics of Maxwell, Hertz, Lorentz, and others still taught today as classical electrodynamics. Their theories annexed other processes. Light, it turned out, was merely a propagating ripple in the electromagnetic field. Although the heat of gases was reduced to random motions of their molecules, heat radiation was found to be just another portion of the electromagnetic spectrum.

With this maturation, the theoretical niche in which speculation about dowsing could flourish was gone. It was no longer plausible that metallic ores or water, buried underground, could exert some force on hazel twigs while evading the now thorough and quantitatively precise measurements of the nineteenth-century physicists. The skeptics, brandishing their mature theory of electrodynamics, were moving from success to success, from strength to strength, whereas the dowsers' theories were successively weakened and in retreat.

3.3. Collapse of the Dowsing Theory

Undeterred, proponents of dowsing continued to urge some sort of electrical or magnetic process as the basis of dowsing. By the later part of the nineteenth century, dowsing had become more prominent as a means of locating underground water. Latimer (1876, 26) claimed that it arose as an electrical effect: ". . . the friction of running waters underground produces an electric current which causes the switch to turn." In evidence, he recounted no exacting measurements, no experiments with running water, and no detailed computation within then developed theories of electromagnetism. Instead, he wore wooden sandals, insulated electrically from the ground by four ink bottles, and attempted to dowse. So insulated, he noted (18) that his dowsing powers were extinguished.

While dowsing proponents persisted in these efforts, they became targets of derision by skeptical scientists. Charles Boys, the English experimental physicist, wrote a scathing review in *Nature* of Tompkins' 1899 volume *The Theory of Water Finding by the Divining Rod: Its History, Method, Utility and Practice*. Tompkins, Boys (1899, 2) reported, attributed the efficacy of dowsing to electrical action and quoted him as asserting the "well-known scientific fact that water is a generator of electricity." Elsewhere, he reported Tompkins asserting that minerals and water emit effluvia. Tompkins followed the tradition of dowsers who claimed that their method could detect much more than metallic ores and water. Their powers of detection extended to precious metals, including gold, boundaries, and murderers. To see whether the rod was detecting gold, one needed only to put gold in each hand, whereupon the motion of the rod ceased. Boys then mocked Tompkins:

> We can only infer that the murderer can be discriminated by putting a murderer in each hand, but this is not stated.

His sobering conclusion:

> But when they [dowsers] put forward preposterous "scientific explanations" such as I have extracted, it makes it very difficult not to come to the almost inevitable conclusion that the water-finder has no case. . . .

An anonymous reviewer of papers on dowsing by William Barrett and T.V. Holmes wasted no words on derision but dismissed without discussion the possibility that successful dowsing results from electrical action. The reviewer (Anonymous 1898, 353) wrote that,

> Moreover, as a physicist, he [Barrett] does not bring to this task any acquired training which is helpful in unravelling the problem; for the only point at which the divining rod touches physics — the assumption that electricity is its motive power — may be dismissed without investigation.

And still the dowsing theorists persisted. A later anonymous reviewer in *Nature* (Anonymous 1940) gave a much more restrained dismissal of

J. Cecil Maby and T. Bedford Franklin's 1939 *The Physics of the Divining Rod*. The authors had attempted to ground dowsing processes in something resembling contemporary physical theory. The reviewer's verdict was dry and devastating:

> The theoretical section, by the second author, postulates some form of cosmic radiation resulting in electromagnetic waves of ten metres wave-length. There seems to be no direct evidence for such waves, and the author's discussion of their polarization cannot be justified on our present knowledge.
>
> In presenting facts and theories to the scientific world, there is a well-accepted and necessary procedure. It is to be regretted that the authors have not followed this procedure, thus making the position of the scientific reviewer impossible.

A convenient marker of the collapse of a physical theory of dowsing was provided by the physicist and psychic researcher William Barrett. He investigated dowsing extensively, convinced himself of its reality, and provided a nonphysical explanation of it in his 1911 volume *Psychical Research*:

> The explanation, I believe, is not physical, but *psychical*. All the evidence points to the fact that the good dowser subconsciously possesses the faculty of clairvoyance, a supersensuous perceptive power such as we have described in a previous chapter. This gives rise to an instinctive, but not conscious, detection of the hidden object for which he is searching. (183; Barrett's emphasis)

The rod, on this account, is then moved by unconscious muscular action.

Today, over a century later, when clairvoyance has secured no scientific credibility, we find this retreat to clairvoyance a damning concession of failure. It would not have been so for Barrett. He was a founder of both the British Society for Psychical Research and the American Society for Psychical Research. They advocated the reality of psychic phenomenon and promoted research on them.

4. The Dispute over Geology

Once the locus of dowsing had moved toward detection of underground water, a new dispute emerged. Just how is the underground water sought by dowsers distributed? Dowsers portrayed the water as commonly residing in flowing streams. The flow of the water, as we saw above, is hypothesized to produce the electricity mediating in its detection. Latimer (1876, 23–24) boasted of his prowess as a dowser in locating a stream of water just ten feet from a well that had run dry and of locating a stream in a yard unfamiliar to him in the dark of night.

These findings are impressive only if the distribution of underground water is sparse and otherwise hard to locate. Critics, however, were quick to dispute this supposition. The anonymous reviewer reported above recorded Holmes, whose work was under review, as making the point clearly:

> He points out, in the first place, that the astonishment caused by the dowser's success is largely due to the fact that the dowser himself, and usually those who employ him, always believe that water-finding is a matter of locating a "spring," which it is possible to miss by a few inches, so that the achievement becomes as wonderful as finding a buried jar of ancient coins. But, as Mr. Holmes points out, while water sometimes runs in underground fissures, water bearing strata usually cover acres or miles, over any point in which a well may be successfully sunk. (Anonymous 1898, 355–56)

Similar points about the ease of finding water are made in an anonymously authored US Geological Survey pamphlet:

> The natural explanation of "successful" water dowsing is that in many areas water would be hard to miss. The dowser commonly implies that the spot indicated by the rod is the *only* one where water could be found, but this is not necessarily true. In a region of adequate rainfall and favorable geology, it is difficult not to drill and find water! (Anonymous 1988, 10; emphasis in the original)

Thomas Riddick (1951) makes the same point and many more in a scathing review of a book written by Kenneth Roberts about the well-established dowser Henry Gross. Riddick, a water-works engineer, decried at length Roberts' "apparent lack of even the most elementary knowledge of the principles of water-works engineering" (62). The title, "Dowsing Is Nonsense," does not hide the fury within the article.

5. Dispute over the Phenomena

5.1. The Early Dispute

Although dowsers maintained a healthy and profitable profession, there are reports from all eras that many in the mining industry itself were skeptical of the reality of the dowsers' powers of detection. Agricola ([1556] 1912, 40) reported it as "in dispute and caus[ing] much dissention amongst miners." Paracelsus was a contemporary of Agricola, both being born in 1493 or 1494. He gave a terse warning:

> You must take particular care, however, not to let yourselves be beguiled by divinations obtained through uncertain arts. These are vain and misleading; and among the first of them are the divining rods, which have deceived many miners.*[5] If they once point out rightly, they deceive ten or twenty times. (as translated in Waite 1894, 185)

The idea that we count both successes and failures in assessing dowsing was later refined greatly and became the basis of the twentieth-century statistical tests of dowsing reported below.

A century later Boyle (1669, 93) noted that "Among the Miners themselves I found some made use of this Wand, and other[s] laughed at it." Even Pryce (1778, 116) had to concede that "many deny, or at least doubt." Coupled with these doubts were strong suspicions that at least some dowsers were frauds and tricksters. Agricola ([1556] 1912, 41) obliquely suggested

5 Editor's footnote here: "Elsewhere Paracelsus says that it is faith which turns and directs the divinatory rod in the hand. — *De Origine Marborum Invisibilium*, Lib. I." I thank Jennifer Whyte for alerting me to Paracelsus' admonition. It must have been written prior to 1541, the year of his death.

deception in calling successful dowsers "cunning manipulators" and pointing out that a forked twig of flexible wood "turns in a circle for any man wherever he stands."

It is also striking that proponents of dowsing relied heavily on anecdotal evidence. Latimer (1876, 10) set out his agenda as "I think I have it in my power to demonstrate to you, principally from my own personal experiences — the relation of which I beg you to accept as strictly accurate. . . ." The demonstration then proceeded through a sequence of boasts of grand dowsing successes from his own professional practice. A favorite anecdote was of Jacques Aymar, who used his dowsing powers in 1692 to solve a notorious murder case in Lyon. The accounts of the episode, though supposedly based on objective contemporary accounts, read like a lurid detective novel, with astonishing moments of high drama. Barrett (1911, 172) included it in his history, favorable to dowsing, but did concede briefly that Aymar was "subsequently somewhat discredited owing to his failure in some tests. . . ." Barin-Gould (1877, 60–78) related the story in all of its lurid details. The account included Aymar's final entrapment in a test that resulted in Aymar being labeled an impostor and sent away "in disgrace" (77). Barin-Gould does not find, however, the exposé to be "conclusive evidence of imposture throughout his career" (78).

At least two commentators were not so credulous. In their colorful exposé of the folly of belief in dowsing, Ozanam and Montucla (1803, 259–67) left no doubt about their skepticism, calling dowsing "illusion, or philosophical quackery" (259–60). Their exposé included the tale of Aymar and suggested that his successful detection depended on ordinary, earlier knowledge of the murders.[6] They concluded their account of his fraud with a lament:

> How could rational minds imagine that an action morally bad, could communicate any physical quality to the authors of it? That the murderer of a human being, or stolen money, should have an effect on the rod, rather than the person who had killed a sheep, or money merely displaced? Those who can believe in such reveries must be exceedingly weak. (263)

6 They reported without giving the reason that "there is reason to think . . ." that Aymar had witnessed the murders. The remark might be more than a rhetorical flourish since they were French authors writing in France closer in time to the events.

5.2. The Modern Dispute

Such weakness persisted. At least as early as the late nineteenth century, proponents of dowsing sought more objective experimental evidence of dowsing. Hansen (1982) reviewed the previous century of experimental research on dowsing. The review provides an extensive synopsis of dowsing-related experiments of various types. For example, the "biophysical" seek to establish a dowser's sensitivity to electric and magnetic fields. The "physiological" seek to establish physiological responses of dowsers. There are many of these tests. The bibliography is over four pages long. However, the results are inconclusive. Hansen says in his final summary that,

> In spite of the large number of investigations made into dowsing, its status remains unclear. This is largely a result of sloppy experimental procedure and or report writing. (362)

It is hard to see how a century of such inconclusive investigation was anything other than a damning indictment of the physical reality of dowsing. It is supposedly an effect so strong that it can break dowsing twigs and lead dowsers to pass out or vomit. Yet a century of careful experimentation failed to establish it. We understand Hansen's curious conclusion best by recalling that the vehicle of publication for his review was the *Journal of the Society for Psychical Research*.

The strongest experimental evidence against dowsing has come in the form of controlled trials, which have occurred sporadically over the past century. Gregory (1929) collected and detailed the tests of dowsing then known to him, many of them unfavorable. Notable among them was a carefully constructed blind test organized by Sir John Cadman of the Anglo-Persian Oil Company (now British Petroleum) at its experimental station at Meadhurst, Sudbury-on-Thames, England, in 1925 (340–43). Dowsers were tested for their abilities to detect various combinations of buried deposits of water, oil, or empty barrels. The result was failure or, to quote Cadman, "a complete fiasco"; "in no case were the diviners able to show any justification for their contention that they could discover such deposits" (341).

In another such test, stage magician and parapsychology debunker James Randi (2020) organized a controlled trial of dowsing in Sydney, Australia, in July 1980. Dowsers were asked to identify which of ten buried pipes contained running water. Despite their confidence, the dowsers performed merely at chance levels.

The largest test of dowsing abilities was conducted in Germany with a grant in 1986 of DM 400,000 from the government ministry Bundesministerium für Forschung und Technologie. It was completed in 1990. Some 500 dowsers were subjected to 10,000 individual tests. Most performed at chance levels. The few (forty-three) who showed more promise were subjected to further tests in a barn, in German *Scheunen*. These tests came to be known as the "*Scheunen* experiment." The dowsers were to locate a position on the barn's second floor directly above a water pipe placed randomly on the floor below. The experimenters proclaimed a successful demonstration of the reality of dowsing. A critic, however, found the experimenters' statistical analysis so flawed as to reverse their conclusion. Enright (1995, 360) concluded that

> A reexamination of the data on which that conclusion was based, however, indicates that no persuasive evidence was obtained for a genuine, reproducible dowsing skill. The absence of reproducibility suggests that the entire research outcome can reasonably be attributed to chance.

The German investigators (Betz et al. 1996) disputed this damning appraisal, and Enright (1996) reaffirmed it.

Although the practice of dowsing and disputes over it persist today, establishment skepticism of it has been unequivocal and well entrenched for over a century. A report in 1917 by the United States Geological Survey (Ellis 1917) responded to the "large number of inquiries received each year by the United States Geological Survey" about the efficacy of dowsing. The "Introductory Note" (5–6) was written by Oscar E. Meinzer, widely recognized as the founding figure of modern groundwater hydrology. His verdict was unequivocal:

> It is doubtful whether so much investigation and discussion have been bestowed on any other subject with such absolute lack of positive results. It is difficult to see how for practical purposes the entire matter could be more thoroughly discredited. . . . (5)

He went on to suggest that part of the dowsing profession is populated by swindlers who deliberately defraud people:

> To all inquirers the United States Geological Survey therefore gives the advice not to expend any money for the services of any "water witch" or for the use or purchase of any machine or instrument devised for locating underground water or other minerals. (6)

6. The Ideo-Motor Principle

This entrenched skeptical conclusion is that there is no real dowsing effect. This presents a problem for the skeptics. Some dowsers, presumably, are frauds and swindlers. However, many sincerely believe that they have the ability and have had the profound experience of their twigs or rods moving as if under the influence of powerful external forces. Why else would these dowsers allow themselves to be subjected to carefully controlled tests?

The skeptical response came in the codification of something long suspected: a sincere dowser might be unconsciously moving the twig. Ellis (1917, 16) noted the idea already advanced in the seventeenth century by Gaspard Schott and Athanasius Kirchner. The modern tradition was initiated by William Carpenter (1852). He argued that muscular motion might occur without one's conscious volition, and he dubbed the effect the "ideo-motor principle." It explains, he assured readers, "numerous phenomena which may have been a source of perplexity. . . ." They include

> . . . the movements of the "divining rod," and the vibration of bodies suspended from the finger; both which have been clearly proved to depend on the state of *expectant attention* on the part of the performer, his Will being temporarily withdrawn from control over his muscles by the state of abstraction to which his mind is given up, and the *anticipation* of a given result being the stimulus which directly and involuntarily prompts the muscular movements that produce it. (153; Carpenter's emphasis)

This possibility had an immediate application in England in the mid-nineteenth century when interest in spiritualism was growing. Participants in séances were startled to find the table under their hands moving even though no one was consciously moving it. Michael Faraday, then an eminent experimental scientist, devised a simple test. He placed stacks of cardboard and other materials under people's hands resting on the table in the séance. The stacks were devised so that they would respond differently according to whether the sitters' hands were moved by the table that moved first or whether their hands moved first and pushed the table. The latter case was demonstrated unequivocally. Faraday reported his results in a letter to the London *Times* on June 30, 1853.[7]

This ideo-motor principle or just the idea of unconscious movement enabled skeptics to account for how sincere dowsers might nonetheless find their twigs moving, as if under some external power. It also explained why sincere dowsers were so successful in controlled trials when they knew where the target was but failed when they did not. Indeed, it could even account for some of the limited successes of dowsers. As has often been noted, there are ordinary clues above ground that a dowser might unwittingly discern. Gregory (1929, 331) concluded that

> Hence a man going over a tract of ground may notice signs of water unconsciously, and some slight mental action may cause the twitching of a finger and a jerk of the rod. While some dowsers may be deliberate frauds, and others may be duped by their vanity, many of the best dowsers probably act by their dissociated mental activities.

The flexibility of the ideo-motor principle also proved to be useful to proponents of dowsing. When it had become increasingly clear that dowsing did not operate by familiar physical processes such as electricity and magnetism, we saw above that Barrett (1911, 183) resorted to clairvoyance as the active mechanism. But how might a clairvoyant thought be known by the dowser's twig? Unconscious muscular movement by the dowser transmits it, Barrett concluded.

---

7 Presumably, he knew of Carpenter's proposal since Faraday was a contributor to the volume of the *Proceedings* in which Carpenter's paper appeared. For an account of the origin and development of the idea of ideo-motor action, see Hyman (1999).

7. The Diverging Inductive Logics

The preceding sections have recounted the dispute among proponents and skeptics of dowsing over which are the facts governing dowsing. According to the material theory of induction, different facts will support different inductive logics. Since these differences among the facts proposed and presumed by each group are large, we should expect and will find these differences reflected in differences in their inductive inferences.

The easiest to see arises from differences in views over the facts of the geological distribution of the water sought by dowsers. If one believes with the dowsers that underground water is distributed sparsely in veins, then one will infer that a dowser's successful prediction of the location of water provides good inductive support for the efficacy of dowsing. Success, if dowsing were ineffective, would be unlikely. If, however, one believes with the skeptics that water is often distributed broadly in readily accessible water tables, then one will find a dowser's successful prediction of the location of water to be evidentially inconsequential. The success is assured independently of any special power of the dowser.

A richer divergence in the inductive logics derives from differences over whether there is a real physical process directly connecting the dowser's target and the movement of the dowser's twig. If one believes with the mainstream of dowsers that there is such a process, then a dowser's success is expected and provides additional support for facts already believed, the efficacy of dowsing. The problematic cases are those in which dowsing fails. In that circumstance, under this logic, we have evidence for a secondary disturbing process or other confounding factor resulting in the failure. The research agenda is to find it. We have seen already that such failures might be explained by proponents of dowsing in a way familiar even to modern parapsychologists — in Agricola's ([1556] 1912, 39) words, "some peculiarity of the individual, which hinders and impedes the power of the veins."

If, however, one believes with the skeptics that no real physical process directly connects the dowser's target and the movement of the dowser's twig, then matters are exactly reversed. The failure of a dowser is expected and provides additional support for facts already believed, the inefficacy of dowsing. The successes are the problematic cases. They are evidence for some secondary process that emulates successful dowsing. The research agenda is to find it. Perhaps the dowser unconsciously reacted to ordinary signs of the target, or success was assured by the prevalence of water, or the reports of success are exaggerated or heavily selected.

These last remarks pertain just to the beliefs of the two sides regarding which are the prevailing facts and thus which are the appropriate inductive inferences. Of course, at most, one of these logics can be applied correctly to dowsing. That one logic is determined by which are the facts actually prevailing over dowsing.

8. Conclusion: The Inductive Instability

I can now summarize the inductive instability that led to the collapse of the credibility of dowsing and the evidential dominance of the skeptics. Initially, when the practice first emerged in the sixteenth century, neither proponents nor skeptics could claim a decisive advantage. If anything, skeptics were at a striking disadvantage, for dowsing was an established practice. Its operation was directly visible in the unambiguous motions of the dowsers' twigs, and there was a financially successful profession of dowsers serving the mining industry. What followed was a steady stream of self-reinforcing victories by the skeptics that so weakened the dowsers' claims that they lost scientific credibility.

As far as the observed reality of the process itself was concerned, the evidential case was unstable, at least in the shorter term. The successes of dowsers strengthened their case and weakened that of the skeptics. Correspondingly, the failures reversed these judgments. These failures were a concern for dowsers from the start, for there were always skeptics who suspected self-deception and even dishonesty by the dowsers. An enduring history of failures is more damaging to the dowsers than to the skeptics, for the dowsers make the positive claim of the existence of a definite process. Yet they are unable to delineate the precise conditions under which that process is guaranteed to appear. Pryce (1778, 116), who championed the efficacy of dowsing, curiously had to concede that he himself was unable to dowse:

> As many deny, or at least doubt, the attributed properties of the divining rod, I shall not take upon me, singly to oppose the general opinion, although I am well convinced of its absolute and improveable virtues. It does not become me to decide upon so controvertible a point; particularly, as from my

natural constitution of mind and body, I am almost incapable of co-operating with its influence; and, therefore, cannot, of my own knowledge and experience, produce satisfactory proofs of its value and excellence.

That is troublesome for an effect supposedly akin to the reliable processes of magnetism and electricity. The persistence of these failures over the centuries must erode the strength of support for dowsing.[8]

The identification of ideo-motor effects in the nineteenth century gave a new advantage to the skeptics at the expense of the dowsers. Pryce had emphasized the honesty and reliability of those giving favorable observational reports of dowsing. He wrote of one,

> . . . my worthy friend Mr. William Cookworthy, of Plymouth, a man, not less esteemed for his refined sense and unimpeachable veracity, than for his chemical abilities. (1778, 116)

Just as the honesty of this observer weighed favorably on Pryce, so too does the sincerity and honesty of at least some of the dowsers who appear to practice dowsing successfully. This part of the case for dowsing was now eliminated. Ideo-motor effects gave skeptics a serviceable account of the illusion of the effectiveness of dowsing. The ideo-motor effects were reproducible reliably. The effect would be present just when the agent knew the targeted answer.

Finally, failures of controlled trials of dowsing completed the experimental side of the skeptics' case.

In parallel with these developments, the strengthening of theories of magnetism, electricity, gravitation, and more left no theoretical niche for the physical processes that would have to mediate in dowsing, if the effect was real. The process unfolded in an instability in which successes by the skeptics strengthened their case while weakening that of the dowsers. That is, as theories of electricity, magnetism, and other physical forces advanced, the theoretical niche available for the physical basis of dowsing contracted. The dowsing theorists were perpetually retreating and shifting their theoretical ground with yet another speculation. Meinzer gave an acerbic appraisal:

---

8 Here we might compare their continuing difficulties with the comparable problem faced by proponents of cold fusion to produce the effect reliably in the laboratory. See Norton (2021, Chapter 4, Section 5).

> A favorite trick for appealing to uneducated persons and yet making specific disproof impossible is to give as the working principle of such a [dowsing] device some newly discovered and vaguely understood phenomenon, as, for example, radioactivity. (Ellis 1917, 5)

Dowsers repeatedly retreated to speculations within existing theories that fell far short of professional standards and then finally to suppositions of psychic effects.

These two observational and theoretical tracks were also mutually reinforcing. When observational or experimental tests fail to manifest an effect, there is always some possibility that a different set of conditions might nonetheless produce it. The skeptics could dismiss this possibility by pointing to the lack of a theoretical niche in known physics for processes that could mediate in dowsing. The skeptical theorists, however, could worry that their theories had failed to probe all of the material processes in their domain of investigation. These theorists could reassure themselves that they had not missed some novel process at work in dowsing by pointing to the failure of objective testing to discover any such process.

In sum, the early viability of both proponents' and skeptics' positions was unstable under further investigation. As those investigations proceeded, on the experimental and theoretical tracks, they favored the skeptics. The investigations reinforced each other, accelerating the skeptics' advantage and leading to their evidential dominance.

REFERENCES

Agricola, Georgius. (1556) 1912. *De re metallica*. Translated by H.C. Hoover and L.H. Hoover. London: Mining Magazine.

Anonymous. 1898. Review of two works on dowsing. *Journal of Mental Science* 44: 352–56.

———. 1940. "The Physics of the Divining Rod." *Nature* 146: 150.

———. 1988. *Water Dowsing*. Washington, DC: US Government Printing Office.

Barin-Gould, Sabine. 1877. *Curious Myths of the Middle Ages*. London: Rivingtons.

Barrett, William F. 1911. *Psychical Research*. New York: Henry Holt and Company.

Betz, Hans-Dieter, et al. 1996. "Dowsing Reviewed — The Effect Persists." *Naturwissenschaften* 83: 272–75.

Boyle, Robert. 1669. *Certain Physiological Essays*. 2nd ed. London: Henry Herringman.

———. 1673. *Of the Strange Subtilty of Effluviums*. London: M. Pitt.

Boys, Charles V. 1899. "Water, Water, Everywhere." *Nature* 61: 1–4.

Bynum, William F., E. Janet Browne, and Roy Porter. 1981. *Dictionary of the History of Science*. Princeton, NJ: Princeton University Press.

Carpenter, William B. 1852. "On the Influence of Suggestion in Modifying and Directing Muscular Movement, Independently of Volition." *Royal Institution of Great Britain (Proceedings)*, March 12, 147–53.

Descartes, Rene. (1644) 1982. *Principles of Philosophy*. Translated by V.L. Miller and R.P. Miller. Dordrecht: Kluwer.

Ellis, Arthur J. 1917. *The Divining Rod: A History of Water Witching*. Washington, DC: Government Printing Office.

Enright, James T. 1995. "Water Dowsing: The *Scheunen* Experiments." *Naturwissenschaften* 82: 360–69.

———. 1996. "Dowsers Lost in a Barn." *Naturwissenschaften* 83: 275–77.

Gilbert, William. (1600) 1893. *De magnete*. Translated by P. Fleury Mottelay. New York: John Wiley and Sons.

Gregory, John W. 1929. "Water Divining." In *Annual Report of the Board of Regents of the Smithsonian Institution, 1928*, 325–48. Washington, DC: Government Printing Office.

Hansen, George P. 1982. "Dowsing: A Review of Experimental Research." *Journal of the Society for Psychical Research* 51: 343–67.

Hyman, Ray. 1999. "The Mischief-Making of Ideomotor Action." *Scientific Review of Alternative Medicine*, fall–winter. https://quackwatch.org/related/ideomotor/.

Latimer, Charles. 1876. *The Divining Rod*. Cleveland: Fairbanks, Benedict, and Company.

May, W.E. 1979. "Garlic and the Magnetic Compass." *Mariner's Mirror* 65: 231–34.

Norton, John D. 2021. *The Material Theory of Induction*. BSPS Open Series. Calgary: University of Calgary Press.

Ozanam, Jacques, and Jean-Étienne Montucla. 1803. *Recreations in Mathematics and Philosophy*. Vol. IV. Translated by Charles Hutton. London: G. Kearsley.

Pryce, William. 1778. *A Treatise on Minerals, Mines and Mining*. London: James Phillips.

Randi, James. 2020. "Australian Skeptics Divining Test." March 29. https://www.skeptics.com.au/resources/articles/australian-skeptics-divining-test/.

Riddick, Thomas M. 1951. "Dowsing Is Nonsense." *Harper's Magazine*, July, 62–68.

Waite, Arthur E. 1894. *The Hermetic and Chemical Writings of Aureolus Philippus Theofrastus Bombast of Hohenheim, Called Paracelsus the Great*. London: James Elliot and Company.

14

Stock Market Prediction: When Inductive Logics Compete

1. Introduction

This chapter continues the investigations of Chapters 4 and 13 of the possibility that a single body of evidence might support competing theories equally well. That possibility is precluded, I argued, by an instability in the competition among rival theories. As long as the evidence is pursued sufficiently, that instability will lead to one theory prevailing over its rival. A small advantage gained from evidence by one theory amplifies its inductive powers at the expense of the rival. This amplification leads to an acceleration of the gains of that theory against its rival and speeds the latter's demise. This process can be completed quickly. The competition between dowsers and their skeptics in the previous chapter was exceptional in its slow pace. The stability of our mature sciences arises from the repeated elimination of rivals by this process. Many outcomes of this process fill most of our present science.

This chapter provides an illustration, occurring now, of an otherwise rarer and enduring competition of theories and their associated inductive logics. The competition has endured over decades and shows no sign of a speedy resolution. It arises through efforts to predict the changes in prices of stocks in the stock market. The competition is relatively easy to assess since the predictions are generally unambiguous and their successes or failures soon evident. Either the stock price went up as predicted, or it did not.

I will describe four systems of prediction. Each is currently in vogue, and each has a history extending over many decades. Each, in effect, is an inductive logic, for each uses past stock performance and related facts to discern which among many future possibilities are more likely. The four systems to be discussed are

- fundamental analysis;
- technical analysis;
- random walk/efficient market analysis; and
- fractal/scale free analysis.

They are sketched in Section 2 below. Since each of these systems has spawned evolving research programs of great complexity, a rudimentary sketch of each is all that is possible here. Each sketch indicates the ideas that motivated the system and its founding hypothesis in its simplest and original form. That, however, will be sufficient for my purposes here. Such sketches provide enough to illustrate the differences between the systems and the dynamics of the competition between them.[1] The mutual incompatibility of the different systems is widely recognized and manifests in repeated attempts by proponents of each system to impugn the others. In Section 3, I collect a representative sample of such cross-system criticism. For my purposes, the important point is that the criticism focuses on proposing facts troublesome for the competition. This is how the material theory of induction dictates that differences among systems are to be resolved — by further factual investigation. A concluding Section 4 summarizes general features of the competition and how the factual investigations proposed could drive the field toward a single inductive logic if only they were pursued.

2. The Systems

Multiple systems of inductive logic are possible, temporarily at least. This is a natural artifact of how these systems are constructed. Each is based on founding propositions that warrant the logic's inferences. We shall see in the examples of stock market prediction below that these founding propositions are introduced initially as hypotheses without full inductive support. The expectation of proponents of each system is that this support will accrue

1 For an engaging historical survey of the development of these systems, written by a philosopher of science, see Weatherall (2013).

eventually. Until this happens, the systems will remain legitimately in conflict while proponents of each seek the strong inductive support needed.

2.1. Fundamental Analysis

This venerable approach is based on a simple idea. Each stock, it is supposed, has an intrinsic value. Often there will be discrepancies between the market price of the stock and its value. These discrepancies will not last. If you can identify a stock whose price is well below its intrinsic value, then it can be purchased with the confidence that the price will rise eventually. Correspondingly, a stock whose price is well above its intrinsic value would be a poor long-term investment since its price will fall eventually. The previous two sentences are predictions inductively supported by the founding

> *Hypothesis of fundamental analysis*. Each stock has an intrinsic value. Discrepancies between the intrinsic value and the market price of a stock will be removed eventually by price moves.

This system has a rich pedigree. The work widely known as the "bible of value investing"[2] is Graham and Dodd (2013). It was first published in 1934 and is now in its sixth edition. In his preface to the latest edition, the legendary investor Warren Buffett endorsed the volume and its approach:

> . . . I studied from *Security Analysis* while I was at Columbia University in 1950 and 1951, when I had the extraordinary good luck to have Ben Graham and Dave Dodd as teachers. Together, the book and the men changed my life.
>
> On the utilitarian side, what I learned then became the bedrock upon which all of my investment and business decisions have been built. . . . (2013, xi)

There is considerably more, of course, to fundamental analysis. Graham and Dodd is a work of 766 pages. Perhaps the most delicate issue is the determination of the intrinsic value of a stock. It cannot be merely the market price on pain of trivializing the whole system of analysis. One important element will be the dividends paid by the stock. Others include less tangible judgments of the stability of the stock's business model and its management's acumen and abilities.

---

2 So reported by Seth Klarman in his preface to Graham and Dodd (2013, xiii).

Fundamentalists make their predictions on the basis of an exhaustive examination of companies behind the stock. In this aspect, fundamental analysis employs a far larger body of evidence than the three remaining approaches discussed below. These latter approaches make their predictions solely on the basis of the history of past stock prices and volumes of trades.

2.2. Technical Analysis ("Chartists")

Technical analysis starts with an observation that can be made by any casual observer of a chart of stock prices over time: the line tracing the prices exhibits all sorts of interesting patterns, some of which appear to be repeated. The core supposition made by technical analysts — "chartists" — is that these patterns are sometimes signals that, properly interpreted, reveal to traders subsequent moves in stock prices. This type of analysis goes back to Charles Dow in the late nineteenth century. This is the same Dow of the Dow Jones Industrial Average. The approach has been refined by many hands. A recent, authoritative exposition is Edwards, Magee, and Bassetti (2019), the eleventh edition of a work first published in 1948.[3]

There are many suppositions underlying that approach. The editor and reviser of the seventh edition attributes to John Magee three principles (Edwards, Magee, and Bassetti 2019, xxxix):

1. Stock prices tend to move in trends.
2. Volume goes with the trends.
3. A trend, once established, tends to continue in force.

A primary goal of technical analysis is the identification in the charts of the signals indicating a reversal of a trend. These signals appear in a bewildering array of patterns in the charts, which are given suggestive names such as "head and shoulders," "symmetrical triangles," "the diamond," and many more.

The existence of these signaling formations is attributed to the behavior of traders reacting to shifts in the market; this behavior, in turn, is explicated by an understanding of the traders' psychology. A simple example is the existence of support and resistance levels, which appear as plateaus of constant price with time in the charts. A support arises when a surge in purchasing

3 Another version of technical analysis is the Elliot wave theory, popularized by Frost and Prechter (2017). It asserts that trader psychology produces nestled waves whose compound action comprises the movements of prices in the market.

forms a plateau that halts a downward trend in prices. A resistance arises when a surge in selling forms a plateau that halts a rising trend in prices.

Following the analysis of Edwards, Magee, and Bassetti (2019, Chapter 13), support and resistance will arise at price levels where, in the past, there was a larger amount of trading. The reason lies with the psychology of the traders involved in these earlier trades. For example, traders might purchase stock at some price level, confident in its price rising. If, instead, the price rises and falls, then traders who have continued to hold the stock might lose confidence in their purchase. When the price rises again and passes through the price at which these traders originally purchased the stock, they would be tempted to sell since then they would have lost nothing on the trade other than the transaction cost. The resulting surge in selling would flood the market and temporarily suppress further price rises. That is, the price would be a resistance level. An inversion of this process could convert the same price level into a support level. If instead the traders become more confident in the wisdom of the purchase, then they might regret not initially purchasing more at the original price. They might be inclined to buy more of the stock when it falls in price to that original level. Then the surge in purchasing forms a support level.

A more elaborate pattern, prominent in technical analysis, is head and shoulders. It consists of three peaks in succession in the charts. In its most characteristic form, the first and third peaks are of the same height, and the second peak is higher. The overall shape is loosely like the silhouette of a person's head and shoulders. Its appearance, we are told by Edwards, Magee, and Bassetti (2019, 44) is common, and it is, they assure us, "by all odds, the most reliable of the Major Reversal Patterns." That is, we can be confident that the stock price will fall once this pattern arises. Their confidence is so high that they later report that, since

> The odds are so overwhelmingly in favor of the downtrend continuing once a Head-and-Shoulders Formation has been confirmed, it pays to believe the evidence of the chart no matter how much it may appear to be out of accord with the prevailing news or market psychology. (48)

As with support and resistance, this head and shoulders formation does not arise by chance. It is a product of the psychology of traders. Edwards, Magee, and Bassetti (2019, 43–44) describe a plausible scenario in which the formation would occur. They imagine a well-financed coterie that has purchased some stock heavily. When it has risen to the price at which they plan to sell, they proceed to sell their holdings hesitantly so as not to precipitate a collapse in the stock's price. In their telling of the scenario, the cautious stopping and starting of the selling happen in just the right way to produce the head and shoulders pattern.

The volume proceeds in this fashion of identifying a prodigious repertoire of patterns for traders to seek and use as signals of reversals in prices. Of course, none of the patterns is infallible. Every few pages, we are warned of "false moves" or "false signals" confounding the technical indicators. The hypothesis that warrants the inferences of this mode of analysis can be summarized as the

> *Hypothesis of technical analysis.* The psychology of market traders leads to trading behavior that imprints distinctive patterns on the changes in time of prices and volumes. The unique association of the earlier and later part of the pattern is strong enough that the presence of the former predicts the coming of the latter.

2.3. Random Walks

The two analytical systems reviewed so far are optimistic. If traders use the right system, each system maintains, then their predictions can lead them to profitable trading. Another approach is pessimistic. Traders, this approach says, are engaged in fierce competition with one another. Any usable indication of a market move is seized and exploited to the full. This happens so rapidly that any actionable indication has already been anticipated, and the move that it foretold is already built into the present price of a stock, at least as far as ordinary investors are concerned. Chance alone governs price movements. It is just self-deception to think that one can beat the averages of market behavior by sophisticated techniques of prediction. The best that one can do is to follow a "buy and hold" strategy that minimizes trading expenses and lets one's fortunes rise with the market as a whole.

Here is how Paul Samuelson (1965, 41) put it, posing it as an enigma that introduced a famous paper:

"In competitive markets, there is a buyer for every seller. If one could be sure that a price will rise, it would have already risen." Arguments like this are used to deduce that competitive prices must display price changes over time, [formula], that perform a random walk with no predictable bias.

The mathematically precise statement of this form of predictive pessimism is the random walk model. It asserts that stock prices meander in a manner akin to the process that Einstein predicted in 1905 for small particles suspended in a liquid. These small particles are affected on all sides by many fluid molecules. The accumulated effect of many of these uncorrelated collisions is the jiggling known as Brownian motion. It is the best-known example in science of a random walk. The proposal is that stock market prices execute a random walk about their mean values. Most importantly, whether the stock will rise or fall momentarily is statistically independent of what it did moments before.

The random walk hypothesis for markets was first proposed by Bachelier (1900) prior to Einstein's work of 1905. A more recent version is elaborated in Fama (1965). The conditions needed for prices to exhibit a random walk are well known. Drawing from Fama (40–41), they are the

Hypothesis of the random walk. Price changes are governed by a probability distribution with a finite mean and variance, and successive price changes are probabilistically independent.

The most significant predictions supported by the random walk model are negative. The best that one can do predictively is to determine the probability distribution of price changes. An examination of the past history of changes in prices, no matter how thorough and extensive, can provide nothing more. It follows that all of the indicators of technical analysis are predictively useless.

Although the random walk model supports few positive predictions, one has proven to be important. The conditions above for a random walk are sufficient to allow the application of the central limit theorem of probability theory to the accumulation of many price changes. That theorem tells us that, if we sum sufficiently many smaller price changes, then the resulting accumulated price change conforms to a Gaussian or normal distribution. Once one knows the standard deviation "σ" of the distribution, the range of probability changes in prices is well circumscribed. They will mass around the mean: 95.4% will lie on average within two standard deviations of the mean. The probability of larger changes diminishes exponentially since the tail of the normal deviation is exponentially thin. Deviations of six sigma, "6σ," or more are vastly improbable. They arise with a probability of about 2 x 10^{-9}. That is, they occur on average once in roughly 500 million changes.[4]

2.4. The Efficient Market Hypothesis

The random walk hypothesis is customarily coupled with what is known as the "efficient market hypothesis." It is the idea sketched above that any usable indication of future price changes has already been reflected fully in the present price. Markets are efficient at exploiting all usable indications immediately so that none is left for ordinary investors to exploit. The efficient market hypothesis is commonly taken to be the grounding of the random walk model. We see it in Samuelson's enigma above. Burton Malkiel (2015), in his successful popularization *Random Walk down Wall Street*, writes favorably (in the preface) of the efficient market hypothesis. However, he also portrays the hypothesis as an "obfuscation" (26) of the random walk hypothesis deployed by academics who attempt to parry critics of the random walk hypothesis.

Malkiel's hesitation is well justified, for the efficient market hypothesis is both imprecisely delimited and weaker logically than the random walk hypothesis. It cannot, by itself, sustain the random walk hypothesis. A significant imprecision lies in a failure to specify just which sorts of information can count as an indication of future price changes. Fama (1970, 383) identifies three candidates. If the information is merely that of the past history of prices, then we have the "weak" form of the hypothesis. If the information includes all publicly available information, then we have the "semi-strong" form. Finally, the "strong" form applies when some monopolistic groups have access to all information relevant to price changes. Fama seeks (384) to give the hypothesis more precise expression in terms of the probabilistic expectations of prices over time. Roughly speaking, it asserts that the expected price of a security at a later time rises just by the increase expected with the best current information. It is immediately clear, as Fama shows (386–87), that a

4 You would be correct to wonder whether this prediction conforms to the stock market's history of rarer but memorable crashes. I will take up this issue in the next section.

condition on probabilistic expectations is weaker than the random walk hypothesis, for this latter hypothesis concerns the full probability distributions and not just their expectations. To his critique, I add that the efficient market hypothesis, as commonly stated, is not necessarily a probabilistic hypothesis at all. It can be expressed for changes, stochastic or otherwise, not governed by a probability distribution.

These last considerations show that an efficient market is not sufficient to produce a random walk. It is also not necessary, for a random walk could also arise if traders were maximally inept and merely traded on idiosyncratic whims.

2.5. Mandelbrot's Fractals

The core supposition of this approach is that the charts recording changes in prices are self-similar under changes of time scale. The program of research associated with it is inseparable from the work of Benoit Mandelbrot, its chief architect and proponent. He is fond of telling heroic tales of his discovery:

> . . . I conceived in the late fifties a tool that was already mentioned, but deserves elaboration. I concluded that much in economics is *self-affine*; a simpler word is *scaling*. This notion is most important, and also most visual (hence closest to being self-explanatory), in the context of the financial charts. Folklore asserts that "all charts look the same." For example, to inspect a chart from close by, then far away, take the whole and diverse pieces of it, and resize each to the same horizontal format known to photographers as "landscape." Two renormalized charts are never identical, of course, but the folklore asserts that they do not differ in kind. The scholarly term for "resize" is to "renormalize" by performing an "affinity," which motivated me in 1977 to coin the term "self-affinity." . . . The scholarly term for "to look alike" is "to remain statistically invariant by dilation or reduction." (1997, 5–6; Mandelbrot's emphasis)

Self-similarity is the defining characteristic of fractal curves, such as the Koch snowflake. Each part is made of smaller parts that are scaled-down versions of the larger part and so on at all levels. Thus, that a curve is self-similar is a powerful constraint. A casual reader, however, might overlook that self-similarity is not quite so restrictive in the financial application. As the remark above allows, the similarity is not exact, as with the Koch snowflake. It is only statistical: that is, there is a similarity in the probabilistic distributions only, not the curve's specific shapes, which means that the curves merely "look alike" at different scales.

We best capture the founding hypothesis by quoting what Mandelbrot calls the "property assumed as 'axiom'" (1997, 2) for him, a collection of his papers in fractal finance:

> *Hypothesis of fractal finance.* "Starting from the rules that govern the variability of price at a certain scale of time, higher-frequency and lower-frequency variation is governed by the same rules, but acting faster or more slowly."

Its implementation is straightforward. Consider the probabilistic distribution of price changes over one day. That distribution is the same distribution that governs prices changes accumulated over a month and again those accumulated over a year. Since the overall magnitude of changes in the periods of a day, a month, and a year is different, we must rescale linearly the distribution in moving between these time periods so that the overall magnitudes align and a sameness of probabilistic distribution is recovered. Here "sameness" means "same analytical formula."

As it happens, just this form of self-similarity is already manifested in the random walk model. Price changes over a large interval of time are just the sums of the changes over the smaller component intervals of time. If price changes in small intervals of time are independent and normally distributed with finite means and variances, then their distribution over the summed time interval will also be normal but with a mean and a variance each of which is the sum of the means and the variances of the distributions in the small time intervals. These distributions scale in the sense that we can map any normal distribution into any other by suitable linear transformation of its variables.

As noted above, the central limit theorem of probability theory tells us that this scaling behavior eventually will emerge as the limiting behavior on sufficiently large time scales even when the probability distributions over the smaller time intervals are not normal. It will happen as long as the probability distributions over the smaller time intervals are independent and have finite means and variances (and, informally speaking, no one time interval makes a disproportionately large contribution to the sum).

The essential observation that Mandelbrot added to this already existing self-similarity is that a Gaussian or normally distributed random walk is not the only distribution satisfying self-similarity. His early paper (1963) outlined a generalization of this self-scaling behavior that arises when the distributions of price changes in the small time intervals are no longer required to have finite means or variances. The most general class of distributions that exhibits the self-similarity under summation of the distributions Mandelbrot called "stable Paretian." That is, if the distribution of price changes in the smaller time intervals is stable Paretian, then so is the distribution of price changes over the summed time interval. These distributions also sustain a generalized version of the central limit theorem. The theorem is as stated above. However, we can drop the requirement that the component distributions have finite means and variances, but we retain their independence. What we are assured to approach in the limit of large sums is a stable Paretian distribution, which includes normal distributions as a special case. So once again we should expect self-similar behavior to be approached over suitably long time periods.[5]

Mandelbrot's contribution was not the identification of this extended class of distributions and the associated extension of the central limit theorem. As Mandelbrot reported, all of this work was already done by the French mathematician Paul Lévy some forty years earlier. Rather, it was to recognize that the nonnormal members of the Paretian class were better suited empirically to market behavior. As we saw above, the normal distribution makes large jumps in prices extremely improbable. Yet such jumps are common in real markets. The nonnormal members of the distribution are distinctive in having "fat tails." That is, they assign considerably larger probabilities than normal distributions to large deviations from the mean. These deviations are the jumps. More specifically, the nonnormal Paretian distributions over some real variable U all approach asymptotically a simple power law distribution for large U. That is, when U is large, the probability of an outcome u greater than U is well approximated by $P(u) = C\,u^{-\alpha}$, for C a constant and $0 < \alpha < 2$. As the variable u increases, any of these power laws decays toward zero slower than the exponential decay of any normal distribution.

5 For a contemporary development of Mandelbrot's analysis, see Fama (1965). A more recent analysis of the generalized central limit theorem is in Ibe (2013, Chapter 8).

Mandelbrot (1997, 29–30) glosses the "scaling" behavior of this tail distribution by noting that, if we were to learn that U must be at least equal to w, then conditioning the original distribution on this fact yields the same power law distribution but now with an altered constant C. This seems to me a weak expression of the scaling behavior, better captured by the generalized central limit theorem. We can forgive Mandelbrot for not giving more mathematical details in a semi-popular presentation since the details become burdensome rapidly. There is no explicit expression for the Paretian class of distributions. They are best characterized by an explicit formula for the characteristic functions of the distributions.

This introduction of Paretian distributions was the first step in a continuing program of research by Mandelbrot. Subsequent work introduced the possibility of various failures of independence of successive price movements while still retaining the statistics of Paretian distributions with their fat tails.

2.6. Random Walkers and Fractals Converge

The random walk theory and the fractal theory might appear to be distinct systems with different logics. That was the view that Mandelbrot urged. He was already describing his work in 1963 as "a radically new approach to the problem of price variation" (395). There were notable differences between his approach and that of the random walk theory at the outset. Mandelbrot denied two of the basic assumptions of the random walk theory: the finite variance of price changes and the independence of subsequent changes. As far as the actual predictive apparatus is concerned, the use of distributions with infinite variance and fat power law tails comprise the main substance of Mandelbrot's deviation from the traditional random walk theory. The scaling hypothesis by itself is not strong enough to preclude the Gaussian random walk theory. Indeed, the introduction of infinite variances and fat-tailed distributions must be supported by observation of the market prices, and those observations might well suffice without the scaling hypothesis if our goal is merely the compact summary of the data.

Viewed more broadly, the random walk theory and the fractal approach agree far more than they disagree. They share a statistical framework that presumes that prices are probabilistically distributed, that market analysis is the mathematical exploration of these distributions, and that these distributions exhaust what the analyst can know. To a chartist, however, whose methods do not include traditional statistical analysis, the differences between the random walk theory and the fractal approach will appear to be mere fine-tuning of details in an analysis remote and alien to them.

More significant for my purposes here, these differences are diminishing. The approaches are converging. In the evolving literature on random walks, empirical investigation is to decide whether the variances are finite and whether there are failures of independence. It now seems to be well established that independence does fail. That recognition is reflected in the provocative title of *A Non-Random Walk down Wall Street* (Lo and MacKinlay 1999). The title is hyperbolic since it turns out that the failures of independence are so slight as not to be serviceable as predictive tools for ordinary traders.

The mainstream of statistical analysts seems to regard Mandelbrot's contribution as mere refinement, as is apparent from the papers collected in Lo and MacKinlay (1999). The word *fractal* appears once (15), and Mandelbrot's work is addressed but treated as an interesting proposal among others for extensions of the probability distributions and dependencies of the mainstream analysis. The word *fractal* and the name Mandelbrot do not appear in Malkiel (2015).

Mandelbrot, for his part, accepts the core lesson of the random walk theory, the unpredictability of price changes. However, he expands this predictive pessimism with a warning that price changes might be far larger than the traditional random walker expects:

> . . . I agree with the orthodox economist that stock prices are probably not predictable in any useful sense of the term. But the risk certainly does follow patterns that can be expressed mathematically and can be modeled on a computer. Thus, my research could help people avoid losing as much money as they do, through foolhardy underestimation of the risk of ruin.[6] (Mandelbrot and Hudson 2004, 6)

3. The Systems Compete

The competition among these systems is unsustainable in the longer term if factual investigations continue and the full import of evidence is respected. The competition might be resolved gently if systems in competition migrate

6 A similar remark is in Mandelbrot (1997, 9).

toward one another. This gentle resolution has brought the random walk theory and fractal analysis into sufficient agreement that they can be regarded as one system. However, if the proponents of competing systems remain intransigent, then, I have argued, a thorough factual investigation will lead at most to one ascending while the others fail.

Proponents of each system do recognize the threat posed by the other systems and have put some effort into impugning their competitors. Here I will collect criticisms levied by proponents of each system against competing systems. The main point for my purposes is that the criticisms all depend on proposing facts whose truth would undermine the competitors' theories. They are most damaging when the proposed facts directly contradict the founding hypotheses of each system. A threat to these founding hypotheses is a threat to the inductive logic and the predictive capacity of the associated view.

This battle of the foundational facts makes clear one of the principal points of this chapter: that the conflict among the systems is to be resolved by factual investigation, as opposed to higher-level examination of abstract principles of inductive inference. Were the facts proposed below by various proponents to be investigated thoroughly and a final decision made on each, that would suffice to leave viable at most one of the systems. The path to this resolution is open. Whether it is taken depends on many factors that go beyond the inductive logic. Is there sufficient motivation by investigators to carry out the requisite studies thoroughly enough to achieve inescapable results? Will proponents of an impugned system accept the results? The persistence of the competing programs indicates that these factors have slowed or even stalled progress toward the final decision.

Below is a sample of the threats mounted against each system.

3.1. Against Fundamental Analysis

Malkiel, the most visible proponent of random walk theory, lists three problems for fundamental analysis:

> Despite its plausibility and scientific appearance, there are three potential flaws in this type of analysis.
>
> *First*, the information and analysis may be incorrect.
>
> *Second*, the security analyst's estimate of "value" may be faulty.

Third, the market may not correct its "mistake," and the stock price may not converge to its value estimate. (2015, 128–29; my emphasis)

Malkiel proceeds to elaborate each. Most striking is his disparaging of the very idea of value:

> It is virtually impossible to translate the specific estimates of growth into a single estimate of intrinsic value. Indeed, attempts to obtain a measure of fundamental value may be an unrewarding search for a will-o'-the-wisp. (129)

Edwards, Magee, and Bassetti, the authoritative source in technical analysis, level similar criticism against fundamental analysis. They reiterate Malkiel's concern about poor information: "The bulk of the statistics the fundamentalists study are past history, already out of date and sterile because the market is not interested in the past or even in the present" (2019, 4). Using an examination of companies listed in the Dow Jones Industrial Average, they also argue that high earnings are a poor indicator of which stock prices will grow most (6). Next they assail the idea of a practically accessible notion of value, urging that ". . . it is futile to assign an intrinsic value to a stock certificate" (4). The claim is reinforced by recounting wild gyrations in the price of a share of US Steel over nearly two decades, from 1929 to 1947. Finally, they doubt that price movements are connected with the factual bases used by fundamentalists to determine value. They assert that "the [fundamental] analyst assumes causality between external events and market movements, a concept which is almost certainly false" (6). Mandelbrot's (1997, 8) critique echoes all of these concerns: "In the real world, causes are usually obscure. Critical information is often unknown or unknowable. . . ."

This combined critique assails the essential elements of the founding hypothesis of fundamental analysis. Intrinsic value is not in practice ascertainable reliably, and market dynamics might not or will not drive prices toward intrinsic value.

The claims of this critique are factual matters. The truth of the founding hypothesis of fundamental analysis can be established empirically. All that fundamental analysts need to display is a successful record of identifying intrinsic values toward which stock prices eventually converge.

3.2. Against Technical Analysis

Of all the approaches, technical analysis has been subject to the most severe criticism, at times bordering on derision.[7] Two factors draw this unflattering appraisal. First, to anyone with a modicum of statistical sophistication, the methods used to ascertain the chartists' patterns are woefully naive. It is all too easy to glance at randomness and see order. We easily see faces in the clouds. In a preface to Edwards, Magee, and Bassetti, Richard McDermott, president of John Magee, Inc., reports the great man's response to this concern:

> To the random walker, who once confronted John [Magee] with the statement that there was no predictable behavior on Wall Street, John's reply was classic. He said, "You fellows rely too heavily on your computers. The best computer ever designed is still the human brain. Theoreticians try to simulate stock market behavior, and, failing to do so with any degree of predictability, declare that a journey through the stock market is a random walk. Isn't it equally possible that the programs simply aren't sensitive enough or the computers strong enough to successfully simulate the thought process of the human brain?" Then John would walk over to his bin of charts, pull out a favorite, and show it to the random walker. There it was — spike up, heavy volume; consolidation, light volume; spike up again, heavy volume. A third time. A fourth time. A beautifully symmetrical chart, moving ahead in a well-defined trend channel, volume moving with price. "Do you really believe that these patterns are random?" John would ask, already knowing the answer. (2019, xxxv)

We would normally pass in silence over such an abysmal display of ignorance of the basics of statistical analysis. However, the second factor that encourages circulation of the unflattering appraisal is that the methods of technical analysis are pervasive in the financial world. Everywhere we find charts annotated in the language of support and resistance levels, breakouts, and more. There is a pretense of learned insight that rests, in practice, on novice

7 Ridicule is a staple in the popular literature. See, for example, Chokkavelu (2010), which opens with the quotation "Stupid is as stupid does" (*Forrest Gump*).

statistical blunders. Yet these instruments are used routinely to make decisions affecting the financial fates of many people. Thus, the long-standing derision is well earned. Long ago, in their original text, Graham and Dodd (1934, 608) reported "many [unnamed] sceptics" who dismiss the analysis as "akin to astrology or necromancy." Mandelbrot (1997, 9) had no need for anonymity and labeled technical analysis "financial astrology."

A footnote in Graham and Dodd's original text also reports one of the earliest versions that I have found of a much-repeated rebuke. The idea is that we can fabricate charts using randomizers that now spuriously manifest the patterns of the technical analysts but without any predictive import. They write that,

> Apropos of this attitude, we refer to a statement made by Frederick R. Macaulay at a meeting of the American Statistical Association in 1925, to the effect that he had plotted the results of tossing a coin several thousand times (heads = "one point up"; tails = "one point down") and had thereby obtained a graph resembling in all respects the typical stock chart — with resistance points, trend lines, double tops, areas of accumulation, etc. Since this graph could not possibly hold any clue as to the future sequence of heads or tails, there was a rather strong inference that stock charts are equally valueless. Mr. Macaulay's remarks were summarized in *Journal of the American Statistical Association*, Vol. 20, p. 248, June 1925.[8] (1934, 608)

The rebuke appears often in later literature. Malkiel (2015, 137–38) reports asking his students to construct such a chart by coin flipping.

Entertaining as such gimmicks might be, they do not really demonstrate the failure of technical analysis. If we are to hold the chartists to a high statistical standard, then we should also apply it to ourselves. To conclude that, *on a superficial scan*, random data might manifest the same patterns as the chartists does not prove them wrong. More cautious analysis is needed. Arditti and McCullough (1978) found that technical analysts could not pick apart real

---

8 The journal article cited is an anonymous report of an April 17, 1925, dinner meeting of the American Statistical Association. Graham and Dodd must be reporting from another source, perhaps their own attendance, since the journal text is briefer and uses dice, not coin tosses, as the randomizers.

from randomly generated charts beyond chance levels in a well-constructed test. However, Hasanhodzic, Lo, and Viola (2010) devised a game in which participants sought to pick real from fabricated charts. The players were given immediate feedback on the correctness of their judgments. The training was effective. They quickly learned to pick the real from the fabricated charts.

All that the examination of fabricated charts can do is cast doubt on the methods that chartists use to arrive at their results. A poor method can still yield a correct result. It might just be that the psychology of traders does imprint identifiable patterns on the charts, as the founding hypothesis asserts. The decisive question to answer is whether the methods work. Here Graham and Dodd (1934, 609) had already leveled a two-part critique. As a historical matter, they reported, the chartists had failed to find a method of prediction that works. "There is no generally known method of chart reading which has been continuously successful for a long period of time." This historical report was coupled with a more principled critique: there can be no such method since it would be self-defeating: "If it were known, it would be speedily adopted by numberless traders. This very following would bring its usefulness to an end."

Here the fundamentalists, Graham and Dodd, offered the same critique as that given later by the random walk proponent, Malkiel. He reported empirical studies showing that the chartists' patterns lack predictive power (e.g., 2015, 114). His principal criticism, however, was the same efficient-market argument as that offered by Graham and Dodd: the chartists' methods cannot work since they undermine themselves.

> Any successful technical scheme must ultimately be self-defeating. The moment I realize that prices will be higher after New Year's Day than they are before Christmas, I will start buying before Christmas ever comes around. If people know a stock will go up tomorrow, you can be sure it will go up today. Any regularity in the stock market that can be discovered and acted upon profitably is bound to destroy itself. This is the fundamental reason why I am convinced that no one will be successful in using technical methods to get above-average returns in the stock market. (Malkiel 2015, 156–57)

As before, the decision on the cogency of the chartists' methods is an empirical matter to be decided by investigations of the market. In principle, arguments such as those against technical analysis are impressive until empirical investigations show their conclusions to be false. Only then do we realize the fragility of assumptions made tacitly in the arguments. Aronson (2007) makes a sustained plea for technical analysts to hold their methods to the standards of routine statistical analysis. Perhaps Graham and Dodd and Malkiel were correct that enough has been done to refute technical analysis. There are dissenters. Lorenzoni et al. (2007) claim that statistical analysis does reveal statistically significant information in two of three patterns: triangle, rectangle, and head and shoulders.

3.3. Against Random Walks

Here I shall construe the random walk theory most broadly as including the possibility of small failures of independence and of distributions with infinite variances. This expanded version includes Mandelbrot's fractal approach. It still retains the main idea that distinguishes the original random walk theory and fractal analysis from other approaches and draws criticism: markets are sufficiently random as to preclude useful prediction of change in prices beyond the broadest averages.

Although this failure of prediction directly contradicts the technical analysts, there is little in the technical analysts' authoritative volume, Edwards, Magee, and Bassetti (2019), to contradict the random walk theory. We have seen Magee's facile response, reported above by McDermott (2019). Otherwise, "random walk" and "efficient market hypothesis" do not appear in the index or, as far as I can tell, in the text. Aronson (2007, 342–55) lays out an extended assault on the efficient market hypothesis. The approach is to undermine what he takes to be the founding assumptions of the hypothesis. For example, he urges that investors are not rational, that their investing errors are not uncorrelated, that arbitrage need not force prices to rational levels, and more. The weakness of the critique is that Aronson does not properly separate the efficient market hypothesis from the hypothesis of a random walk. However, important for my purposes here is that all of the objections depend on factual matters, such as those just listed, and their truth can be ascertained by empirical investigations.

Buffett gave the authoritative response from the fundamentalists to random walk theory. His extraordinary record of profitable investing alone indicates that an astute analyst can make successful predictions over sustained periods. His "Superinvestors of Graham-and-Doddsville" (1984) makes the case against the impossibility of predicting the market in a direct way. He reports nine successful investment funds that exceeded market averages in their returns by wide margins and did so over long periods. The longest of them was 1956 to 1984.

This behavior contradicts the unpredictability of markets central to the random walk theory. More specifically, when the prices of undervalued stocks eventually rise assuredly to their true values, the sequence of upward changes in prices contradicts the independence or near independence of the price changes hypothesized in random walk theory.

The obvious random walk theorist's response is that, in any large economy with many such funds, there will always be outliers that perform well merely by chance. Buffett goes to some pains to answer this objection. The funds on which he reports were selected prior to their successes. As he puts it, "these winners were all well known to me and pre-identified as superior investors, the most identification occurring over 15 years ago" (1984, 4). Buffett also stresses the many differences between the funds while retaining the major common factor: they all follow the Graham and Dodd policy of investing when prices and values are mismatched. This common factor, we are to believe, is responsible for their successes.

There is also a casual rebuttal of the efficient market hypothesis, memorable because of the credentials of its source:

> I'm convinced that there is much inefficiency in the market. These Graham-and-Doddsville investors have successfully exploited gaps between price and value. When the price of a stock can be influenced by a "herd" on Wall Street with prices set at the margin by the most emotional person, or the greediest person, or the most depressed person, it is hard to argue that the market always prices rationally. In fact market prices are frequently nonsensical. (Buffett 1984, 13)

Once again Buffett's argument is a direct challenge to the founding hypothesis of the random walk theory and its embellished versions. The basis of the challenge is empirical. If it is an empirical fact that a particular sort of investment strategy leads to long-term profits, well in excess of market averages, then the unpredictability of the market has been refuted.

4. Conclusion: The Instability of Competing Systems

Competing systems arise when analysts proceed from different, mutually incompatible hypotheses. The competition should be transient while we await further evidential scrutiny that will decide which, if any, of the hypotheses is well supported. As the full import of the existing evidence and that of new evidence is brought to bear, we have seen two ways that the competition could be resolved.

4.1. The Gentle Way: Convergence

In the gentler way, one or more of the systems in competition alter their founding hypotheses to accommodate evidential pressures. If this process of adaptation proceeds far enough, then competing systems might converge. This convergence has happened in the case of the random walk theory and fractal analysis. While the systems might first appear to be very different, they agree on so much at the outset that convergence was easily attained. Both adopt an essentially probabilistic outlook using the standard statistical methods of analysis. They differ only in smaller matters that can be settled by smaller empirical analysis. Are the variances of the probability distributions of price changes finite or infinite? What are the extent and nature of any probabilistic dependence among successive price changes? Insofar as proponents of the approaches accept the results of empirical studies, and if the statistical approach is viable in the first place, then the convergence was inevitable.

In principle, a convergence of this generalized random walk theory and technical analysis is also possible. It would be inevitable if chartists would heed Aronson's (2007) urging of the use of sound statistical methodology. Either the statistical studies will show a correlation between the head and shoulders formation and a subsequent decline in prices, or they will not. Once both groups of theorists accept these statistical methods, agreement on the efficacy or otherwise of these chartists' signals is inevitable if only the empirical studies are pursued thoroughly. The losing approach then would need to adapt its founding hypotheses accordingly. Or both might adapt to some compromise account containing elements of both original approaches.

4.2. The Severe Way: Elimination

The more severe path to a unique logic arises when proponents of each competing logic are intransigent and refuse to adapt their logic to emerging evidence, for the competition is unstable. Evidence that turns out to support one system's founding hypothesis will strengthen that system while weakening those that disagree with it. A stronger system can infer to still more that strengthens it further while weakening the competition. The process is akin to the instability of a pencil balanced on its tip. Once the pencil starts to fall to one side, the forces pulling it to that side are strengthened, and the fall accelerates.

The competition between random walk theorists and chartists illustrates this instability. The generalized random walk theory depends essentially on the independence or meager dependence of the probability distributions of successive price changes. This meager dependence needs to be demonstrated, in principle, for each stock or each stock sector index. Each success would detract from the prospects of the chartists, whose theories depend essentially on a failure of independence. Their head and shoulders formation can be a reliable indicator of a coming reversal only if there is a strong correlation between it and subsequent price changes.

As this independence is established for more individual stocks or indices, each success provides indirect support for independence among untested stocks or indices. This last inference is supported by a warranting hypothesis that the mechanisms governing price moves are much the same across the market. These successes form a cascade of continuing successes, each amplifying the strength of support of the random walk theory's claims elsewhere. Each also brings the corresponding collapse of the competing chartists' system. This is a cycle of positive reinforcement that would terminate in the elimination of technical analysis.

The reverse process would arise if, instead, chartists were able to demonstrate with statistical rigor the efficacy of one of their formations as a signal for future price movements. Such a success would contradict the very limited dependence among successive price changes that the random walk theory is prepared to accept. The assumption that the mechanisms moving prices are much the same across the market would support an inference that similar signals are possible elsewhere. As their successes mount, the prospects for the limited dependencies allowed by the random walk theory would narrow.

Continuing successes eventually would end in the demise of the random walk theory.

As we saw above, the fundamentalists' challenge to the other systems is laid out most cogently by Buffett (1984). Using the evidence of several successful investment funds, he claims that pursuit of value-price discrepancies led them to purchase stocks whose long-term price gains greatly exceed market averages. He argues that the only common factor among them is their focus on value. He insists that the successful funds dismissed daily price movements as meaningless distractions. If they prove demonstrably to be correct about daily price movements, then the basic supposition of technical analysis would be refuted. If the success of value investing persists and is sustainable under careful statistical analysis, then random walk theorists who respect statistical methods must accept the fundamentalist approach. Conversely, if statistical analysis reveals their successes to be merely the luck of a few, then fundamentalists would have to retreat. With each new report of a successful value investor, the fundamentalist approach would be strengthened, once again under the assumption that the mechanisms moving prices are much the same across the market. The random walk theory would be weakened, for it would be harder to dismiss these successes as mere chance.

4.3. Multiple Systems Are Possible if They Do Not Compete

The processes assuring ascendance of one dominant logic at most arise only when the systems truly conflict. In the earlier chapter, I raised the possibility of multiple systems coexisting if the domains could be divided so that each logic would apply in its partition only. Such a possibility could be realized in principle here. Fundamental analysis draws from a different body of evidence from the other three systems and makes predictions over a longer time span. We might divide the field of stock market prediction into two partitions.

The evidence base for the first is the detailed compilation of facts about all aspects of the companies associated with each stock, and the time scale for predictions is some suitably chosen longer term. Fundamental analysis would apply in this partition.

The evidence base for the second partition is restricted to the past history of stock prices and volumes traded. Predictions would be made over the shorter term. Each of the remaining systems has aspirations in this partition.

Although such a partition is possible in principle, fundamental analysts and those of the other systems do regard themselves as being in competition. Each does seek to impugn the basic suppositions of the others.

4.4. Principle and Practice

The processes sketched above map out how, in principle, suitable empirical investigations can and should dissolve eventually the competition among the logics. Convergence to a single logic, then, awaits only analysts willing to undertake the investigations and proponents of the systems willing to accept the results. In practice, however, the differing systems persist, and there is little hope that this circumstance will change. We can speculate about why this is so. Perhaps the continuing infusion of new traders into the stock market replenishes the pool of novice enthusiasts, well informed on just one system. Perhaps there is too much inertia among proponents of each competing system. The chartists are too wedded to their charts, the random walk theorists are too wedded to their theorems, and the value investors are too wedded to company balance sheets. Whatever the reasons, this persistence reveals little of the applicable inductive logic and more of the contingent social factors.

4.5. Material and Formal Approaches

How can competition among different inductive logics in some domain be resolved? These examples display how a material approach to inductive inference succeeds in answering easily where a purely formal approach cannot. According to the material theory, facts warrant inductive inferences. Hence, a local resolution is possible merely through investigations that establish which are the facts of the domain. Such investigations have been the substance of the dispute among the systems discussed here.

If instead we were to conceive of inductive logics as governed by universally applicable formal schemas, then no such easy resolution would be possible. A dispute over which is the right logic must proceed at the remotest level of generality, separated from any considerations specific to the domain. No such domain-specific considerations can enter, tempting as they would be. To say that this logic is better adapted to this domain and that logic better adapted to that domain is to give up the universal applicability of the formal schemas. It is tacitly to become a material theorist who looks to facts of each domain to decide which inductive logic applies.

For example, a probabilist might argue for the probabilistic methods of random walk theory on the supposition that all uncertainties everywhere are probabilistic. This is a supposition at the highest level of generality that, as I have argued in *The Material Theory of Induction* (Norton 2021), is unsustainable. A more realistic probabilist might argue merely that the sorts of uncertainties in stock prices are factually of a type to which probability theory applies. To do that is just to adopt the core idea of the material theory of induction: facts in the domain warrant the inductive logic applicable.

REFERENCES

Arditti, Fred D., and W. Andrew McCullough. 1978. "Can Analysts Distinguish between Real and Randomly Generated Stock Prices?" *Financial Analysts Journal* 34: 70–74.

Aronson, David R. 2007. *Evidence-Based Technical Analysis: Applying the Scientific Method and Statistical Inference to Trading Signals*. Hoboken, NJ: John Wiley and Sons.

Bachelier, Louis. 1900. "Théorie de la spéculation." *Annales scientifiques de l'É.N.S.*, 3rd Series, 17: 21–86.

Buffett, Warren. 1984. "The Superinvestors of Graham-and-Doddsville." *Hermes: The Columbia Business School Magazine*, fall, 4–15.

———. 2013. Preface. In *Security Analysis: Principles and Technique*, by Benjamin Graham and David Dodd, 6th ed., xi–xii. New York: McGraw-Hill.

Chokkavelu, Anand. 2010. "Technical Analysis Is Stupid." Motley Fool, April 30. Archived at https://web.archive.org/web/20100503010345/http://www.fool.com:80/investing/value/2010/04/30/technical-analysis-is-stupid.aspx.

Edwards, Robert D., John Magee, and W.H.C. Bassetti. 2019. *Technical Analysis of Stock Trends*. 11th ed. New York: Routledge.

Einstein, Albert. 1905. "Über die von der molekularkinetischen Theorie der Wärme geforderte Bewegung von in ruhenden Flüssigkeiten suspendierten Teilchen." *Annalen der Physik* 17: 549–60.

Fama, Eugene F. 1965. "The Behavior of Stock-Market Prices." *Journal of Business* 38: 34–105.

———. 1970. "Efficient Capital Markets: A Review of Theory and Empirical Work." *Journal of Finance* 25: 383–417.

Frost, Alfred J., and Robert R. Prechter. 2017. *Elliott Wave Principle: Key to Market Behavior*. 11th ed. Gainesville, GA: New Classics Library.

Graham, Benjamin, and David Dodd. 1934. *Security Analysis*. 1st ed. New York: McGraw-Hill.

———. 2013. *Security Analysis: Principles and Technique*. 6th ed. New York: McGraw-Hill.

Hasanhodzic, Jasmina, Andrew W. Lo, and Emanuele Viola. 2010. "Is It Real, or Is It Randomized? A Financial Turing Test." https://arxiv.org/abs/1002.4592v1.

Ibe, Oliver. 2013. *Elements of Random Walk and Diffusion Processes*. Hoboken, NJ: Wiley.

Klarman, Seth. 2013. Preface. In *Security Analysis: Principles and Technique*, by Benjamin Graham and David Dodd, 6th ed., xiii–xl. New York: McGraw-Hill.

Lo, Andrew W., and A. Craig MacKinlay. 1999. *A Non-Random Walk down Wall Street*. Princeton, NJ: Princeton University Press.

Lorenzoni, Giuliano, Adrian Pizzinga, Rodrigo Atherino, Cristiano Fernandes, and Rosane Riera Freire. 2007. "On the Statistical Validation of Technical Analysis." *Brazilian Review of Finance, Brazilian Society of Finance* 5: 3–28.

Malkiel, Burton. 2015. *A Random Walk down Wall Street*. Rev. ed. New York: Norton.

Mandelbrot, Benoit B. 1963. "The Variation of Certain Speculative Prices." *Journal of Business* 36: 394–419.

———. 1997. *Fractals and Scaling in Finance: Discontinuity, Concentration, Risk*. New York: Springer.

Mandelbrot, Benoit B., and Richard L. Hudson. 2004. *The (Mis)behavior of Markets: A Fractal View of Risk, Ruin, and Reward*. New York: Basic Books.

McDermott, Richard. 2019. Preface. In *Technical Analysis of Stock Trends*, by Robert D. Edwards, John Magee, and W.H.C. Bassetti, 11th ed., xxxv–xxxvi. New York: Routledge.

Norton, John D. 2021. *The Material Theory of Induction*. BSPS Open Series. Calgary: University of Calgary Press.

Samuelson, Paul A. 1965. "Proof that Properly Anticipated Prices Fluctuate Randomly." *Industrial Management Review* 6: 41–49.

Weatherall, James O. 2013. *The Physics of Wall Street: A Brief History of Predicting the Unpredictable*. Boston: Houghton Mifflin Harcourt.

Afterword

In this volume, I have sought to describe the large-scale structure of the inductive inferences and the inductive relations of support in science. Although I am satisfied that the many chapters devoted to this task have made considerable progress in delineating that structure, I am sure that there is much more to be done. The research that led to this volume has been research in the history and philosophy of science. As I noted in the introduction, that research involves a continuing exchange between the philosophy of science and the history of science. One component of the exchange needs to be emphasized. I have found that a major source of theses in the philosophy of science lies in the study of the history of science. That history recounts the many examples of scientists who grappled with inductive problems of great difficulty and overcame them with inductive maneuvers of still greater ingenuity. Time spent studying the history is philosophically fertile in a way that armchair reflection is not. Armchair reflection can only return what each of us can think up ourselves. A study of the history of science can draw from the ingenuity of generations of the cleverest minds at their moments of greatest achievements. It provides an endlessly fertile repository of inductive ideas in the philosophy of science for those willing to explore it. This volume explores only a tiny portion of this repository. Much remains to be found. My hope is that this volume will encourage others to enter this repository and see what marvels they can find.

Index

Page numbers followed by f, t, and n denote figures, tables, and notes, respectively.

FSC
www.fsc.org
MIX
Papier | Fördert
gute Waldnutzung
FSC® C083411

Kolibri 360 GmbH
Ferdinand-Jühlke-Straße 7
99095 Erfurt, Deutschland
produktsicherheit@kolibri360.de